Graduate Texts in Mathematics 233

Graduate Texts in Mathematics

(continued after index)

Fernando Albiac and Nigel J. Kalton

Topics in Banach Space Theory

 Springer

Fernando Albiac
Department of Mathematics
University of Missouri
Columbia, Missouri 65211
USA
albiac@math.missouri.edu.

Nigel J. Kalton
Department of Mathematics
University of Missouri
Columbia, Missouri 65211
USA
nigel@math.missouri.edu

Mathematics Subject Classification (2000): 46B25

Library of Congress Cataloging in Publication Data:2005933143

ISBN10: 0-387-28141-X
ISBN13: 978-0387-28141-4

Printed on acid-free paper.

Printed in the United States of America.   (MP)

9 8 7 6 5 4 3 2 1   SPIN 10951439

Springer-Verlag is a part of Springer Science+Business Media

springeronline.com

Preface

This book grew out of a one-semester course given by the second author in 2001 and a subsequent two-semester course in 2004-2005, both at the University of Missouri-Columbia. The text is intended for a graduate student who has already had a basic introduction to functional analysis; the aim is to give a reasonably brief and self-contained introduction to classical Banach space theory.

Banach space theory has advanced dramatically in the last 50 years and we believe that the techniques that have been developed are very powerful and should be widely disseminated amongst analysts in general and not restricted to a small group of specialists. Therefore we hope that this book will also prove of interest to an audience who may not wish to pursue research in this area but still would like to understand what is known about the structure of the classical spaces.

Classical Banach space theory developed as an attempt to answer very natural questions on the structure of Banach spaces; many of these questions date back to the work of Banach and his school in Lvov. It enjoyed, perhaps, its golden period between 1950 and 1980, culminating in the definitive books by Lindenstrauss and Tzafriri [138] and [139], in 1977 and 1979 respectively. The subject is still very much alive but the reader will see that much of the basic groundwork was done in this period.

We will be interested specifically in questions of the following type: given two Banach spaces X and Y, when can we say that they are linearly isomorphic, or that X is linearly isomorphic to a subspace of Y? Such questions date back to Banach's book in 1932 [8] where they are treated as *problems of linear dimension*. We want to study these questions particularly for the classical Banach spaces, that is, the spaces c_0, ℓ_p $(1 \leq p \leq \infty)$, spaces $\mathcal{C}(K)$ of continuous functions, and the Lebesgue spaces L_p, for $1 \leq p \leq \infty$.

At the same time, our aim is to introduce the student to the fundamental techniques available to a Banach space theorist. As an example, we spend much of the early chapters discussing the use of Schauder bases and basic sequences in the theory. The simple idea of extracting basic sequences in order to understand subspace structure has become second-nature in the subject, and so the importance of this notion is too easily overlooked.

It should be pointed out that this book is intended as a text for graduate students, not as a reference work, and we have selected material with an eye to what we feel can be appreciated relatively easily in a quite leisurely two-semester course. Two of the most spectacular discoveries in this area during the last 50 years are Enflo's solution of the basis problem [54] and the Gowers-Maurey solution of the unconditional basic sequence problem [71]. The reader will find discussion of these results but no presentation. Our feeling, based on experience, is that detouring from the development of the theory to present lengthy and complicated counterexamples tends to break up the flow of the course. We prefer therefore to present only relatively simple and easily appreciated counterexamples such as the James space and Tsirelson's space. We also decided, to avoid disruption, that some counterexamples of intermediate difficulty should be presented only in the last optional chapter and not in the main body of the text.

Let us describe the contents of the book in more detail. Chapters 1-3 are intended to introduce the reader to the methods of bases and basic sequences and to study the structure of the sequence spaces ℓ_p for $1 \leq p < \infty$ and c_0. We then turn to the structure of the classical function spaces. Chapters 4 and 5 concentrate on $\mathcal{C}(K)$-spaces and $L_1(\mu)$-spaces; much of the material in these chapters is very classical indeed. However, we do include Miljutin's theorem that all $\mathcal{C}(K)$-spaces for K uncountable compact metric are linearly isomorphic in Chapter 4; this section (Section 4.4) and the following one (Section 4.5) on $\mathcal{C}(K)$-spaces for K countable can be skipped if the reader is more interested in the L_p-spaces, as they are not used again. Chapters 6 and 7 deal with the basic theory of L_p-spaces. In Chapter 6 we introduce the notions of type and cotype. In Chapter 7 we present the fundamental ideas of Maurey-Nikishin factorization theory. This leads into the Grothendieck theory of absolutely summing operators in Chapter 8. Chapter 9 is devoted to problems associated with the existence of certain types of bases. In Chapter 10 we introduce Ramsey theory and prove Rosenthal's ℓ_1-theorem; we also cover Tsirelson space, which shows that not every Banach space contains a copy of ℓ_p for some p, $1 \leq p < \infty$, or c_0. Chapters 11 and 12 introduce the reader to local theory from two different directions. In Chapter 11 we use Ramsey theory and infinite-dimensional methods to prove Krivine's theorem and Dvoretzky's theorem, while in Chapter 12 we use computational methods and the concentration of measure phenomenon to prove again Dvoretzky's theorem. Finally Chapter 13 covers, as already noted, some important examples which we removed from the main body of the text.

The reader will find all the prerequisites we assume (without proofs) in the Appendices. In order to make the text flow rather more easily we decided to make a default assumption that all Banach spaces are real. That is, unless otherwise stated, we treat only real scalars. In practice, almost all the results in the book are equally valid for real or complex scalars, but we leave to the reader the extension to the complex case when needed.

There are several books which cover some of the same material from somewhat different viewpoints. Perhaps the closest relatives are the books by Diestel [39] and Wojtaszczyk [221], both of which share some common themes. Two very recent books, namely, Carothers [23] and Li and Queffélec [126], also cover some similar topics. We feel that the student will find it instructive to compare the treatments in these books. Some other texts which are highly relevant are [10], [78], [149], and [56]. If, as we hope, the reader is inspired to learn more about some of the topics, a good place to start is the *Handbook of the Geometry of Banach Spaces*, edited by Johnson and Lindenstrauss [90, 92] which is a collection of articles on the development of the theory; this has the advantage of being (almost) up to date at the turn of the century. Included is an article by the editors [91] which gives a condensed summary of the basic theory.

The first author gratefully acknowledges Gobierno de Navarra for funding, and wants to express his deep gratitude to Sheila Johnson for all her patience and unconditional support for the duration of this project. The second author acknowledges support from the National Science Foundation and wishes to thank his wife Jennifer for her tolerance while he was working on this project.

Columbia, Missouri,   *Fernando Albiac*
November 2005   *Nigel Kalton*

Contents

1

Bases and Basic Sequences

In this chapter we are going to introduce the fundamental notion of a Schauder basis of a Banach space and the corresponding notion of a basic sequence. One of the key ideas in the isomorphic theory of Banach spaces is to use the properties of bases and basic sequences as a tool to understanding the differences and similarities between spaces. The systematic use of basic sequence arguments also turns out to simplify some classical theorems and we illustrate this with the Eberlein-Šmulian theorem on weakly compact subsets of a Banach space.

Before proceeding let us remind the reader that our convention will be that all Banach spaces are real, unless otherwise stated. In fact there is very little change in the theory in switching to complex scalars, but to avoid keeping track of minor notational changes it is convenient to restrict ourselves to the real case. Occasionally, we will give proofs in the complex case when it appears to be useful to do so. In other cases the reader is invited to convince himself that he can obtain the same result in the complex case.

1.1 Schauder bases

The basic idea of functional analysis is to combine the techniques of linear algebra with topological considerations of convergence. It is therefore very natural to look for a concept to extend the notion of a basis of a finite dimensional vector space.

In the context of Hilbert spaces orthonormal bases have proved a very useful tool in many areas of analysis. We recall that if $(e_n)_{n=1}^{\infty}$ is an orthonormal basis of a Hilbert space H, then for every $x \in H$ there is a unique sequence of scalars $(a_n)_{n=1}^{\infty}$ given by $a_n = \langle x, e_n \rangle$ such that

$$x = \sum_{n=1}^{\infty} a_n e_n.$$

The usefulness of orthonormal bases stems partly from the fact that they are relatively easy to find; indeed, every separable Hilbert space has an orthonormal basis. Procedures such as the Gram-Schmidt process allow very easy constructions of new orthonormal bases.

There are several possible extensions of the basis concept to Banach spaces, but the following definition is the most useful.

Definition 1.1.1. A sequence of elements $(e_n)_{n=1}^{\infty}$ in an infinite-dimensional Banach space X is said to be a *basis* of X if for each $x \in X$ there is a *unique* sequence of scalars $(a_n)_{n=1}^{\infty}$ such that

$$x = \sum_{n=1}^{\infty} a_n e_n.$$

This means that we require that the sequence $(\sum_{n=1}^{N} a_n e_n)_{N=1}^{\infty}$ converges to x in the norm topology of X.

It is clear from the definition that a basis consists of linearly independent, and in particular nonzero, vectors. If X has a basis $(e_n)_{n=1}^{\infty}$ then its closed linear span, $[e_n]$, coincides with X and therefore X is separable (the rational finite linear combinations of (e_n) will be dense in X). Let us stress that the order of the basis is important; if we permute the elements of the basis then the new sequence can very easily fail to be a basis. We will discuss this phenomenon in much greater detail later, in Chapter 3.

The reader should not confuse the notion of basis in an infinite-dimensional Banach space with the purely algebraic concept of Hamel basis or vector space basis. A *Hamel basis* $(e_i)_{i \in \mathcal{I}}$ for X is a collection of linearly independent vectors in X such that each x in X is uniquely representable as a *finite* linear combination of e_i. From the Baire Category theorem it is easy to deduce that if $(e_i)_{i \in \mathcal{I}}$ is a Hamel basis for an infinite-dimensional Banach space X then $(e_i)_{i \in \mathcal{I}}$ must be uncountable. Henceforth, whenever we refer to a basis for an infinite-dimensional Banach space X it will be in the sense of Definition 1.1.1.

We also note that if $(e_n)_{n=1}^{\infty}$ is a basis of a Banach space X, the maps $x \mapsto a_n$ are linear functionals on X. Let us write, for the time being, $e_n^{\#}(x) = a_n$. However, it is by no means immediate that the linear functionals $(e_n^{\#})_{n=1}^{\infty}$ are actually continuous. Let us make the following definition:

Definition 1.1.2. Let $(e_n)_{n=1}^{\infty}$ be a sequence in a Banach space X. Suppose there is a sequence $(e_n^*)_{n=1}^{\infty}$ in X^* such that

(i) $e_k^*(e_j) = 1$ if $j = k$, and $e_k^*(e_j) = 0$ otherwise, for any k and j in $\mathbb{N}$,
(ii) $x = \sum_{n=1}^{\infty} e_n^*(x) e_n$ for each $x \in X$.

Then $(e_n)_{n=1}^{\infty}$ is called a *Schauder basis* for X and the functionals $(e_n^*)_{n=1}^{\infty}$ are called the *biorthogonal functionals* associated with $(e_n)_{n=1}^{\infty}$.

If $(e_n)_{n=1}^{\infty}$ is a Schauder basis for X and $x = \sum_{n=1}^{\infty} e_n^*(x)e_n \in X$, the *support of x* is the subset of integers n such that $e_n^*(x) \neq 0$. We denote it by supp (x). If $|\text{supp }(x)| < \infty$ we say that x is *finitely supported*.

The name *Schauder* in the previous definition is in honor of J. Schauder, who first introduced the concept of a basis in 1927 [203]. In practice, nevertheless, every basis of a Banach space is a Schauder basis, and the concepts are not distinct (the distinction is important, however, in more general locally convex spaces).

The proof of the equivalence between the concepts of basis and Schauder basis is an early application of the Closed Graph theorem ([8], p. 111). Although this result is a very nice use of some of the basic principles of functional analysis, it has to be conceded that it is essentially useless in the sense that in all practical situations we are only able to prove that $(e_n)_{n=1}^{\infty}$ is a basis by showing the formally stronger conclusion that it is already a Schauder basis. Thus the reader can safely skip the next theorem.

Theorem 1.1.3. *Let X be a (separable) Banach space. A sequence $(e_n)_{n=1}^{\infty}$ in X is a Schauder basis for X if and only if $(e_n)_{n=1}^{\infty}$ is a basis for X.*

Proof. Let us assume that $(e_n)_{n=1}^{\infty}$ is a basis for X and introduce the *partial sum projections* $(S_n)_{n=0}^{\infty}$ associated to $(e_n)_{n=1}^{\infty}$ defined by $S_0 = 0$ and for $n \geq 1$,

$$S_n(x) = \sum_{k=1}^{n} e_k^{\#}(x)e_k.$$

Of course, we do not yet know that these operators are bounded! Let us consider a new norm on X defined by the formula

$$|||x||| = \sup_{n \geq 1} \|S_n x\|.$$

Since $\lim_{n \to \infty} \|x - S_n x\| = 0$ for each $x \in X$, it follows that $||| \cdot ||| \geq \| \cdot \|$. We will show that $(X, ||| \cdot |||)$ is complete.

Suppose that $(x_n)_{n=1}^{\infty}$ is a Cauchy sequence in $(X, ||| \cdot |||)$. $(x_n)_{n=1}^{\infty}$ is indeed convergent to some $x \in X$ for the original norm. Our goal is to prove that $\lim_{n \to \infty} |||x_n - x||| = 0$.

Notice that for each fixed k the sequence $(S_k x_n)_{n=1}^{\infty}$ is convergent in the original norm to some $y_k \in X$, and note also that $(S_k x_n)_{n=1}^{\infty}$ is contained in the finite-dimensional subspace $[e_1, \ldots, e_k]$. Certainly, the functionals $e_j^{\#}$ are continuous on any finite-dimensional subspace; hence if $1 \leq j \leq k$ we have

$$\lim_{n \to \infty} e_j^{\#}(x_n) = e_j^{\#}(y_k) := a_j.$$

Next we argue that $\sum_{j=1}^{\infty} a_j e_j = x$ for the original norm.

Given $\epsilon > 0$, pick an integer n so that if $m \geq n$ then $|||x_m - x_n||| \leq \frac{1}{3}\epsilon$, and take k_0 so that $k \geq k_0$ implies $\|x_n - S_k x_n\| \leq \frac{1}{3}\epsilon$. Then for $k \geq k_0$ we have

$$\|y_k - x\| \leq \lim_{m\to\infty} \|S_k x_m - S_k x_n\| + \|S_k x_n - x_n\| + \lim_{m\to\infty} \|x_m - x_n\| \leq \epsilon.$$

Thus $\lim_{k\to\infty} \|y_k - x\| = 0$ and, by the uniqueness of the expansion of x with respect to the basis, $S_k x = y_k$.

Now,

$$\||x_n - x\|| = \sup_{k\geq 1} \|S_k x_n - S_k x\| \leq \limsup_{m\to\infty} \sup_{k\geq 1} \|S_k x_n - S_k x_m\|,$$

so $\lim_{n\to\infty} \||x_n - x\|| = 0$ and $(X, \||\cdot\||)$ is complete.

By the Closed Graph theorem (or the Open Mapping theorem), the identity map $\iota : (X, \|\cdot\|) \to (X, \||\cdot\||)$ is bounded, i.e., there exists K so that $\||x\|| \leq K\|x\|$ for $x \in X$. This implies that

$$\|S_n x\| \leq K\|x\|, \qquad x \in X, \ n \in \mathbb{N}.$$

In particular,

$$|e_n^{\#}(x)|\|e_n\| = \|S_n x - S_{n-1} x\| \leq 2K\|x\|,$$

hence $e_n^{\#} \in X^*$ and $\|e_n^{\#}\| \leq 2K\|e_n\|^{-1}$.

$\square$

Let $(e_n)_{n=1}^{\infty}$ be a basis for a Banach space X. The preceding theorem tells us that $(e_n)_{n=1}^{\infty}$ is actually a Schauder basis, hence we use $(e_n^*)_{n=1}^{\infty}$ for the biorthogonal functionals.

As above, we consider the partial sum operators $S_n : X \to X$, given by $S_0 = 0$ and, for $n \geq 1$,

$$S_n \left(\sum_{k=1}^{\infty} e_k^*(x) e_k \right) = \sum_{k=1}^{n} e_k^*(x) e_k.$$

S_n is a continuous linear operator since each e_k^* is continuous. That the operators $(S_n)_{n=1}^{\infty}$ are uniformly bounded was already proved in Theorem 1.1.3, but we note it for further reference:

Proposition 1.1.4. *Let $(e_n)_{n=1}^{\infty}$ be a Schauder basis for a Banach space X and $(S_n)_{n=1}^{\infty}$ the natural projections associated with it. Then*

$$\sup_n \|S_n\| < \infty.$$

Proof. For a Schauder basis the operators $(S_n)_{n=1}^{\infty}$ are bounded *a priori*. Since $S_n(x) \to x$ for every $x \in X$ we have $\sup_n \|S_n(x)\| < \infty$ for each $x \in X$. Then the Uniform Boundedness principle yields that $\sup_n \|S_n\| < \infty$.

$\square$

Definition 1.1.5. If $(e_n)_{n=1}^{\infty}$ is a basis for a Banach space X then the number $K = \sup_n \|S_n\|$ is called the *basis constant*. In the optimal case that $K = 1$ the basis $(e_n)_{n=1}^{\infty}$ is said to be *monotone*.

Remark 1.1.6. We can always renorm a Banach space X with a basis in such a way that the given basis is monotone. Just put

$$|||x||| = \sup_{n \geq 1} \|S_n x\|.$$

Then $\|x\| \leq |||x||| \leq K\|x\|$, so the new norm is equivalent to the old one and it is quickly verified that $|||S_n||| = 1$ for $n \in \mathbb{N}$.

The next result establishes a method for constructing a basis for a Banach space X, provided we have a family of projections enjoying the properties of the partial sum operators.

Proposition 1.1.7. *Suppose $S_n : X \to X$, $n \in \mathbb{N}$, is a sequence of bounded linear projections on a Banach space X such that*

(i) dim $S_n(X) = n$ for each n;
(ii) $S_n S_m = S_m S_n = S_{\min\{m,n\}}$, for any integers m and n; and
(iii) $S_n(x) \to x$ for every $x \in X$.

Then any nonzero sequence of vectors $(e_k)_{k=1}^{\infty}$ in X chosen inductively so that $e_1 \in S_1(X)$, and $e_k \in S_k(X) \cap S_{k-1}^{-1}(0)$ if $k \geq 2$ is a basis for X with partial sum projections $(S_n)_{n=1}^{\infty}$.

Proof. Let $0 \neq e_1 \in S_1(X)$ and define $e_1^* : X \to \mathbb{R}$ by $e_1^*(x)e_1 = S_1(x)$. Next we pick $0 \neq e_2 \in S_2(X) \cap S_1^{-1}(0)$ and define the functional $e_2^* : X \to \mathbb{R}$ by $e_2^*(x)e_2 = S_2(x) - S_1(x)$. This gives us by induction the procedure to extract the basis and its biorthogonal functionals: for each integer n, we pick $0 \neq e_n \in S_n(X) \cap S_{n-1}^{-1}(0)$ and define $e_n^* : X \to \mathbb{R}$ by $e_n^*(x)e_n = S_n(x) - S_{n-1}(x)$. Then

$$|e_n^*(x)| = \|S_n(x) - S_{n-1}(x)\| \|e_n\|^{-1} \leq 2\sup_n \|S_n\| \|e_n\|^{-1} \|x\|,$$

hence $e_n^* \in X^*$. It is immediate to check that $e_k^*(e_j) = \delta_{kj}$ for any two integers k, j.

On the other hand, if we let $S_0(x) = 0$ for all x, we can write

$$S_n(x) = \sum_{k=1}^{n}(S_k(x) - S_{k-1}(x)) = \sum_{k=1}^{n} e_k^*(x)e_k,$$

which, by *(iii)* in the hypothesis, converges to x for every $x \in X$. Therefore, the sequence $(e_n)_{n=1}^{\infty}$ is a basis and $(S_n)_{n=1}^{\infty}$ its natural projections. $\square$

In the next definition we relax the assumption that a basis must span the entire space.

Definition 1.1.8. A sequence $(e_k)_{k=1}^{\infty}$ in a Banach space X is called a *basic sequence* if it is a basis for $[e_k]$, the closed linear span of $(e_k)_{k=1}^{\infty}$.

As the reader will quickly realize, basic sequences are of fundamental importance in the theory of Banach spaces and will be exploited throughout this volume. To recognize a sequence of elements in a Banach space as a basic sequence we use the following test, also known as *Grunblum's criterion* [77]:

Proposition 1.1.9. *A sequence* $(e_k)_{k=1}^{\infty}$ *of nonzero elements of a Banach space X is basic if and only if there is a positive constant K such that*

$$\left\| \sum_{k=1}^{m} a_k e_k \right\| \le K \left\| \sum_{k=1}^{n} a_k e_k \right\| \tag{1.1}$$

for any sequence of scalars (a_k) and any integers m, n such that $m \le n$.

Proof. Assume $(e_k)_{k=1}^{\infty}$ is basic, and let $S_N : [e_k] \to [e_k]$, $N = 1, 2, \ldots$, be its partial sum projections. Then, if $m \le n$ we have

$$\left\| \sum_{k=1}^{m} a_k e_k \right\| = \left\| S_m \left(\sum_{k=1}^{n} a_k e_k \right) \right\| \le \sup_m \| S_m \| \left\| \sum_{k=1}^{n} a_k e_k \right\|,$$

so (1.1) holds with $K = \sup_m \| S_m \|$.

For the converse, let E be the linear span of $(e_k)_{k=1}^{\infty}$ and $s_m : E \to [e_k]_{k=1}^{m}$ be the finite-rank operator defined by

$$s_m \left(\sum_{k=1}^{n} a_j e_j \right) = \sum_{k=1}^{\min(m,n)} a_k e_k, \qquad m, n \in \mathbb{N}.$$

By density each s_m extends to $S_m : [e_k] \to [e_k]_{k=1}^{m}$ with $\| S_m \| = \| s_m \| \le K$. Notice that for each $x \in E$ we have

$$S_n S_m(x) = S_m S_n(x) = S_{\min(m,n)}(x), \qquad m, m \in \mathbb{N}, \tag{1.2}$$

so, by density, (1.2) holds for all $x \in [e_n]$.

$S_n x \to x$ for all $x \in [e_n]$ since the set $\{ x \in [e_n] : S_m(x) \to x \}$ is closed (see D.14 in the Appendix) and contains E, which is dense in $[e_n]$. Proposition 1.1.7 yields that (e_k) is a basis for $[e_k]$ with partial sum projections (S_m). $\qquad \square$

1.2 Examples: Fourier series

Some of the classical Banach spaces come with a naturally given basis. For example, in the spaces ℓ_p for $1 \le p < \infty$ and c_0 there is a canonical basis given by the sequence $e_n = (0, \ldots, 0, 1, 0, \ldots)$, where the only nonzero entry is in the nth coordinate. We leave the verification of these simple facts to the reader. In this section we will discuss an example from Fourier analysis and also Schauder's original construction of a basis in $\mathcal{C}[0,1]$.

Let $\mathbb{T}$ be the unit circle $\{z \in \mathbb{C} : |z| = 1\}$. We denote a typical element of $\mathbb{T}$ by $e^{i\theta}$ and then we can identify the space $\mathcal{C}_{\mathbb{C}}(\mathbb{T})$ of continuous complex-valued functions on $\mathbb{T}$ with the space of continuous 2π-periodic functions on $\mathbb{R}$. Let us note that in the context of Fourier series it is more natural to consider complex function spaces than real spaces.

For every $n \in \mathbb{Z}$ let $e_n \in \mathcal{C}_{\mathbb{C}}(\mathbb{T})$ be the function such that $e_n(\theta) = e^{in\theta}$. The question we wish to tackle is whether the sequence $(e_0, e_1, e_{-1}, e_2, e_{-2}, \dots)$ (in this particular order) is a basis of $\mathcal{C}_{\mathbb{C}}(\mathbb{T})$. In fact, we shall see that it is not. This is a classical result in Fourier analysis (a good reference is Katznelson [108]) which is equivalent to the statement that there is a continuous function f whose Fourier series does not converge uniformly. The stronger statement that there is a continuous function whose Fourier series does not converge at some point is due to Du Bois-Reymond and a nice treatment can be found in Körner [117]; we shall prove this below.

That $[e_n]_{n \in \mathbb{Z}} = \mathcal{C}_{\mathbb{C}}(\mathbb{T})$ follows directly from the Stone-Weierstrass theorem, but we shall also prove this directly.

The *Fourier coefficients* of $f \in \mathcal{C}_{\mathbb{C}}(\mathbb{T})$ are defined by the formula

$$\hat{f}(n) = \int_{-\pi}^{\pi} f(t) e^{-int} \frac{dt}{2\pi}, \qquad n \in \mathbb{Z}.$$

The linear functionals

$$e_n^* : \mathcal{C}_{\mathbb{C}}(\mathbb{T}) \to \mathbb{C}, \quad f \mapsto e_n^*(f) = \hat{f}(n)$$

are biorthogonal to the sequence $(e_n)_{n \in \mathbb{Z}}$.

The *Fourier series* of f is the formal series

$$\sum_{-\infty}^{\infty} \hat{f}(n) e^{in\theta}.$$

For each integer n let $T_n : \mathcal{C}_{\mathbb{C}}(\mathbb{T}) \to \mathcal{C}_{\mathbb{C}}(\mathbb{T})$ be the operator

$$T_n(f) = \sum_{k=-n}^{n} \hat{f}(k) e_k,$$

which gives us the nth partial sum of the Fourier series of f. Then

$$T_n(f)(\theta) = \sum_{k=-n}^{n} \int_{\theta-\pi}^{\theta+\pi} f(t) e^{ik(\theta-t)} \frac{dt}{2\pi}$$

$$= \int_{-\pi}^{\pi} f(\theta - t) \sum_{k=-n}^{n} e^{ikt} \frac{dt}{2\pi}$$

$$= \int_{-\pi}^{\pi} f(\theta - t) \frac{\sin(n + \frac{1}{2})t}{\sin \frac{t}{2}} \frac{dt}{2\pi}.$$

The function

$$D_n(t) = \frac{\sin(n + \frac{1}{2})t}{\sin \frac{t}{2}}$$

is known as the *Dirichlet kernel.*

Let us also consider the operators

$$A_n = \frac{1}{n}(T_0 + \cdots + T_{n-1}), \quad n = 2, 3, \ldots.$$

Then

$$A_n f(\theta) = \frac{1}{n} \int_{-\pi}^{\pi} f(\theta - t) \sum_{k=0}^{n-1} \frac{\sin(k + \frac{1}{2})t}{\sin \frac{t}{2}} \frac{dt}{2\pi}$$

$$= \frac{1}{n} \int_{-\pi}^{\pi} f(\theta - t) \left(\frac{\sin(\frac{nt}{2})}{\sin \frac{t}{2}} \right)^2 \frac{dt}{2\pi}.$$

The function

$$F_n(t) = \frac{1}{n} \left(\frac{\sin(\frac{nt}{2})}{\sin \frac{t}{2}} \right)^2$$

is called the *Fejer kernel.* Note that

$$\int_{-\pi}^{\pi} D_n(t) \frac{dt}{2\pi} = \int_{-\pi}^{\pi} F_n(t) \frac{dt}{2\pi} = 1.$$

Nevertheless, a crucial difference is that F_n is a positive function whereas D_n is not.

Let us now show that if $f \in C_{\mathbb{C}}(\mathbb{T})$ then $\|A_n f - f\| \to 0$. Since f is uniformly continuous, given $\epsilon > 0$ we can find $0 < \delta < \pi$ so that $|\theta - \theta'| < \delta$ implies $|f(\theta) - f(\theta')| \le \epsilon$. Then for any θ we have

$$A_n f(\theta) - f(\theta) = \int_{-\pi}^{\pi} F_n(t)(f(\theta - t) - f(\theta)) \frac{dt}{2\pi}.$$

Hence

$$\|A_n f - f\| \le \|f\| \int_{\delta < |t| \le \pi} F_n(t) \frac{dt}{2\pi} + \epsilon \int_{-\delta}^{\delta} F_n(t) \frac{dt}{2\pi}.$$

Now

$$\int_{\delta < |t| \le \pi} F_n(t) \frac{dt}{2\pi} \le \frac{1}{n} \sin^{-2}(\delta/2)$$

and so

$$\limsup \|A_n f - f\| \le \epsilon.$$

This shows that $[e_n]_{n \in \mathbb{Z}} = C_{\mathbb{C}}(\mathbb{T})$.

Since the biorthogonal functionals are given by the Fourier coefficients, it follows that if $(e_0, e_1, e_{-1}, \ldots)$ is a basis then the partial sum operators (S_n)

satisfy $S_{2n+1} = T_n$ for all n. To show that it is not a basis it therefore suffices to show that the sequence of operators $(T_n)_{n=1}^{\infty}$ is not uniformly bounded.

Let $\varphi \in \mathcal{C}_{\mathbb{C}}(\mathbb{T})^*$ be given by

$$\varphi(f) = f(0).$$

Then

$$\varphi(T_n f) = \int_{-\pi}^{\pi} D_n(t) f(-t) \frac{dt}{2\pi},$$

hence

$$\|T_n^* \varphi\| = \int_{-\pi}^{\pi} |D_n(t)| \frac{dt}{2\pi}.$$

Thus, since $|\sin x| \le |x|$ for all real x,

$$\|T_n\| \ge \int_{-\pi}^{\pi} |D_n(t)| \frac{dt}{2\pi}$$

$$= \frac{1}{\pi} \int_0^{\pi} \left| \frac{\sin\left(n + \frac{1}{2}\right)t}{\sin \frac{t}{2}} \right| dt$$

$$\ge \frac{2}{\pi} \int_0^{(n+1/2)\pi} \left| \frac{\sin t}{\sin \frac{t}{2n+1}} \right| \frac{dt}{2n+1}$$

$$\ge \frac{2}{\pi} \int_0^{(n+1/2)\pi} \frac{|\sin t|}{t} \, dt.$$

By Fatou's lemma

$$\liminf_{n \to \infty} \|T_n\| \ge \frac{2}{\pi} \int_0^{\infty} \frac{|\sin x|}{x} \, dx = \infty.$$

Let us remark that we have actually proved that $\sup_n \|T_n^* \varphi\| = \infty$; therefore by the Uniform Boundedness principle there must exist $f \in \mathcal{C}_{\mathbb{C}}(\mathbb{T})$ such that $(T_n f(0))_{n=1}^{\infty}$ is unbounded. Notice also that this is not an explicit example; see [117] for such an example.

If we prefer to deal with the space of continuous real-valued functions $\mathcal{C}(\mathbb{T})$, exactly the same calculations show that the trigonometric system $\{1, \cos \theta, \sin \theta, \cos 2\theta, \sin 2\theta, \dots \}$ fails to be a basis. Indeed, the operators (T_n) are unbounded on the space $\mathcal{C}(\mathbb{T})$ and correspond to the partial sum operators (S_{2n+1}) as before.

However, $\mathcal{C}(\mathbb{T})$ and $\mathcal{C}_{\mathbb{C}}(\mathbb{T})$ do have a basis. This can easily be shown in a very similar way to Schauder's original construction of a basis in $\mathcal{C}[0,1]$, which we now describe. Let $(q_n)_{n=1}^{\infty}$ be a sequence which is dense in $[0,1]$ and such that $q_1 = 0$ and $q_2 = 1$. We construct inductively a sequence of operators $(S_n)_{n=1}^{\infty}$, defined on $\mathcal{C}[0,1]$, by $S_1 f(t) = f(q_1)$ for $0 \le t \le 1$ and subsequently $S_n f$ is the piecewise linear function defined by $S_n f(q_k) = f(q_k)$ for $1 \le k \le n$ and linear on all the intervals of $[0,1] \backslash \{q_1, \dots, q_n\}$. It is then easy to see that $\|S_n\| = 1$ for all n and that the assumptions of Proposition 1.1.7 are verified. In this way we obtain a monotone basis for $\mathcal{C}[0, 1]$. The basis elements are given by $e_1(t) = 1$ for all t and then e_n is defined recursively by $e_n(q_n) = 1$, $e_n(q_k) = 0$ for $1 \leq k \leq n - 1$ and e_n is linear on each interval in $[0, 1] \setminus \{q_1, \ldots, q_n\}$.

To modify this for the case of the circle we identify $\mathcal{C}(\mathbb{T})$ [respectively, $\mathcal{C}_{\mathbb{C}}(\mathbb{T})$] with the functions in $\mathcal{C}[0, 2\pi]$ [respectively, $\mathcal{C}_{\mathbb{C}}[0, 2\pi]$] such that $f(0) = f(2\pi)$. Let $q_1 = 0$ and suppose $(q_n)_{n=1}^{\infty}$ is dense in $[0, 2\pi)$. Then $S_n f$ for $n > 1$ is defined by $S_n f(q_k) = f(q_k)$ for $1 \leq k \leq n$ and $S_n f(2\pi) = f(q_1)$ and to be affine on each interval in $[0, 2\pi) \setminus \{q_1, \ldots, q_n\}$.

In both cases this procedure constructs a monotone basis. To summarize we have:

Theorem 1.2.1. *The spaces $\mathcal{C}[0, 1]$, $\mathcal{C}_{\mathbb{C}}(\mathbb{T})$ both have a monotone basis. The exponential system $(1, e^{i\theta}, e^{-i\theta}, \ldots)$ fails to be a basis of $\mathcal{C}_{\mathbb{C}}(\mathbb{T})$.*

1.3 Equivalence of bases and basic sequences

If we select a basis in a finite-dimensional vector space then we are, in effect, selecting a system of coordinates. Bases in infinite-dimensional Banach spaces play the same role. Thus, if we have a basis $(e_n)_{n=1}^{\infty}$ of X then we can specify $x \in X$ by its coordinates $(e_n^*(x))_{n=1}^{\infty}$. Of course, it is not true that every scalar sequence $(a_n)_{n=1}^{\infty}$ defines an element of X. Thus X is coordinatized by a certain sequence space, i.e., a linear subspace of the vector space of all sequences. This leads us naturally to the following definition.

Definition 1.3.1. Two bases (or basic sequences) $(x_n)_{n=1}^{\infty}$ and $(y_n)_{n=1}^{\infty}$ in the respective Banach spaces X and Y are *equivalent*, and we write $(x_n)_{n=1}^{\infty} \sim (y_n)_{n=1}^{\infty}$, if whenever we take a sequence of scalars $(a_n)_{n=1}^{\infty}$, then $\sum_{n=1}^{\infty} a_n x_n$ converges if and only if $\sum_{n=1}^{\infty} a_n y_n$ converges.

Hence if the bases $(x_n)_{n=1}^{\infty}$ and $(y_n)_{n=1}^{\infty}$ are equivalent then the corresponding sequence spaces associated to X by $(x_n)_{n=1}^{\infty}$ and to Y by $(y_n)_{n=1}^{\infty}$ coincide. It is an easy consequence of the Closed Graph theorem that if $(x_n)_{n=1}^{\infty}$ and $(y_n)_{n=1}^{\infty}$ are equivalent then the spaces X and Y must be isomorphic. More precisely, we have:

Theorem 1.3.2. *Two bases (or basic sequences) $(x_n)_{n=1}^{\infty}$ and $(y_n)_{n=1}^{\infty}$ are equivalent if and only if there is an isomorphism $T : [x_n] \to [y_n]$ such that $Tx_n = y_n$ for each n .*

Proof. Let $X = [x_n]$ and $Y = [y_n]$. It is obvious that $(x_n)_{n=1}^{\infty}$ and $(y_n)_{n=1}^{\infty}$ are equivalent if there is an isomorphism T from X onto Y such that $Tx_n = y_n$ for each n.

Suppose conversely that $(x_n)_{n=1}^{\infty}$ and $(y_n)_{n=1}^{\infty}$ are equivalent. Let us define $T : X \to Y$ by $T(\sum_{n=1}^{\infty} a_n x_n) = \sum_{n=1}^{\infty} a_n y_n$. T is one-to-one and onto.

To prove that T is continuous we use the Closed Graph theorem. Suppose $(u_j)_{j=1}^{\infty}$ is a sequence such that $u_j \to u$ in X and $Tu_j \to v$ in Y. Let us write $u_j = \sum_{n=1}^{\infty} x_n^*(u_j)x_n$ and $u = \sum_{n=1}^{\infty} x_n^*(u)x_n$. It follows from the continuity of the biorthogonal functionals associated respectively with $(x_n)_{n=1}^{\infty}$ and $(y_n)_{n=1}^{\infty}$ that $x_n^*(u_j) \to x_n^*(u)$ and $y_n^*(Tu_j) = x_n^*(u_j) \to y_n^*(v)$ for all n. By the uniqueness of limit, $x_n^*(u) = y_n^*(v)$ for all n. Therefore $Tu = v$ and so T is continuous.

$\square$

Corollary 1.3.3. *Let $(x_n)_{n=1}^{\infty}$ and $(y_n)_{n=1}^{\infty}$ be two bases for the Banach spaces X and Y respectively. Then $(x_n)_{n=1}^{\infty} \sim (y_n)_{n=1}^{\infty}$ if and only if there exists a constant $C > 0$ such that for all finitely nonzero sequences of scalars $(a_i)_{i=1}^{\infty}$ we have*

$$C^{-1}\left\|\sum_{i=1}^{\infty} a_i y_i\right\| \le \left\|\sum_{i=1}^{\infty} a_i x_i\right\| \le C\left\|\sum_{i=1}^{\infty} a_i y_i\right\|. \tag{1.3}$$

If $C = 1$ in (1.3) then the basic sequences $(x_n)_{n=1}^{\infty}$ and $(y_n)_{n=1}^{\infty}$ are said to be *isometrically equivalent*.

Equivalence of basic sequences (and in particular of bases) will become a powerful technique for studying the isomorphic structure of Banach spaces.

Let us now introduce a special type of basic sequence:

Definition 1.3.4. Let $(e_n)_{n=1}^{\infty}$ be a basis for a Banach space X. Suppose that $(p_n)_{n=1}^{\infty}$ is a strictly increasing sequence of integers with $p_0 = 0$ and that $(a_n)_{n=1}^{\infty}$ are scalars. Then a sequence of nonzero vectors $(u_n)_{n=1}^{\infty}$ in X of the form

$$u_n = \sum_{j=p_{n-1}+1}^{p_n} a_j e_j$$

is called a *block basic sequence* of $(e_n)_{n=1}^{\infty}$.

Lemma 1.3.5. *Suppose $(e_n)_{n=1}^{\infty}$ is a basis for the Banach space X with basis constant K. Let $(u_k)_{k=1}^{\infty}$ be a block basic sequence of $(e_n)_{n=1}^{\infty}$. Then $(u_k)_{k=1}^{\infty}$ is a basic sequence with basis constant less than or equal to K.*

Proof. Suppose that $u_k = \sum_{j=p_{k-1}+1}^{p_k} a_j e_j$, $k \in \mathbb{N}$, is a block basic sequence of $(e_n)_{n=1}^{\infty}$. Then, for any scalars (b_k) and integers m, n with $m \le n$ we have

$$\left\|\sum_{k=1}^{m} b_k u_k\right\| = \left\|\sum_{k=1}^{m} b_k \sum_{j=p_{k-1}+1}^{p_k} a_j e_j\right\|$$

$$= \left\|\sum_{k=1}^{m} \sum_{j=p_{k-1}+1}^{p_k} b_k a_j e_j\right\|$$

$$= \left\|\sum_{j=1}^{p_m} c_j e_j\right\|, \text{ where } c_j = a_j b_k \text{ if } p_{k-1}+1 \le j \le p_k$$

$$\leq K \left\| \sum_{j=1}^{p_n} c_j e_j \right\|$$

$$= K \left\| \sum_{k=1}^{n} b_k u_k \right\|.$$

That is, (u_k) satisfies Grunblum's condition (Proposition 1.1.9), therefore (u_k) is a basic sequence with basis constant at most K.

□

Definition 1.3.6. A basic sequence $(x_n)_{n=1}^{\infty}$ in X is *complemented* if $[x_n]$ is a complemented subspace of X.

Remark 1.3.7. Suppose $(x_n)_{n=1}^{\infty}$ is a complemented basic sequence in a Banach space X. Let $Y = [x_n]$ and $P : X \to Y$ be a projection. If $(x_n^*)_{n=1}^{\infty} \subset Y^*$ are the biorthogonal functionals associated to $(x_n)_{n=1}^{\infty}$, using the Hahn-Banach theorem we can obtain a biorthogonal sequence $(\hat{x}_n^*)_{n=1}^{\infty} \subset X^*$ such that each $\hat{x}_n^*$ is an extension of x_n^* to X with preservation of norm. But since we have a projection, P, we can also extend each x_n^* to the whole of X by putting $u_n^* = x_n^* \circ P$. Then for $x \in X$, we will have

$$\sum_{n=1}^{\infty} u_n^*(x) x_n = P(x).$$

Conversely, if we can make a sequence $(u_n^*)_{n=1}^{\infty} \subset X^*$ such that $u_n^*(x_m) = \delta_{nm}$ and the series $\sum_{n=1}^{\infty} u_n^*(x) x_n$ converges for all $x \in X$, then the subspace $[x_n]$ is complemented by the projection $X \to [x_n]$, $x \mapsto \sum_{n=1}^{\infty} u_n^*(x) x_n$.

Definition 1.3.8. Let X and Y be Banach spaces. We say that two sequences $(x_n)_{n=1}^{\infty} \subset X$ and $(y_n)_{n=1}^{\infty} \subset Y$ are *congruent with respect to* (X, Y) if there is an invertible operator $T : X \to Y$ such that $T(x_n) = y_n$ for all $n \in \mathbb{N}$. When (x_n) and (y_n) satisfy this condition in the particular case that $X = Y$ we will simply say that they are *congruent*.

Let us suppose that the sequences $(x_n)_{n=1}^{\infty}$ in X and $(y_n)_{n=1}^{\infty}$ in Y are congruent with respect to (X, Y). The operator T of X onto Y that exists by the previous definition preserves any isomorphic property of $(x_n)_{n=1}^{\infty}$. For example if $(x_n)_{n=1}^{\infty}$ is a basis of X then $(y_n)_{n=1}^{\infty}$ is a basis of Y; if K is the basis constant of $(x_n)_{n=1}^{\infty}$ then the basis constant of $(y_n)_{n=1}^{\infty}$ is at most $K \|T\| \|T^{-1}\|$.

The following stability result dates back to 1940 [118]. It says, roughly speaking, that if $(x_n)_{n=1}^{\infty}$ is a basic sequence in a Banach space X and $(y_n)_{n=1}^{\infty}$ is another sequence in X so that $\|x_n - y_n\| \to 0$ fast enough then $(y_n)_{n=1}^{\infty}$ and $(x_n)_{n=1}^{\infty}$ are congruent.

Theorem 1.3.9 (Principle of small perturbations). *Let* $(x_n)_{n=1}^{\infty}$ *be a basic sequence in a Banach space X with basis constant K. If $(y_n)_{n=1}^{\infty}$ is a sequence in X such that*

$$2K \sum_{n=1}^{\infty} \frac{\|x_n - y_n\|}{\|x_n\|} = \theta < 1,$$

then $(x_n)_{n=1}^{\infty}$ and $(y_n)_{n=1}^{\infty}$ are congruent. In particular:

(i) If $(x_n)_{n=1}^{\infty}$ is a basis, so is $(y_n)_{n=1}^{\infty}$ (in which case the basis constant of $(y_n)_{n=1}^{\infty}$ is at most $K(1+\theta)(1-\theta)^{-1}$),
(ii) $(y_n)_{n=1}^{\infty}$ is a basic sequence (with basis constant at most $K(1+\theta)(1-\theta)^{-1}$),
(iii) If $[x_n]$ is complemented then $[y_n]$ is complemented.

Proof. For every $n \geq 2$ and any $x \in [x_n]$ we have

$$x_n^*(x)x_n = \sum_{k=1}^{n} x_k^*(x)x_k - \sum_{k=1}^{n-1} x_k^*(x)x_k,$$

where $(x_n^*) \subset [x_n]^*$ are the biorthogonal functionals of (x_n). Then $\|x_n^*(x)x_n\| \leq 2K\|x\|$ and so $\|x_n^*\|\|x_n\| \leq 2K$. For $n = 1$ it is clear that $\|x_1^*\|\|x_1\| \leq K$. These inequalities still hold if we replace x_n^* by its Hahn-Banach extension to X, $\hat{x}_n^*$.

For each $x \in X$ put

$$A(x) = x + \sum_{n=1}^{\infty} \hat{x}_n^*(x)(y_n - x_n).$$

A is a bounded operator from X to X with $A(x_n) = y_n$ and with norm

$$\|A\| \leq 1 + \sum_{n=1}^{\infty} \|\hat{x}_n^*\|\|y_n - x_n\|$$

$$\leq 1 + 2K \sum_{n=1}^{\infty} \frac{\|y_n - x_n\|}{\|x_n\|}$$

$$= 1 + \theta.$$

Moreover,

$$\|A - I\| \leq \sum_{n=1}^{\infty} \|\hat{x}_n^*\|\|y_n - x_n\| = \theta < 1,$$

which implies that A is invertible and $\|A^{-1}\| \leq (1 - \theta)^{-1}$.

□

As an application we obtain the following result known as the *Bessaga-Pełczyński Selection Principle*. It was first formulated in [12]. The technique used in its proof has come to be called the "gliding hump" (or "sliding hump") argument; the reader will see this type of argument in other contexts.

Proposition 1.3.10 (The Bessaga-Pełczyński Selection Principle). *Let* $(e_n)_{n=1}^{\infty}$ *be a basis for a Banach space* X *with basis constant* K *and dual functionals* $(e_n^*)_{n=1}^{\infty}$. *Suppose* $(x_n)_{n=1}^{\infty}$ *is a sequence in* X *such that*

(i) $\inf_n \|x_n\| > 0$, *but*
(ii) $\lim_{n\to\infty} e_k^*(x_n) = 0$ *for all* $k \in \mathbb{N}$.

Then $(x_n)_{n=1}^{\infty}$ *contains a subsequence* $(x_{n_k})_{k=1}^{\infty}$ *which is congruent to some block basic sequence* $(y_k)_{k=1}^{\infty}$ *of* $(e_n)_{n=1}^{\infty}$. *Furthermore, for every* $\epsilon > 0$ *it is possible to choose* $(n_k)_{k=1}^{\infty}$ *so that* $(x_{n_k})_{k=1}^{\infty}$ *has basis constant at most* $K + \epsilon$. *In particular the same result holds if* $(x_n)_{n=1}^{\infty}$ *converges to* 0 *weakly but not in the norm topology.*

Proof. Let $\alpha = \inf_n \|x_n\| > 0$ and let K be the basis constant of $(e_n)_{n=1}^{\infty}$. Suppose $0 < \nu < \frac{1}{4}$.

Pick $n_1 = 1$, $r_0 = 0$. There exists $r_1 \in \mathbb{N}$ such that

$$\|x_{n_1} - S_{r_1}x_{n_1}\| < \frac{\nu\alpha}{2K}.$$

Here, as usual, S_m denotes the mth-partial sum operator with respect to the basis $(e_n)_{n=1}^{\infty}$. We know that $\lim_{n\to\infty}\|S_{r_1}x_n\| = 0$, therefore there is $n_2 > n_1$ such that

$$\|S_{r_1}x_{n_2}\| < \frac{\nu^2\alpha}{2K}.$$

Pick $r_2 > r_1$ such that

$$\|x_{n_2} - S_{r_2}x_{n_2}\| < \frac{\nu^2\alpha}{2K}.$$

Again, since $\lim_{n\to\infty}\|S_{r_2}x_n\| = 0$, there exists $n_3 > n_2$ so that

$$\|S_{r_2}x_{n_3}\| < \frac{\nu^3\alpha}{2K}.$$

In this way, we get a sequence $(x_{n_k})_{k=1}^{\infty} \subset X$ and a sequence of integers $(r_k)_{k=0}^{\infty}$ with $r_0 = 0$, such that

$$\|S_{r_{k-1}}x_{n_k}\| < \frac{\nu^k\alpha}{2K}, \qquad \|x_{n_k} - S_{r_k}x_{n_k}\| < \frac{\nu^k\alpha}{2K}.$$

For each $k \in \mathbb{N}$, let $y_k = S_{r_k}x_{n_k} - S_{r_{k-1}}x_{n_k}$. (y_k) is a block basic sequence of the basis (e_n). Hence, by Lemma 1.3.5, (y_k) is a basic sequence with basis constant less than K.

Notice that for each k

$$\|y_k - x_{n_k}\| < \frac{\nu^k\alpha}{K},$$

hence,

$$\|y_k\| > \alpha - \frac{\nu\alpha}{K} \geq (1 - \nu)\alpha.$$

Then

$$2K \sum_{k=1}^{\infty} \frac{\|y_k - x_{n_k}\|}{\|y_k\|} < 2(1 - \nu)^{-1} \sum_{k=1}^{\infty} \nu^k = 2\nu(1 - \nu)^{-2} < \frac{8}{9}.$$

By Theorem 1.3.9, (x_{n_k}) is a basic sequence equivalent to (y_k). Since ν can be made arbitrarily small, we can arrange the basis constant for (x_{n_k}) to be as close to K as we wish. Moreover, if (y_k) is complemented in X so is (x_{n_k}).
□

1.4 Bases and basic sequences: discussion

The abstract concept of a Banach space grew very naturally from work in the early part of the twentieth century by Fredholm, Hilbert, F. Riesz, and others on concrete function spaces such as $\mathcal{C}[0, 1]$ and L_p for $1 \leq p < \infty$. The original motivation of these authors was to study linear differential and integral equations by using the methods of linear algebra with analysis. By the end of the First World War the definition of a Banach space was almost demanding to be made and it is therefore not surprising that it was independently discovered by Norbert Wiener and Stefan Banach around the same time. The axioms for a Banach space were introduced in Banach's thesis (1920), published in *Fundamenta Mathematicae* in 1922 in French.

The initial results of functional analysis are the underlying principles (Uniform Boundedness, Closed Graph and Open Mapping theorems and the Hahn-Banach theorem) which crystallized the common theme in so many arguments in analysis of the early twentieth century. However, after this, it was Banach and the school (Steinhaus, Mazur, Orlicz, Schauder, Ulam, etc.) in Lvov (then in Poland but now in the Ukraine) that developed the program of studying the isomorphic theory of Banach spaces. This school flourished until the time of the Second World War. In 1939, under the terms of the Nazi-Soviet pact, shortly after Germany invaded Poland, the Soviet Union occupied eastern Poland, including Lvov. After the Soviet invasion Banach was able to continue working, but the German invasion of 1941 effectively and tragically ended the work of his group. Banach himself suffered great hardship during the German occupation and died shortly after the end of the war, in 1945.

Given two classical Banach spaces X and Y one can ask questions such as whether X is isomorphic to Y, or whether X is isomorphic to a [complemented] subspace of Y. For these sort of questions, bases and basic sequences are an invaluable tool.

In 1932 Banach formulated in his book ([8], p. 111) the following:

The basis problem: *Does every separable Banach space have a basis?*

This problem motivated a great deal of research over the next forty years. Undoubtedly, the Lvov school knew much more about this problem than was ever published but, unfortunately, their research came to an untimely end with the German invasion of the Soviet Union in 1941. In particular, Mazur in the Scottish Book (an informal collection of problems kept in Lvov) formulated a very closely related problem which has come to be known as the *Approximation Problem*. Both problems were eventually solved by Per Enflo in 1973 [54], when he gave an example of a separable Banach space failing to have the Approximation Property and hence also failing to have a basis. This solution is beyond the scope of this book (see [138]), but we can at least present two facts that were known to Banach: Theorem 1.4.3 and Theorem 1.4.4. To that end, let us first record the following lemma, which will be required many times.

Lemma 1.4.1. *Let X be a Banach space.*

(i) If X is separable then the closed unit ball of X^*, B_{X^*}, is (compact and) metrizable for the weak* topology.

(ii) Suppose X^* contains a separating (or total) sequence $(x_n^*)_{n=1}^\infty$ for X; that is, $x_n^*(x) = 0$ for all $n \in \mathbb{N}$ implies that $x = 0$. Then any weakly compact subset of X is metrizable for the weak topology.

The conditions of (ii) hold when X is separable.

Proof. The proofs of both (i) and (ii) rely on the following simple observation. If K is a compact set for some topology τ, and τ' is any Hausdorff topology on K which is weaker than τ, then τ and τ' coincide. Indeed, suppose A is a τ-closed subset of K. Then A is τ-compact and so its continuous image in (K, τ') under the mapping $id_K : (K, \tau) \to (K, \tau')$ is also compact, i.e., A is τ'-compact. Since τ' is Hausdorff, A is τ'-closed.

For (i), let us take $(x_n)_{n=1}^\infty$ dense in the unit ball B_X of X. We define the topology ρ induced on X^* by convergence on each x_n. Precisely, a base of neighborhoods for ρ at a point $x_0^* \in X^*$ is given by sets of the form

$$V_\epsilon(x_0^*; x_1, \dots, x_N) = \left\{ x^* \in X^* : |x^*(x_n) - x_0^*(x_n)| < \epsilon, \; n = 1, \dots, N \right\},$$

where $\epsilon > 0$ and $N \in \mathbb{N}$. This topology is metrizable, and a metric inducing ρ may be defined by

$$d(x^*, y^*) = \sum_{n=1}^\infty 2^{-n} \min(1, |x^*(x_n) - y^*(x_n)|), \qquad x^*, y^* \in X^*.$$

ρ is Hausdorff and weaker than the weak* topology, so it coincides with the weak* topology on the weak* compact set B_{X^*}.

To prove (ii) we choose for ρ the topology on X induced by convergence in each x_n^*. The details are very similar; the point separation property is equivalent to ρ being Hausdorff.

Finally, if X is separable let $(x_n)_{n=1}^\infty$ be a sequence of nonzero vectors which is dense in X. For each n, using the Hahn-Banach theorem pick $x_n^* \in X^*$ so that $x_n^*(x_n) = \|x_n\|$ and $\|x_n^*\| = 1$. Suppose $x_n^*(x) = 0$ for all n. Then if $\epsilon > 0$ there exists $m \in \mathbb{N}$ so that $\|x - x_m\| < \epsilon$. Thus $\|x_m\| = x_m^*(x_m) < \epsilon$ and so $\|x\| < 2\epsilon$. Since $\epsilon > 0$ is arbitrary we have $x = 0$.

$\square$

Remark 1.4.2. (a) Note that if $X = \ell_\infty$ then the conditions of (ii) in the lemma hold (use the coordinate functionals) but X is not separable. Thus, every weakly compact subset of ℓ_∞ is metrizable.

(b) Let us observe as well that if X is separable then not only is the sequence $(x_n^*)_{n=1}^\infty$ in (ii) separating for X but it is also *norming* in X. That is, the norm of any $x \in X$ is completely determined by this numerable set of functionals:

$$\|x\| = \sup_n |x_n^*(x)|, \qquad x \in X.$$

The next theorem is in [8], p. 185. The proof uses the Cantor set and some of its topological properties.

By the *Cantor set*[1], Δ, we mean the topological space $\{0,1\}^\mathbb{N}$, the countable product of the two-point space $\{0,1\}$, endowed with the product topology[2].

Among the features of the Cantor set we single out the following:

• Δ *embeds homeomorphically as a closed subspace of* $[0,1]$.
The map

$$\Delta \to [0,1], \qquad (t_n) \mapsto \sum_{n=1}^\infty \frac{2t_n}{3^n}$$

does the job.

• $[0,1]$ *is the continuous image of* Δ.
Indeed, the function $\varphi : \Delta \to [0,1]$ defined by $\varphi((t_n)_{n=1}^\infty) = \sum_{n=1}^\infty t_n/2^n$ is continuous and surjective (but not one-to-one).

• Δ *is homeomorphic to the countable product of Cantor sets,* $\Delta^\mathbb{N}$.
This follows from the fact that if $(A_i, \tau_i)_{i \in \mathbb{N}}$ is a countable family of topological spaces each of which is homeomorphic to the countable product of two-point spaces, $\{0,1\}^\mathbb{N}$, then the topological product space $\prod_{i \in \mathbb{N}} A_i$ is homeomorphic to $\{0,1\}^\mathbb{N}$.

---

[1] On the other hand, the *Cantor middle third set*, $\mathcal{C}$, consists of all those real numbers x in $[0,1]$ so that when we write x in ternary form $x = \sum_{i=1}^\infty a_i/3^i$, then none of the numbers $a_1, a_2, \ldots$ equals 1 (i.e., either $a_i = 0$ or $a_i = 2$). Actually, the ternary correspondence from $\mathcal{C}$ onto Δ, $\sum_{i=1}^\infty a_i/3^i \mapsto (a_1/2, a_2/2, \ldots)$ is a homeomorphism.

[2] Sometimes, for convenience, we will equivalently realize the Cantor set as $\Delta = \{-1,1\}^\mathbb{N}$.

- $[0,1]^{\mathbb{N}}$ *is the continuous image of* Δ.

Since Δ is homeomorphic to $\Delta^{\mathbb{N}}$, a point in Δ can be assumed to be of the form $(x_1, x_2, \ldots)$, where $x_i \in \Delta$ for each i. If $\varphi : \Delta \to [0,1]$ is a continuous surjection, then $\psi : \Delta \to [0,1]^{\mathbb{N}}$ defined by $\psi(x_1, x_2, \ldots) = (\varphi(x_1), \varphi(x_2), \ldots)$ is continuous and surjective as well.

Theorem 1.4.3 (The Banach-Mazur Theorem). *If X is a separable Banach space then X embeds isometrically into $\mathcal{C}[0,1]$ (and hence embeds isometrically in a space with a monotone basis).*

Proof. The proof will be a direct consequence of the following two Facts:

Fact 1. *If X is a separable Banach space, then there exists a compact, Hausdorff, metrizable space K such that X embeds isometrically into $\mathcal{C}(K)$.*

Indeed, take $K = B_{X^*}$ with the relative weak* topology. If X is separable then B_{X^*} is compact and metrizable as we saw in Lemma 1.4.1. The isometric embedding of X into $\mathcal{C}(B_{X^*})$ is easily checked to be achieved by the mapping $x \to f_x$ where $f_x(x^*) = x^*(x)$ for all $x^* \in B_{X^*}$.

Fact 2. *If K is a compact metrizable space then $\mathcal{C}(K)$ embeds isometrically into $\mathcal{C}[0,1]$.*

We split the proof of this statement into some steps:

- *If K is a compact metrizable space, then K embeds homeomorphically into* $[0,1]^{\mathbb{N}}$. Being compact and metrizable, K contains a countable dense set, $(s_n)_{n=1}^{\infty}$. Let ρ be a metric on K inducing its topology. Without loss of generality we can assume that $0 \le \rho \le 1$. Now we define $\theta : K \to [0,1]^{\mathbb{N}}$ by $\theta(x) = (\rho(x, s_n))_{n=1}^{\infty}$.

θ is continuous since the mapping $x \mapsto \rho(x, s_n)$ is continuous for each n. θ is injective because if x and y are two different points in K then there exists some s_n such that $\rho(x, s_n) < \rho(y, s_n)$ (or the other way round) and, therefore, $\theta(x)$ and $\theta(y)$ will differ in the nth-coordinate.

Since K is compact and $[0,1]^{\mathbb{N}}$ is Hausdorff, it follows that θ maps K homeomorphically into its image.

- *If E is a closed subset of $[0,1]$, then $\mathcal{C}(E)$ embeds isometrically into $\mathcal{C}[0,1]$.* To show this, we need only define a norm-one extension operator $A : \mathcal{C}(E) \to \mathcal{C}[0,1]$, i.e., a norm-one linear map so that $Af|_E = f$ for all $f \in \mathcal{C}(E)$. Notice that $[0,1] \setminus E$ is a countable disjoint union of relatively open intervals; thus, we may extend f to be affine on each such interval interior to $[0,1]$ and to be constant on any such interval containing an endpoint of $[0,1]$. This procedure clearly gives a linear extension operator.

We are ready now to complete the proof of Fact 2 and, therefore, of the theorem. Let $\psi : \Delta \to [0,1]^{\mathbb{N}}$ be a continuous surjection and let us consider K as a closed subset of $[0,1]^{\mathbb{N}}$. It follows that if $E = \psi^{-1}(K)$, then E is homeomorphic to a (closed) subset of $[0,1]$. Then $\mathcal{C}(E)$ embeds isometrically into $\mathcal{C}[0, 1]$. Finally, $f \to f \circ \psi$ embeds $\mathcal{C}(K)$ isometrically into $\mathcal{C}(E)$ and, therefore, $\mathcal{C}(K)$ embeds isometrically into $\mathcal{C}[0, 1]$.

□

Theorem 1.4.4 was also known to Banach's school in their approach to tackle the *basis problem* and it is mentioned without proof by Banach in [8], p. 238. Several proofs have been given ever since; for example a proof due to Mazur is presented on p.4 of [138] and we shall revisit this theorem in the next section (Corollary 1.5.3). The proof we include here is due to Bessaga and Pełczyński [12].

Theorem 1.4.4. *Every separable, infinite-dimensional Banach space contains a basic sequence (i.e., a closed infinite-dimensional subspace with a basis). Furthermore if $\epsilon > 0$ we may find a basic sequence with basis constant at most $1 + \epsilon$.*

Proof. By the Banach-Mazur theorem (Theorem 1.4.3) we can consider the case when the separable Banach space X is a closed subspace of $\mathcal{C}[0, 1]$. Let $(e_n)_{n=1}^{\infty}$ be a monotone basis for $\mathcal{C}[0, 1]$ with biorthogonal functionals $(e_n^*)_{n=1}^{\infty}$. Since X is infinite-dimensional we may pick a sequence $(f_n)_{n=1}^{\infty}$ in X with $\|f_n\| = 1$ and $e_k^*(f_n) = 0$ for $1 \le k \le n$. By Proposition 1.3.10 we can find a subsequence $(f_{n_k})_{k=1}^{\infty}$ which is basic with constant at most $1 + \epsilon$.

□

1.5 Constructing basic sequences

The study of the isomorphic theory of Banach spaces went into retreat after the Second World War and was revived with the emergence of a new Polish school in Warsaw around 1958. There were some profound advances in Banach space theory between 1941 and 1958 (for example, the work of James and Grothendieck) but it seems that only after 1958 was there a concerted attack on problems of isomorphic structure. The prime mover in this direction was Pełczyński. Pełczyński, together with his collaborators, developed the theory of bases and basic sequences into a subtle and effective tool in Banach space theory. One nice aspect of the new theory was that basic sequences could be used to establish some classical results. In this section we are going to look deeper into the problem of constructing basic sequences and then show in the next section how this theory gives a nice and quite brief proof of the Eberlein-Šmulian theorem on weakly compact sets.

We will now present a refinement of the Mazur method for constructing basic sequences. We work in the dual X^* of a Banach space for purely technical reasons; ultimately we will apply Lemma 1.5.1 and Theorem 1.5.2 to X^{**}.

Lemma 1.5.1. *Suppose that S is a subset of X^* such that $0 \in \overline{S}^{weak^*}$ but $0 \notin \overline{S}^{\|\cdot\|}$. Let E be a finite-dimensional subspace of X^*. Then given $\epsilon > 0$ there exists $x^* \in S$ such that*

$$\|e^* + \lambda x^*\| \geq (1 - \epsilon) \|e^*\|$$

for all $e^* \in E$ and $\lambda \in \mathbb{R}$.

Proof. Let us notice that such a set S exists because the weak* topology and the norm topology of an infinite-dimensional Banach space do not coincide. $0 \notin \overline{S}^{\|\cdot\|}$ implies $\alpha \leq \|x^*\|$ for all $x^* \in S$, for some $0 < \alpha < \infty$. Given $\epsilon > 0$ put

$$\overline{\epsilon} = \frac{\alpha \epsilon}{2(1 + \alpha)}.$$

Let $U_E = \{e^* \in E : \|e^*\| = 1\}$. Since E is finite-dimensional U_E is norm-compact. Take $y_1^*, y_2^*, \ldots, y_N^* \in U_E$ such that whenever $e^* \in U_E$ then $\|e^* - y_k^*\| < \overline{\epsilon}$ for some $k = 1, \ldots, N$; for each $k = 1, \ldots, N$ pick $x_k \in B_X$ so that $y_k^*(x_k) > 1 - \overline{\epsilon}$.

Since $0 \in \overline{S}^{weak^*}$ each neighborhood of 0 in the weak* topology of X^* contains at least one point of S distinct from 0. In particular there is $x^* \in S$ such that $|x^*(x_k)| < \overline{\epsilon}$ for each $k = 1, \ldots, N$.

If $e^* \in U_E$ and $|\lambda| \geq \frac{2}{\alpha}$ we have

$$\|e^* + \lambda x^*\| \geq |\lambda|\alpha - 1 \geq 1.$$

If $|\lambda| < \frac{2}{\alpha}$ we pick y_k^* such that $\|e^* - y_k^*\| < \overline{\epsilon}$. Then

$$\begin{aligned}
\|y_k^* + \lambda x^*\| &\geq y_k^*(x_k) + \lambda x^*(x_k) \\
&> (1 - \overline{\epsilon}) + \lambda x^*(x_k) \\
&\geq (1 - \overline{\epsilon}) - |\lambda|\overline{\epsilon} \\
&\geq \left(1 - (1 + \frac{2}{\alpha})\overline{\epsilon}\right)
\end{aligned}$$

and, therefore,

$$\begin{aligned}
\|e^* + \lambda x^*\| &\geq \left| \|e^* - y_k^*\| - \|y_k^* + \lambda x^*\| \right| \\
&\geq 1 - \left(1 + \frac{2}{\alpha}\right)\overline{\epsilon} - \overline{\epsilon} \\
&= 1 - \epsilon.
\end{aligned}$$

$\square$

Theorem 1.5.2. *Suppose that S is a subset of X^* such that $0 \in \overline{S}^{weak^*}$ but $0 \notin \overline{S}^{\|\cdot\|}$. Then for any $\epsilon > 0$, S contains a basic sequence with basis constant less than $1 + \epsilon$.*

Proof. Fix a decreasing sequence of positive numbers $(\epsilon_n)_{n=1}^{\infty}$ such that $\sum_{n=1}^{\infty} \epsilon_n < \infty$ and so that $\prod_{n=1}^{\infty}(1 - \epsilon_n) > (1 + \epsilon)^{-1}$.

Pick $x_1^* \in S$ and consider the 1-dimensional space $E_1 = [x_1^*]$. By Lemma 1.5.1 there is $x_2^* \in S$ such that

$$\|e^* + \lambda x_2^*\| \geq (1 - \epsilon_1)\|e^*\|$$

for all $e^* \in E_1$ and $\lambda \in \mathbb{R}$.

Now let E_2 be the 2-dimensional space generated by x_1^*, x_2^*, $E_2 = [x_1^*, x_2^*]$. Lemma 1.5.1 yields $x_3^* \in S$ such that

$$\|e^* + \lambda x_3^*\| \geq (1 - \epsilon_2)\|e^*\|$$

for all $e^* \in E_2$ and $\lambda \in \mathbb{R}$.

Repeating this process we produce a sequence $(x_n^*)_{n=1}^\infty$ in S such that for each $n \in \mathbb{N}$ and any scalars (a_k),

$$\left\| \sum_{k=1}^{n+1} a_k x_k^* \right\| \geq (1 - \epsilon_n) \left\| \sum_{k=1}^{n} a_k x_k^* \right\|.$$

Therefore given any integers m, n with $m \leq n$ we have

$$\left\| \sum_{k=1}^{m} a_k x_k^* \right\| \leq \frac{1}{\displaystyle\prod_{j=1}^{n-1}(1 - \epsilon_j)} \left\| \sum_{k=1}^{n} a_k x_k^* \right\|.$$

Applying the Grunblum condition (Proposition 1.1.9) we conclude that $(x_n^*)_{n=1}^\infty$ is a basic sequence with basis constant at most $1 + \epsilon$.

$\square$

Corollary 1.5.3. *Every infinite-dimensional Banach space contains, for $\epsilon > 0$, a basic sequence with basis constant less than $1 + \epsilon$.*

Proof. Let X be an infinite-dimensional Banach space. Consider $S = \partial B_X = \{x \in X : \|x\| = 1\}$. We claim that 0 belongs to the weak closure of S, therefore it belongs to the weak* closure of S as a subspace of X^{**}.

If our claim fails then there exist some $\epsilon > 0$ and linear functionals $x_1^*, \ldots, x_n^*$ in X^* such that the weak neighborhood of 0

$$V = \{x \in X : |x_k^*(x)| < \epsilon, \text{ for } k = 1, \ldots, n\}$$

satisfies $V \cap S = \emptyset$. This is impossible because the intersection of the null subspaces of the x_k^*'s is a nontrivial subspace of X contained in V with points in S.

Now Theorem 1.5.2 yields the existence of a basic sequence (x_n) in S with basis constant as close to 1 as we wish.

$\square$

The following proposition is often stated as a special case of Theorem 1.5.2. It may also be deduced equally easily using Theorem 1.4.4.

Proposition 1.5.4. *If $(x_n)_{n=1}^{\infty}$ is a weakly null sequence in an infinite-dimensional Banach space X such that $\inf_n \|x_n\| > 0$ then, for $\epsilon > 0$, $(x_n)_{n=1}^{\infty}$ contains a basic subsequence with basis constant less than $1 + \epsilon$.*

Proof. Consider $S = \{x_n : n \in \mathbb{N}\}$. Since $(x_n)_{n=1}^{\infty}$ is weakly convergent, the set S is norm bounded. Furthermore $0 \in \overline{S}^{\text{weak}}$ hence, by Theorem 1.5.2, S contains a basic sequence with basis constant at most $1+\epsilon$. To finish the proof we just have to prune this basic sequence by extracting terms in increasing order and we obtain a basic subsequence of $(x_n)_{n=1}^{\infty}$.

□

The next technical lemma will be required for our main result on basic sequences.

Lemma 1.5.5. *Let $(x_n)_{n=1}^{\infty}$ be a basic sequence in X. Suppose that there exists a linear functional $x^* \in X^*$ such that $x^*(x_n) = 1$ for all $n \in \mathbb{N}$. If $u \notin [x_n]$ then the sequence $(x_n + u)_{n=1}^{\infty}$ is basic.*

Proof. Since $u \notin [x_n]$, without loss of generality we can assume $x^*(u) = 0$. Let $T : X \to X$ be the operator given by $T(x) = x^*(x)u$. Then $I_X + T$ is invertible with inverse $I_X - T$. Since $(I_X + T)(x_n) = x_n + u$, the sequences $(x_n)_{n=1}^{\infty}$ and $(x_n + u)_{n=1}^{\infty}$ are congruent, hence $(x_n + u)_{n=1}^{\infty}$ is basic.

□

We are now ready to give a criterion for a subset of a Banach space to contain a basic sequence. This criterion is due to Kadets and Pełczyński (1965) [99].

Theorem 1.5.6. *Let S be a bounded subset of a Banach space X such that $0 \notin \overline{S}^{\|\cdot\|}$. Then the following are equivalent:*

(i) S fails to contain a basic sequence,

(ii) $\overline{S}^{\text{weak}}$ is weakly compact and fails to contain 0.

Proof. $(ii) \Rightarrow (i)$. Suppose $(x_n)_{n=1}^{\infty} \subset S$ is a basic sequence. Since $\overline{S}^{\text{weak}}$ is weakly compact, $(x_n)_{n=1}^{\infty}$ has a weak cluster point, x, in $\overline{S}^{\text{weak}}$. By Mazur's theorem, x belongs to $[x_n]$, so we can write $x = \sum_{n=1}^{\infty} x_n^*(x)x_n$.

By the continuity of the coefficient functionals $(x_n^*)_{n=1}^{\infty}$, it follows that for each n, $x_n^*(x)$ is a cluster point of the scalar sequence $(x_n^*(x_m))_{m=1}^{\infty}$, which converges to 0. Therefore, $x_n^*(x) = 0$ for all n and, as a consequence, $x = 0$. This contradicts the hypothesis, so S contains no basic sequences.

For the forward implication, $(i) \Rightarrow (ii)$, assume S contains no basic sequences. We can apply Theorem 1.5.2 to S considered as a subset of X^{**} with the weak* topology and we conclude that 0 cannot be a weak closure point of S. It remains to show that S is relatively weakly compact. To achieve this we simply need to show that any weak* cluster point of S in X^{**} is already contained in X. Let us suppose x^{**} is a weak* cluster point of S and that $x^{**} \in X^{**} \setminus X$. Consider the set $S - x^{**} = \{s - x^{**} : s \in S\}$ in X^{**}. By

Theorem 1.5.2 there exists $(x_n)_{n=1}^{\infty}$ in S such that the sequence $(x_n - x^{**})_{n=1}^{\infty}$ is basic. We can suppose that $x^{**} \notin [x_n - x^{**} : n \geq 1]$ because it is certainly true that $x^{**} \notin [x_n - x^{**} : n \geq N]$ for some choice of N. By the Hahn-Banach theorem there exists $x^{***} \in X^{***}$ so that $x^{***} \in X^{\perp}$ and $x^{***}(x^{**}) = -1$. This implies that $x^{***}(x_n - x^{**}) = 1$ for all $n \in \mathbb{N}$. Now Lemma 1.5.5 applies and we deduce that $(x_n)_{n=1}^{\infty}$ is also basic, contrary to our assumption on S.

$\square$

1.6 The Eberlein-Šmulian Theorem

Let M be a topological space and A be a subset of M. Let us recall that A is said to be *sequentially compact* [respectively, *relatively sequentially compact*] if every sequence in A has a subsequence convergent to a point in A [respectively, to a point in M] and that A is *countably compact* [respectively, *relatively countably compact*] if every sequence in A has a cluster point in A [respectively, in M].

Countable compactness is implied by both compactness and sequential compactness. If M is a metrizable topological space these three concepts certainly coincide but if M is instead a general topological space these equivalences are no longer valid. The easiest counterexample is obtained by considering $B_{\ell_{\infty}^*}$, the unit ball in ℓ_{∞}^* with the weak* topology. $B_{\ell_{\infty}^*}$ is, of course, weak* compact but fails to be weak* sequentially compact: the sequence of functionals (e_n^*) given by $e_n^*(\xi) = \xi(n)$ has no weak* convergent subsequence.

In this section we will prove the Eberlein-Šmulian theorem, which asserts that in a Banach space the weak topology behaves like a metrizable topology in this respect although it need not be metrizable even on compact sets (except in the case of separable Banach space, see Lemma 1.4.1). That weak compactness implies weak sequential compactness was discovered by Šmulian in 1940 [207]; the more difficult converse direction is due to Eberlein (1947) [51]. This result is rather hard and the original proof did not use the concept of a basic sequence, as the result predates the development of basic sequence techniques. The proof via basic sequences is due to Pełczyński [172]. Basic sequences seem to provide a conceptual simplification of the idea of the proof.

The lemmas we will need are the following:

Lemma 1.6.1. *If $(x_n)_{n=1}^{\infty}$ is a basic sequence in a Banach space and x is a weak cluster point of $(x_n)_{n=1}^{\infty}$ then $x = 0$.*

Proof. Since x is in the weak closure of the convex set $\langle x_n : n \in \mathbb{N} \rangle$ (the linear span of the sequence (x_n)), Mazur's theorem yields that x belongs to the norm-closed linear span, $[x_n]$, of (x_n). Hence $x = \sum_{n=1}^{\infty} x_n^*(x)x_n$, where (x_n^*) are the biorthogonal functionals of (x_n). Now, for each n, $x_n^*(x)$ is a cluster point of $(x_n^*(x_m))_{m=1}^{\infty}$ and is, therefore, forced to be zero. Thus $x = 0$.

$\square$

Lemma 1.6.2. *Let A be a relatively weakly countably compact subset of a Banach space X. Suppose that $x \in X$ is the only weak cluster point of the sequence $(x_n)_{n=1}^{\infty} \subset A$. Then $(x_n)_{n=1}^{\infty}$ converges weakly to x.*

Proof. Assume that (x_n) does not converge weakly to x. Then for some $x^* \in X^*$ the sequence $(x^*(x_n))_{n=1}^{\infty}$ fails to converge to $x^*(x)$, hence we may pick a subsequence $(x_{n_k})_{k=1}^{\infty}$ of (x_n) such that $\inf_k |x^*(x) - x^*(x_{n_k})| > 0$. But this prevents x from being a weak cluster point of (x_{n_k}), contradicting the hypothesis.

$\square$

Theorem 1.6.3 (The Eberlein-Šmulian Theorem). *Let A be a subset of a Banach space X. The following are equivalent:*

(i) A is [relatively] weakly compact,
(ii) A is [relatively] weakly sequentially compact,
(iii) A is [relatively] weakly countably compact.

Proof. Since (i) and (ii) both imply (iii) we need only show that (iii) implies both (ii) and (i). We will prove the relativized versions; minor modifications can be made to prove the nonrelativized versions. Note that each of the statements of the theorem implies that A is bounded.

Let us first do the case (iii) implies (ii). Suppose $(x_n)_{n=1}^{\infty}$ is any sequence in A. Then, by hypothesis, there is a weak cluster point x of $(x_n)_{n=1}^{\infty}$. If x is a point in the norm-closure of the set $\{x_n\}_{n=1}^{\infty}$, then there is a subsequence which converges in norm and we are done. If not, using Theorem 1.5.6, we can extract a subsequence $(y_n)_{n=1}^{\infty}$ of (x_n) so that $(y_n - x)_{n=1}^{\infty}$ is a basic sequence. But $(y_n)_{n=1}^{\infty}$ has a weak cluster point, y, hence $y - x$ is a weak cluster point of the basic sequence $(y_n - x)_{n=1}^{\infty}$. By Lemma 1.6.1 we have $y = x$. Thus x is the *only* weak cluster point of $(y_n)_{n=1}^{\infty}$. Then $(y_n)_{n=1}^{\infty}$ converges weakly to x by Lemma 1.6.2.

Let us turn to the case (iii) implies (i). Suppose A fails to be relatively weakly compact. Since the weak* closure W of A in X^{**} is necessarily weak* compact by Banach-Alaoglu's theorem, we conclude that this set cannot be contained in X. Thus there exists $x^{**} \in W \backslash X$. Pick $x^* \in X^*$ so that $x^{**}(x^*) > 1$. Then consider the set $A_0 = \{x \in A : x^*(x) > 1\}$. The set A_0 is not relatively weakly compact since x^{**} is in its weak* closure. Theorem 1.5.6 gives us a basic sequence $(x_n)_{n=1}^{\infty}$ contained in A_0. Appealing to countable compactness, $(x_n)_{n=1}^{\infty}$ has a weak cluster point, x, which by Lemma 1.6.1 must be $x = 0$. This is a contradiction since, by construction, $x^*(x) \geq 1$.

$\square$

Combining Theorem 1.6.3 with Proposition G.2 we obtain:

Corollary 1.6.4. *A Banach space X is reflexive if and only if every bounded sequence has a weakly convergent subsequence.*

The Eberlein-Šmulian theorem was probably the deepest result of earlier (pre-1950) Banach space theory. Not surprisingly it inspired more examination and it is far from the end of the story. In [74] the Eberlein-Šmulian theorem is extended to bounded subsets of $\mathcal{C}(K)$ (K a compact Hausdorff space) with the weak topology replaced by the topology of pointwise convergence. This does not follow from basic sequence techniques because it is no longer true that a cluster point of a basic sequence for pointwise convergence is necessarily zero. Later, Bourgain, Fremlin, and Talagrand [16] proved similar results for subsets of the Baire class one functions on a compact metric space. A function is of *Baire class one* if it is a pointwise limit of a sequence of continuous functions.

Problems

1.1. Mazur's Weak Basis Theorem.
A sequence $(e_n)_{n=1}^{\infty}$ is called a *weak basis* of a Banach space X if for each $x \in X$ there is a unique sequence of scalars $(a_n)_{n=1}^{\infty}$ such that $x = \sum_{n=1}^{\infty} a_n x_n$ in the weak topology. Show that every weak basis is a basis. [*Hint*: Try to imitate Theorem 1.1.3.]

1.2. Krein-Milman-Rutman Theorem.
Let X be a Banach space with a basis and D be a dense subset of X. Show that D contains a basis for X.

1.3. Let (e_n) be a normalized basis for a Banach space X and suppose there exists $x^* \in X^*$ with $x^*(e_n) = 1$ for all n. Show that the sequence $(e_n - e_{n-1})_{n=1}^{\infty}$ is also a basis for X (we let $e_0 = 0$ in this definition).

1.4. The Bounded Approximation Property.
A separable Banach space X has the *bounded approximation property* (BAP) if there is a sequence (T_n) of finite-rank operators so that

$$\lim_{n \to \infty} \|x - T_n x\| = 0, \qquad x \in X. \tag{1.4}$$

(a) Show (1.4) implies $\sup_n \|T_n\| < \infty$ and, hence, (BAP) implies the approximation property.

(b) Show that every complemented subspace of a space with a basis has (BAP).

1.5. Let X be a Banach space and $A : X \to X$ a finite-rank operator. Show that for $\epsilon > 0$ there is a finite sequence of rank-one operators $(B_n)_{n=1}^{N}$ so that $A = B_1 + \cdots + B_N$ and

$$\sup_{1 \le n \le N} \left\| \sum_{k=1}^{n} B_k \right\| < \|A\| + \epsilon.$$

1.6. Show that if X has (BAP) then there is a sequence of rank-one operators $(B_n)_{n=1}^\infty$ so that $x = \sum_{n=1}^\infty B_n x$ for each $x \in X$. [*Hint*: Apply Problem 1.5 to $A = T_1$ and $A = T_n - T_{n-1}$ for $n = 2, 3, \ldots$.]

1.7. If X has (BAP) let $(B_n)_{n=1}^\infty$ be the sequence of rank-one operators given in Problem 1.6. Let $B_n x = x_n^*(x) x_n$ where $x_n^* \in X^*$ and $x_n \in X$. Define Y to be the space of all sequences $\xi = (\xi(n))_{n=1}^\infty$ so that $\sum_{n=1}^\infty \xi(n) x_n$ converges under the norm

$$\|\xi\|_Y = \sup_n \left\| \sum_{k=1}^n \xi(k) x_k \right\|.$$

(a) Show that $(Y, \|\cdot\|_Y)$ is a Banach space and that the canonical basis vectors $(e_n)_{n=1}^\infty$ form a basis of Y.

(b) Show further that X is isomorphic to a complemented subspace of Y.

Thus X has (BAP) if and only if it is isomorphic to a complemented subspace of a space with a basis. This is due independently to Johnson, Rosenthal, and Zippin [94] and Pełczyński [175]. In 1987 Szarek [212] gave an example to show that not every space with (BAP) has a basis; this is very difficult! We refer to [24] for a full discussion of the problems associated with the bounded approximation property. See also Chapter 13 for the construction of Pełczyński's universal basis space U.

1.8. Suppose X is a separable Banach space with the property that there is a sequence of finite-rank operators (T_n) such that $\lim_{n\to\infty}\langle T_n x, x^*\rangle = \langle x, x^*\rangle$ for all $x \in X$, $x^* \in X^*$. Show that X has the (BAP).

1.9. Suppose that X is a Banach space and that $(T_n)_{n=1}^\infty$ is a sequence of finite-rank operators such that $\lim_{n\to\infty}\langle T_n^* x^*, x^{**}\rangle = \langle x^*, x^{**}\rangle$ for every $x^* \in X^*$, $x^* \in X^*$.

(a) Show that $(T_n)_{n=1}^\infty$ is a weakly Cauchy sequence in the space $\mathcal{K}(X)$ of compact operators on X and that $(T_n)_{n=1}^\infty$ converges weak* to an element $\chi \in \mathcal{K}(X)^{**}$ where $\|\chi\| = 1$. [*Hint:* Consider B_{X^*} and $B_{X^{**}}$ with their respective weak* topologies. Embed $\mathcal{K}(X)$ into $\mathcal{C}(B_{X^*} \times B_{X^{**}})$ via the embedding $T \to f_T$ where $f_T(x^*, x^{**}) = \langle T^* x^*, x^{**}\rangle$.]

(b) Using Goldstine's theorem deduce the existence of a sequence of finite-rank operators $(S_n)_{n=1}^\infty$ so that $\lim_{n\to\infty} \|S_n\| = 1$ and $\lim_{n\to\infty} \|S_n x - x\| = 0$ for $x \in X$. [*Hint:* Choose each S_n as a convex combination of $\{T_n, T_{n+1}, \ldots\}$.]

Thus if X is reflexive and has (BAP) we can choose the operators T_n to have $\|T_n\| \leq 1$; thus X has the *metric approximation property* (MAP).

1.10. Consider $\mathbb{T}$ with the normalized measure $\frac{d\theta}{2\pi}$.

(a) Show that the exponentials $(e_0, e_1, e_{-1}, \ldots)$ (see Section 1.2) do not form a basis of the complex space $L_1(\mathbb{T})$. [*Hint*: Prove that the partial sum operators $S_n f = \sum_{k=-n}^n \hat{f}(k) e_k$ are not uniformly bounded.]

(b) Show that if $1 < p < \infty$, $(e_0, e_1, e_{-1}, \dots)$ form a basis of $L_p(\mathbb{T})$. (You may assume that the *Riesz projection* is bounded on $L_p(\mathbb{T})$, i.e., there is a bounded linear operator $R : L_p \to L_p$ such that $Re_k = 0$ when $k \leq 0$ and $Re_k = e_k$ for $k \geq 0$. This is equivalent to the boundedness of the Hilbert transform; see for example Theorem 1.8, p. 68, of [108].)

2

The Classical Sequence Spaces

We now turn to the classical sequence spaces ℓ_p for $1 \leq p < \infty$ and c_0. The techniques developed in the previous chapter will prove very useful in this context. These Banach spaces are, in a sense, the simplest of all Banach spaces and their structure has been well understood for many years. However, if $p \neq 2$, there can still be surprises and there remain intriguing open questions.

To avoid some complicated notation we will write a typical element of ℓ_p or c_0 as $\xi = (\xi(n))_{n=1}^{\infty}$. Let us note at once that the spaces ℓ_p and c_0 are equipped with a canonical monotone Schauder basis $(e_n)_{n=1}^{\infty}$ given by $e_n(k) = 1$ if $k = n$ and 0 otherwise. It is useful, and now fairly standard, to use c_{00} to denote the subspace of all sequences of scalars $\xi = (\xi(n))_{n=1}^{\infty}$ such that $\xi(n) = 0$ except for finitely many n.

One feature of the canonical basis of the ℓ_p-spaces and c_0 that is useful to know is that $(e_n)_{n=1}^{\infty}$ is equivalent to the basis $(a_n e_n)_{n=1}^{\infty}$ whenever $0 < \inf_n |a_n| \leq \sup_n |a_n| < \infty$. This property is equivalent to the *unconditionality* of the basis, but we will not formally introduce this concept until the next chapter.

2.1 The isomorphic structure of the ℓ_p-spaces and c_0

We first ask ourselves a very simple question: are the ℓ_p-spaces distinct (i.e., mutually nonisomorphic) Banach spaces? This question may seem absurd because they look different, but recall that $L_2[0,1]$ and ℓ_2 are actually the same space in two different disguises. We can observe, for instance, that c_0 and ℓ_1 are nonreflexive while the spaces ℓ_p for $1 < p < \infty$ are reflexive; further the dual of c_0 (i.e., ℓ_1) is separable but the dual of ℓ_1 (i.e., ℓ_∞) is nonseparable.

To help answer our question we need the following lemma:

Lemma 2.1.1. *Let $(u_n)_{n=1}^{\infty}$ be a normalized block basic sequence in c_0 or in ℓ_p for some $1 \leq p < \infty$. Then $(u_n)_{n=1}^{\infty}$ is isometrically equivalent to the canonical basis of the space and $[u_n]$ is the range of a contractive projection.*

Proof. Let us treat the case when (u_n) is a block basic sequence in ℓ_p for $1 \leq p < \infty$ and leave the modifications for the c_0 case to the reader. Let us suppose that

$$u_k = \sum_{j=r_{k-1}+1}^{r_k} a_j e_j, \qquad k \in \mathbb{N},$$

where $0 = r_0 < r_1 < r_2 < \ldots$ are positive integers and $(a_j)_{j=1}^{\infty}$ are scalars such that

$$\|u_k\|^p = \sum_{j=r_{k-1}+1}^{r_k} |a_j|^p = 1, \qquad k \in \mathbb{N}.$$

Then, given any $m \in \mathbb{N}$ and any scalars $b_1, \ldots, b_m$ we have

$$\left\| \sum_{k=1}^{m} b_k u_k \right\| = \left\| \sum_{k=1}^{m} \sum_{j=r_{k-1}+1}^{r_k} b_k a_j e_j \right\|$$

$$= \left(\sum_{k=1}^{m} |b_k|^p \sum_{j=r_{k-1}+1}^{r_k} |a_j|^p \right)^{1/p}$$

$$= \left(\sum_{k=1}^{m} |b_k|^p \right)^{1/p}.$$

This establishes isometric equivalence.

We shall construct a contractive projection onto $[u_n]_{n=1}^{\infty}$. Here we suppose $1 < p < \infty$ and leave both cases c_0 and ℓ_1 to the reader. For each k we select scalars $(b_j)_{j=r_{k-1}+1}^{r_k}$ so that

$$\sum_{j=r_{k-1}+1}^{r_k} |b_j|^q = 1$$

and

$$\sum_{j=r_{k-1}+1}^{r_k} b_j a_j = 1.$$

Put

$$u_k^* = \sum_{j=r_{k-1}+1}^{r_k} b_j e_j^*.$$

Clearly, $(u_n^*)_{n=1}^{\infty}$ is biorthogonal to $(u_n)_{n=1}^{\infty}$ and $\|u_n^*\| = \|u_n\| = 1$. Our aim is to see that the operator

$$P(\xi) = \sum_{k=1}^{\infty} u_k^*(\xi) u_k, \qquad \xi \in \ell_p,$$

defines a norm-one projection from ℓ_p onto $[u_k]$. We will show that $\|P\xi\| \leq \|\xi\|$ when $\xi \in c_{00}$ and then observe that P extends by density to a contractive projection.

For each $\xi \in c_{00}$,

$$|u_k^*(\xi)| = \left| \sum_{j=r_{k-1}+1}^{r_k} b_j \xi(j) \right|$$

$$\leq \left(\sum_{j=r_{k-1}+1}^{r_k} |b_j|^q \right)^{\frac{1}{q}} \left(\sum_{j=r_{k-1}+1}^{r_k} |\xi(j)|^p \right)^{\frac{1}{p}}$$

$$= \left(\sum_{j=r_{k-1}+1}^{r_k} |\xi(j)|^p \right)^{\frac{1}{p}}.$$

Then, using the isometric equivalence of $(u_n)_{n=1}^\infty$ and $(e_n)_{n=1}^\infty$, we have

$$\|P(\xi)\| = \left(\sum_{k=1}^\infty |u_k^*(\xi)|^p \right)^{\frac{1}{p}}$$

$$\leq \left(\sum_{k=1}^\infty \sum_{j=r_{k-1}+1}^{r_k} |\xi(j)|^p \right)^{\frac{1}{p}}$$

$$= \|\xi\|.$$

$\square$

Remark 2.1.2. Notice that if (u_n) is not normalized but satisfies instead an inequality

$$0 < a \leq \|u_n\| \leq b < \infty, \quad n \in \mathbb{N},$$

for some constants a, b (in which case (u_n) is said to be *seminormalized*), then we can apply the previous lemma to $(u_n/\|u_n\|)$ and we obtain that $(u_n)_{n=1}^\infty$ is equivalent to $(e_n)_{n=1}^\infty$ (but not isometrically) and $[u_n]$ is complemented by a contractive projection.

Although the preceding lemma was quite simple it already leads to a powerful conclusion:

Proposition 2.1.3. *Let $(x_n)_{n=1}^\infty$ be a normalized sequence in ℓ_p for $1 \leq p < \infty$ [respectively, c_0] such that for each $j \in \mathbb{N}$ we have $\lim_{n\to\infty} x_n(j) = 0$ (for example suppose $(x_n)_{n=1}^\infty$ is weakly null). Then there is a subsequence $(x_{n_k})_{k=1}^\infty$ which is a basic sequence equivalent to the canonical basis of ℓ_p and such that $[x_{n_k}]_{k=1}^\infty$ is complemented in ℓ_p [respectively, c_0].*

Proof. Proposition 1.3.10 (using the "gliding hump" technique) yields a subsequence $(x_{n_k})_{k=1}^\infty$ and a block basic sequence $(u_k)_{k=1}^\infty$ of $(e_n)_{n=1}^\infty$ such that $(x_{n_k})_{k=1}^\infty$ is basic, equivalent to $(u_k)_{k=1}^\infty$ and such that $[x_{n_k}]_{k=1}^\infty$ is complemented whenever $[u_k]_{k=1}^\infty$ is. By Lemma 2.1.1 we are done.

$\square$

Now let us prove a classical result from the 1930s (Pitt [189]).

Theorem 2.1.4 (Pitt's Theorem). *Suppose $1 \leq p < r < \infty$. If X is a closed subspace of ℓ_r and $T : X \to \ell_p$ is a bounded operator then T is compact.*

Proof. ℓ_r is reflexive, hence X is reflexive and so B_X is weakly compact. Therefore in order to prove that T is compact it suffices to show that $T|_{B_X}$ is weak-to-norm continuous. Since the weak topology of X restricted to B_X is metrizable (Lemma 1.4.1 (ii)) it suffices to see that whenever $(x_n)_{n=1}^{\infty} \subset B_X$ is weakly convergent to some x in B_X then $(T(x_n))_{n=1}^{\infty}$ converges in norm to Tx.

We need only show that if $(x_n)_{n=1}^{\infty}$ is a weakly null sequence in X then $\lim_{n\to\infty} \|Tx_n\| = 0$. If this fails, we may suppose the existence of a weakly null sequence $(x_n)_{n=1}^{\infty}$ with $\|x_n\| = 1$ such that $\|Tx_n\| \geq \delta > 0$ for all n. By passing to a subsequence we may suppose that $(x_n)_{n=1}^{\infty}$ is a basic sequence equivalent to the canonical ℓ_r-basis (Proposition 2.1.3). But then, since $(Tx_n)_{n=1}^{\infty}$ is also weakly null, by passing to a further subsequence we may suppose that $(Tx_n/\|Tx_n\|)_{n=1}^{\infty}$, and hence $(Tx_n)_{n=1}^{\infty}$, is basic and equivalent to the canonical ℓ_p-basis. Since T is bounded we have effectively shown that the identity map $\iota : \ell_r \to \ell_p$ is bounded, which is absurd. Or, alternatively, there exist constants C_1 and C_2 such that the following inequalities hold simultaneously for all n:

$$\left\|\sum_{k=1}^{n} x_k\right\|_r \leq C_1 n^{\frac{1}{r}} \text{ and } \left\|\sum_{k=1}^{n} Tx_k\right\|_p \geq C_2 n^{\frac{1}{p}},$$

which contradicts the boundedness of T. Thus the theorem is proved.   □

Remark 2.1.5. (a) Essentially the same proof works with c_0 replacing ℓ_r; although c_0 is nonreflexive, Lemma 1.4.1 can still be used to show that B_X is at least weakly metrizable, and the weak-to-norm continuity of $T|_{B_X}$ is enough to show that the image is relatively norm-compact.

(b) We would like to single out the following crucial ingredient in the proof of Pitt's theorem. *Suppose $T : \ell_r \to \ell_p$ is a bounded operator with $1 \leq p < r < \infty$. Then whenever (x_n) is a weakly null sequence in ℓ_r we have $\|Tx_n\|_p \to 0$. In particular $\|Te_n\|_p \to 0$. The same is true for any operator $T : c_0 \to \ell_p$.*

Corollary 2.1.6. *The spaces of the set $\{c_0\} \cup \{\ell_p : 1 \leq p < \infty\}$ are mutually nonisomorphic. In fact, if X is an infinite-dimensional subspace of one of the spaces $\{c_0\} \cup \{\ell_p : 1 \leq p < \infty\}$, then it is not isomorphic to a subspace of any other.*

This suggests the following definition:

Definition 2.1.7. Two infinite-dimensional Banach spaces X, Y are said to be *totally incomparable* if they have no infinite-dimensional subspaces in common (up to isomorphism).

What can be said for bounded operators $T : \ell_p \longrightarrow \ell_r$ for $p < r$? First, notice that in this case Pitt's theorem is not true. Take, for example, the natural inclusion $\iota : \ell_p \hookrightarrow \ell_r$. ι is a norm-one operator which is not compact since the image of the canonical basis of ℓ_p is a sequence contained in $\iota(B_{\ell_p})$ with no convergent subsequences.

Definition 2.1.8. A bounded operator T from a Banach space X into a Banach space Y is *strictly singular* if there is no infinite-dimensional subspace $E \subset X$ such that $T|_E$ is an isomorphism onto its range.

Theorem 2.1.9. *If $p < r$, every $T : \ell_p \longrightarrow \ell_r$ is strictly singular.*

Proof. This is immediate from Corollary 2.1.6.

$\square$

2.2 Complemented subspaces of ℓ_p $(1 \leq p < \infty)$ and c_0

The results of this section are due to Pełczyński (1960) [169]; they demonstrate the power of basic sequence techniques.

Proposition 2.2.1. *Every infinite-dimensional closed subspace Y of ℓ_p $(1 \leq p < \infty)$ [respectively, c_0] contains a closed subspace Z such that Z is isomorphic to ℓ_p [respectively, c_0] and complemented in ℓ_p [respectively, c_0].*

Proof. Since Y is infinite-dimensional, for every n there is $y_n \in Y$, $\|y_n\| = 1$, such that $e_k^*(y_n) = 0$ for $1 \leq k \leq n$. If not, for some $N \in \mathbb{N}$ the projection $S_N(\sum_{n=1}^{\infty} a_n e_n) = \sum_{n=1}^{N} a_n e_n$ restricted to Y would be injective (since $0 \neq y \in Y$ would imply $S_N(y) \neq 0$) and so $S_N|_Y$ would be an isomorphism onto its image, which is impossible because Y is infinite-dimensional. By Proposition 2.1.3 the sequence $(y_n)_{n=1}^{\infty}$ has a subsequence $(y_{n_k})_{k=1}^{\infty}$ which is basic, equivalent to the canonical basis of the space and such that the subspace $Z = [y_{n_k}]$ is complemented.

$\square$

Since c_0 and ℓ_1 are nonreflexive and every closed subspace of a reflexive space is reflexive, using Proposition 2.2.1 we obtain:

Proposition 2.2.2. *Let Y be an infinite-dimensional closed subspace of either c_0 or ℓ_1. Then Y is not reflexive.*

Suppose now that Y is itself complemented in ℓ_p $(1 \leq p < \infty)$ [respectively, c_0]. Proposition 2.2.1 certainly tells us that Y contains a complemented copy of ℓ_p [respectively, c_0]. Can we say more? Remarkably, Pełczyński discovered a trick which enables us, by rather "soft" arguments, to do quite a bit better. This trick is nowadays known as the *Pełczyński decomposition technique* and has proved very useful in different contexts.

The situation is: we have two Banach spaces X and Y so that Y is isomorphic to a complemented subspace of X and X is isomorphic to a complemented subspace of Y. We would like to deduce that X and Y are isomorphic. This is known (by analogy with a similar result for cardinals as the *Schroeder-Bernstein problem* for Banach spaces. The next theorem gives two criteria where the Schroeder-Bernstein problem has a positive solution. To this end we need to introduce the spaces $\ell_p(X)$ for $1 \leq p < \infty$ and $c_0(X)$, where X is a given Banach space.

For $1 \leq p < \infty$, the space $\ell_p(X) = (X \oplus X \oplus \ldots)_p$ called the *infinite direct sum of X in the sense of ℓ_p*, consists of all sequences $x = (x(n))_{n=1}^{\infty}$ with values in X so that $(\|x(n)\|)_{n=1}^{\infty} \in \ell_p$, with the norm

$$\|x\| = \|(\|x(n)\|)_{n=1}^{\infty}\|_p.$$

Similarly, the *infinite direct sum of X in the sense of c_0*, $c_0(X) = (X \oplus X \oplus \ldots)_{c_0}$ is the space of X-valued sequences $x = (x(n))_{n=1}^{\infty}$ so that $\lim_{n \to \infty} \|x(n)\| = 0$ under the norm

$$\|x\| = \max_{1 \leq n < \infty} \|x(n)\|.$$

Notice that $\ell_p(\ell_p)$ can be identified with $\ell_p(\mathbb{N} \times \mathbb{N})$ and hence is isometric to ℓ_p. Analogously, $c_0(c_0)$ is isometric to c_0.

Theorem 2.2.3 (The Pełczyński decomposition technique [169]). *Let X and Y be Banach spaces so that X is isomorphic to a complemented subspace of Y and Y is isomorphic to a complemented subspace of X. Suppose further that either:*
(a) $X \approx X^2 = X \oplus X$ and $Y \approx Y^2$, or
(b) $X \approx c_0(X)$ or $X \approx \ell_p(X)$ for some $1 \leq p < \infty$.
Then X is isomorphic to Y.

Proof. Let us put $X \approx Y \oplus E$ and $X \approx Y \oplus F$. If (a) holds then we have

$$X \approx Y \oplus Y \oplus E \approx Y \oplus X,$$

and by a symmetrical argument $Y \approx X \oplus Y$. Hence $Y \approx X$.

If X satisfies (b) in particular we have $X \approx X^2$ so as in part (a) we obtain $Y \approx X \oplus Y$. On the other hand,

$$\ell_p(X) \approx \ell_p(Y \oplus E) \approx \ell_p(Y) \oplus \ell_p(E).$$

Hence if $X \approx \ell_p(X)$,

$$X \approx Y \oplus \ell_p(Y) \oplus \ell_p(E) \approx Y \oplus \ell_p(X) \approx Y \oplus X.$$

The proof is analogous if $X \approx c_0(X)$.

□

We are ready to prove a beautiful theorem due to Pełczyński (1960) [169] which had a profound influence on the development of Banach space theory.

Theorem 2.2.4. *Suppose Y is a complemented infinite-dimensional subspace of ℓ_p where $1 \leq p < \infty$ [respectively, c_0]. Then Y is isomorphic to ℓ_p [respectively, c_0].*

Proof. Proposition 2.2.1 gives an infinite-dimensional subspace Z of Y such that Z is isomorphic to ℓ_p [respectively, c_0] and Z is complemented in ℓ_p [respectively, c_0]. Obviously Z is also complemented in Y, therefore ℓ_p [respectively, c_0] is (isomorphic to) a complemented subspace in Y. Since $\ell_p(\ell_p) = \ell_p$ [respectively, $c_0(c_0) = c_0$], (b) of Theorem 2.2.3 applies and we are done.

□

At this point let us discuss where this theorem leads. First, the alert reader may ask whether it is true that *every* subspace of ℓ_p is actually complemented. Certainly this is true when $p = 2$! This is a special case of:

The complemented subspace problem. *If X is a Banach space such that every closed subspace is complemented, is X isomorphic to a Hilbert space?*

This problem was settled positively by Lindenstrauss and Tzafriri in 1971 [135]. We will later discuss its general solution but, at the moment, let us point out that it is not so easy to demonstrate the answer even for the ℓ_p-spaces when $p \neq 2$. In this chapter we will show that ℓ_1 has an uncomplemented subspace.

Another way to approach the complemented subspace problem is to demonstrate that ℓ_p has a subspace which is not isomorphic to the whole space. Here we meet another question dating back to Banach:

The homogeneous space problem. *Let X be a Banach space which is isomorphic to every one of its infinite-dimensional closed subspaces. Is X isomorphic to a Hilbert space?*

This problem was finally solved, again positively, by Komorowski and Tomczak-Jaegermann [115] in 1996 (using an important ingredient by Gowers [70]).

Oddly enough, the ℓ_p-spaces for $p \neq 2$ are not as regular as one would expect. In fact, for every $p \neq 2$, ℓ_p contains a subspace without a basis. For $p > 2$ this was proved by Davie in 1973 [34]; for general p it was obtained by Szankowski [211] a few years later. However, the construction of such subspaces is far from easy and will not be covered in this book. Notice that this provides an example of a separable Banach space without a basis.

One natural idea that comes out of Theorem 2.2.4 is the notion that the ℓ_p-spaces and c_0 are the building blocks from which Banach spaces are constructed; by analogy they might play the role of primes in number theory. This thinking is behind the following definition:

Definition 2.2.5. A Banach space X is called *prime* if every complemented infinite-dimensional subspace of X is isomorphic to X.

Thus the ℓ_p-spaces and c_0 are prime. Are there other primes? One may immediately ask about ℓ_∞ and, indeed, this is a (nonseparable) prime space as was shown by Lindenstrauss in 1967 [129]; we will show this later. The quest for other prime spaces has proved difficult, some candidates have been found but in general it is very hard to prove that a particular space is prime. Eventually another prime space was found by Gowers and Maurey [72] but the construction is very involved and the space is far from being "natural." In fact the Gowers-Maurey prime space has the property that the only complemented subspaces of infinite dimension are of *finite* codimension. One can say that this space is prime only because it has very few complemented subspaces at all!

2.3 The space ℓ_1

The space ℓ_1 has a special role in Banach space theory. In this section we develop some of its elementary properties. We start by proving a universal property of ℓ_1 with respect to separable spaces due to Banach and Mazur [9] from 1933.

Theorem 2.3.1. *If X is a separable Banach space then there exists a continuous operator $Q : \ell_1 \to X$ from ℓ_1 onto X.*

Proof. It suffices to show that X admits of a continuous operator $Q : \ell_1 \to X$ such that $Q\{\xi \in \ell_1 : \|\xi\|_1 < 1\} = \{x \in X : \|x\| < 1\}$.

Let $(x_n)_{n=1}^\infty$ be a dense sequence in B_X and define $Q : \ell_1 \to X$ by $Q(\xi) = \sum_{n=1}^\infty \xi(n)x_n$. Notice that Q is well defined: for every $\xi = (\xi(n)) \in \ell_1$ the series $\sum_{n=1}^\infty \xi(n)x_n$ is absolutely convergent in X. Q is clearly linear and has norm one since

$$\|Q(\xi)\| = \left\| \sum_{n=1}^\infty \xi(n)x_n \right\| \leq \sum_{n=1}^\infty |\xi(n)| = \|(\xi(n))\|_1.$$

$Q(B_{\ell_1})$ is a dense subset of B_X, hence given $x \in B_X$ and $0 < \epsilon < 1$ there exists $\xi_1 \in B_{\ell_1}$ such that $\|x - T\xi_1\| < \epsilon$. Next we find $\xi_2' \in B_{\ell_1}$ such that $\|\frac{1}{\epsilon}(x - Q\xi_1) - Q\xi_2'\| < \epsilon$. If we let $\xi_2 = \epsilon\xi_2'$ we obtain

$$\|x - Q(\xi_1 + \xi_2)\| < \epsilon^2.$$

Iterating we find a sequence (ξ_n) in B_{ℓ_1} satisfying $\|\xi_n\|_1 < \epsilon^{n-1}$ and $\|x - Q(\xi_1 + \cdots + \xi_n)\| < \epsilon^n$. Let $\xi = \sum_{n=1}^\infty \xi_n$. Then $\|\xi\|_1 \leq (1-\epsilon)^{-1}$ and $Q\xi = x$. Since $0 < \epsilon < 1$ is arbitrary, by scaling we deduce that $Q\{\xi \in \ell_1 : \|\xi\|_1 < 1\} = \{x \in X : \|x\| < 1\}$. $\qquad\square$

Corollary 2.3.2. *If X is a separable Banach space then X is isometrically isomorphic to a quotient of ℓ_1.*

Proof. Let $Q : \ell_1 \to X$ be the quotient map in the proof of Theorem 2.3.1. Then it follows that $\ell_1/\ker Q$ is isometrically isomorphic to X. $\square$

Corollary 2.3.3. *ℓ_1 has an uncomplemented closed subspace.*

Proof. Take X a separable Banach space which is not isomorphic to ℓ_1. Theorem 2.3.1 yields an operator Q from ℓ_1 onto X whose kernel is a closed subspace of ℓ_1. If $\ker Q$ were complemented in ℓ_1 then we would have $\ell_1 = \ker Q \oplus M$ for some closed subspace M of ℓ_1 and therefore

$$X = \ell_1/\ker Q \approx M.$$

But this can only occur if X is isomorphic to ℓ_1 by Theorem 2.2.4. $\square$

Definition 2.3.4. A Banach space X has the *Schur property* (or X is a *Schur space*) if weak and norm sequential convergence coincide in X, i.e., a sequence $(x_n)_{n=1}^\infty$ in X converges to 0 weakly if and only if $(x_n)_{n=1}^\infty$ converges to 0 in norm.

Example 2.3.5. Neither of the spaces ℓ_p for $1 < p < \infty$ nor c_0 have the Schur property since the canonical basis is weakly null but cannot converge to 0 in norm.

The next result was discovered in an equivalent form by Schur in 1920 [205].

Theorem 2.3.6. *ℓ_1 has the Schur property.*

Proof. Suppose (x_n) is a weakly null sequence in ℓ_1 that does not converge to 0 in norm. Using Proposition 2.1.3, (x_n) contains a subsequence which is basic and equivalent to the canonical basis; this gives a contradiction because the canonical basis of ℓ_1 is clearly not weakly null. $\square$

Theorem 2.3.7. *Let X be a Banach space with the Schur property. Then a subset W of X is weakly compact if and only if W is norm compact.*

Proof. Suppose W is weakly compact and consider a sequence $(x_n)_{n=1}^\infty$ in W. By the Eberlein-Šmulian theorem W is weakly sequentially compact, so $(x_n)_{n=1}^\infty$ has a subsequence $(x_{n_k})_{k=1}^\infty$ that converges weakly to some $x \in W$. Since X has the Schur property, $(x_{n_k})_{k=1}^\infty$ converges to x in norm as well. Therefore W is compact for the norm topology. $\square$

Corollary 2.3.8. *If X is a reflexive Banach space with the Schur property then X is finite-dimensional.*

Proof. If a reflexive Banach space X has the Schur property then its unit ball is norm-compact by Theorem 2.3.7 and so X is finite-dimensional. □

Definition 2.3.9. A sequence $(x_n)_{n=1}^{\infty}$ in a Banach space X is *weakly Cauchy* if $\lim_{n\to\infty} x^*(x_n)$ exists for every x^* in X^*.

Any weakly Cauchy sequence $(x_n)_{n=1}^{\infty}$ in a Banach space X is norm-bounded by the Uniform Boundedness principle. If X is reflexive, by Corollary 1.6.4, $(x_n)_{n=1}^{\infty}$ will have a weak cluster point, x, and so $(x_n)_{n=1}^{\infty}$ will converge weakly to x. If X is nonreflexive, however, there may be sequences which are weakly Cauchy but not weakly convergent.

Definition 2.3.10. A Banach space X is said to be *weakly sequentially complete* (wsc) if every weakly Cauchy sequence in X converges weakly.

Example 2.3.11. In the space c_0 consider the sequence $x_n = e_1 + \cdots + e_n$, where (e_n) is the unit vector basis. $(x_n)_{n=1}^{\infty}$ is obviously weakly Cauchy but it does not converge weakly in c_0. $(x_n)_{n=1}^{\infty}$ converges weak* in the bidual, ℓ_{∞}, to the element $(1, 1, \ldots, 1, \ldots)$. Thus c_0 is not weakly sequentially complete.

Proposition 2.3.12. *Any Banach space with the Schur property (in particular ℓ_1) is weakly sequentially complete.*

Proof. Suppose $(x_n)_{n=1}^{\infty}$ is weakly Cauchy. Then for any two strictly increasing sequences of integers $(n_k)_{k=1}^{\infty}, (m_k)_{k=1}^{\infty}$ the sequence $(x_{m_k} - x_{n_k})_{k=1}^{\infty}$ is weakly null and so $\lim_{k\to\infty} \|x_{m_k} - x_{n_k}\| = 0$. Thus, being norm-Cauchy, $(x_n)_{n=1}^{\infty}$ is norm-convergent and hence weak-convergent. □

2.4 Convergence of series

Definition 2.4.1. Let $(x_n)_{n=1}^{\infty}$ be a sequence in a Banach space X. A (formal) series $\sum_{n=1}^{\infty} x_n$ in X is said to be *unconditionally convergent* if $\sum_{n=1}^{\infty} x_{\pi(n)}$ converges for every permutation π of $\mathbb{N}$.

We will see in Chapter 8 that except in finite-dimensional spaces, unconditional convergence is weaker than *absolute convergence*, i.e., convergence of $\sum_{n=1}^{\infty} \|x_n\|$.

Lemma 2.4.2. *Given a series $\sum_{n=1}^{\infty} x_n$ in a Banach space X, the following are equivalent:*

(a) $\sum_{n=1}^{\infty} x_n$ is unconditionally convergent;

(b) The series $\sum_{k=1}^{\infty} x_{n_k}$ converges for every increasing sequence of integers $(n_k)_{k=1}^{\infty}$;
(c) The series $\sum_{n=1}^{\infty} \epsilon_n x_n$ converges for every choice of signs (ϵ_n);
(d) For every $\epsilon > 0$ there exists an n so that if F is any finite subset of $\{n+1, n+2, \ldots\}$ then

$$\left\| \sum_{j \in F} x_j \right\| < \epsilon.$$

Proof. We will establish only $(a) \Rightarrow (d)$ and leave the other easier implications to the reader. Suppose that (d) fails. Then there exists $\epsilon > 0$ so that for every n we can find a finite subset F_n of $\{n+1, \ldots\}$ with

$$\left\| \sum_{j \in F_n} x_j \right\| \geq \epsilon.$$

We will build a permutation π of $\mathbb{N}$ so that $\sum_{n=1}^{\infty} x_{\pi(n)}$ diverges.
     Take $n_1 = 1$ and let $A_1 = F_{n_1}$. Next pick $n_2 = \max A_1$ and let $B_1 = \{n_1 + 1, \ldots, n_2\} \setminus A_1$. Now repeat the process taking $A_2 = F_{n_2}$, $n_3 = \max A_2$ and $B_2 = \{n_2 + 1, \ldots, n_3\} \setminus A_2$. Iterating we generate a sequence $(n_k)_{k=1}^{\infty}$ and a partition $\{n_k + 1, \ldots, n_{k+1}\} = A_k \cup B_k$. Define π so that π permutes the elements of $\{n_k + 1, \ldots, n_{k+1}\}$ in such a way that A_k precedes B_k. Then the series $\sum_{n=1}^{\infty} x_{\pi(n)}$ is divergent because the Cauchy condition fails. $\qquad \square$

Definition 2.4.3. A (formal) series $\sum_{n=1}^{\infty} x_n$ in a Banach space X is *weakly unconditionally Cauchy* (WUC) or *weakly unconditionally convergent* if for every $x^* \in X^*$ $\sum_{n=1}^{\infty} |x^*(x_n)| < \infty$.

Proposition 2.4.4. *Suppose the series $\sum_{n=1}^{\infty} x_n$ converges unconditionally to some x in a Banach space X. Then*

(i) $\sum_{n=1}^{\infty} x_{\pi(n)} = x$ for every permutation π.
(ii) $\sum_{n \in \mathbb{A}} x_n$ converges unconditionally for every infinite subset $\mathbb{A}$ of $\mathbb{N}$.
(iii) $\sum_{n=1}^{\infty} x_n$ is WUC.

Proof. Parts (i) and (ii) are immediate. For (iii), given $x^* \in X^*$ the scalar series $\sum_{n=1}^{\infty} x^*(x_{\pi(n)})$ converges for every permutation π. It is a classical theorem of Riemann that for scalar sequences the series $\sum_{n=1}^{\infty} a_n$ converges unconditionally if and only if it converges absolutely, i.e., $\sum_{n=1}^{\infty} |a_n| < \infty$. Thus we have $\sum_{n=1}^{\infty} |x^*(x_n)| < \infty$. $\qquad \square$

     Let us notice that the name "weakly unconditionally convergent" series can be misleading because such series need not be weakly convergent; we will therefore use the term weakly unconditionally Cauchy or more usually its abbreviation (WUC).

Example 2.4.5. The series $\sum_{n=1}^{\infty} e_n$ in c_0, where $(e_n)_{n=1}^{\infty}$ is the canonical basis of the space, is WUC but fails to converge weakly (and so it cannot converge unconditionally). In fact, this is in a certain sense the only counterexample as we shall see.

In Proposition 2.4.7 we shall prove that WUC series are in a very natural correspondence with bounded operators on c_0. Let us first see a lemma.

Lemma 2.4.6. Let $\sum_{n=1}^{\infty} x_n$ be a formal series in a Banach space X. Then the following are equivalent:

(i) $\sum_{n=1}^{\infty} x_n$ is WUC.

(ii) There exists $C > 0$ such that for all $(\xi(n)) \in c_{00}$ we have

$$\left\| \sum_{n=1}^{\infty} \xi(n)x_n \right\| \leq C \max_n |\xi(n)|.$$

(iii) There exists $C' > 0$ such that

$$\left\| \sum_{n \in F} \epsilon_n x_n \right\| \leq C'$$

for any finite subset F of $\mathbb{N}$ and all $\epsilon_n = \pm 1$.

Proof. $(i) \Rightarrow (ii)$. Put

$$S = \left\{ \sum_{n=1}^{\infty} \xi(n)x_n \in X : \xi = (\xi(n)) \in c_{00}, \|\xi\|_\infty \leq 1 \right\}.$$

The WUC property implies that S is weakly bounded, therefore it is norm-bounded by the Uniform Boundedness principle.

Obviously, (ii) implies (iii). For $(iii) \Rightarrow (i)$, given $x^* \in X^*$ let $\epsilon_n = \text{sgn } x^*(x_n)$. Then for each integer N we have

$$\sum_{n=1}^{N} |x^*(x_n)| = \left| x^* \left(\sum_{n=1}^{N} \epsilon_n x_n \right) \right| \leq C \|x^*\|$$

and therefore the series $\sum_{n=1}^{\infty} |x^*(x_n)|$ converges.

$\square$

Proposition 2.4.7. Let $\sum_{n=1}^{\infty} x_n$ be a series in a Banach space X. Then $\sum_{n=1}^{\infty} x_n$ is WUC if and only if there is a bounded operator $T : c_0 \to X$ with $Te_n = x_n$.

Proof. If $\sum_{n=1}^{\infty} x_n$ is WUC then the operator $T : c_{00} \to X$ defined by $T\xi = \sum_{n=1}^{\infty} \xi(n)x_n$ is bounded for the c_0-norm by Lemma 2.4.6. By density T extends to a bounded operator $T : c_0 \to X$.

For the converse, let $T : c_0 \to X$ be a bounded operator with $Te_n = x_n$ for all n. For each $x^* \in X^*$ we have

$$\sum_{n=1}^{\infty} |x^*(x_n)| = \sum_{n=1}^{\infty} |x^*(Te_n)| = \sum_{n=1}^{\infty} |T^*(x^*)(e_n)|,$$

which is finite since $\sum_{n=1}^{\infty} e_n$ is WUC.

$\square$

Proposition 2.4.8. *Let $\sum_{n=1}^{\infty} x_n$ be a WUC series in a Banach space X. Then $\sum_{n=1}^{\infty} x_n$ converges unconditionally in X if and only if the operator $T : c_0 \to X$ such that $Te_n = x_n$ is compact.*

Proof. Suppose $\sum_{n=1}^{\infty} x_n$ is unconditionally convergent. We will show that $\lim_{n\to\infty} \|T - TS_n\| = 0$, where $(S_n)_{n=1}^{\infty}$ are the partial sum projections associated to the canonical basis (e_n) of c_0. Thus, being a uniform limit of finite-rank operators, T will be compact.

Given $\epsilon > 0$ we use Lemma 2.4.2 to find $n = n(\epsilon)$ so that if F is a finite subset of $\{n + 1, n + 2, \ldots\}$ then $\|\sum_{j\in F} x_j\| \le \epsilon/2$. For every $x^* \in X^*$ with $\|x^*\| \le 1$ we have

$$\sum_{\{j\in F \,:\, x^*(x_j)\ge 0\}} x^*(x_j) \le \frac{\epsilon}{2},$$

therefore

$$\sum_{j\in F} |x^*(x_j)| \le \epsilon.$$

Hence if $\xi \in c_{00}$ with $\|\xi\|_\infty \le 1$ it follows that $|x^*(T - TS_m)\xi| \le \epsilon$ for $m \ge n$ and all $x^* \in X^*$. By density we conclude that $\|T - TS_m\| \le \epsilon$.

Assume, conversely, that T is compact. Let us consider

$$T^{**} : c_0^{**} = \ell_\infty \longrightarrow X \subset X^{**}.$$

The restriction of T^{**} to B_{ℓ_∞} is weak*-to-norm continuous because on a norm compact set the weak* topology agrees with the norm topology. Since $\sum_{n=1}^{\infty} e_{\pi(n)}$ converges weak* in ℓ_∞ for every permutation π, $\sum_{n=1}^{\infty} x_n$ also converges unconditionally in X.

$\square$

Note that the above argument also implies the following stability property of unconditionally convergent series with respect to the multiplication by bounded sequences. The proof is left as an exercise.

Proposition 2.4.9. *A series $\sum_{n=1}^{\infty} x_n$ in a Banach space X is unconditionally convergent if and only if $\sum_{n=1}^{\infty} t_n x_n$ converges (unconditionally) for all $(t_n) \in \ell_\infty$.*

The next theorem and its consequences are essentially due to Bessaga and Pełczyński in their 1958 paper [12] and represent some of the earliest applications of the basic sequence methods.

Theorem 2.4.10. *Suppose $T : c_0 \to X$ is a bounded operator. Then the following conditions on T are equivalent:*

(i) T is compact,
(ii) T is weakly compact,
(iii) T is strictly singular.

Proof. $(i) \Rightarrow (ii)$ is obvious. For $(ii) \Rightarrow (iii)$, let us suppose that T fails to be strictly singular. Then there exists an infinite-dimensional subspace Y of c_0 such that $T|_Y$ is an isomorphism onto its range. If T is weakly compact this forces Y to be reflexive, contradicting Proposition 2.2.2.

We now consider $(iii) \Rightarrow (i)$. Assume that T fails to be compact. Then, by Proposition 2.4.8, $\sum_{n=1}^{\infty} Te_n$ does not converge unconditionally so, by Lemma 2.4.2, there exists $\epsilon > 0$ and a sequence of disjoint finite subsets of integers $(F_n)_{n=1}^{\infty}$ so that $\left\| \sum_{k \in F_n} Te_k \right\| \geq \epsilon$ for every n. Let $x_n = \sum_{k \in F_n} Te_k$. $(x_n)_{n=1}^{\infty}$ is weakly null in X since $\sum_{k \in F_n} e_k$ is weakly null in c_0. Using Proposition 1.3.10 we can, by passing to a subsequence of $(x_n)_{n=1}^{\infty}$, assume it is basic in X with basis constant K, say. Then for $\xi = (\xi(n))_{n=1}^{\infty} \in c_{00}$,

$$\left\| \sum_{n=1}^{\infty} \xi(n) x_n \right\| = \left\| T\left(\sum_{n=1}^{\infty} \xi(n) \sum_{k \in F_n} e_k \right) \right\| \leq \|T\| \max_{n \in \mathbb{N}} |\xi(n)|.$$

On the other hand,

$$\max_{n \in \mathbb{N}} |\xi(n)| \leq 2K \left\| \sum_{n=1}^{\infty} \xi(n) x_n \right\|.$$

Thus $(x_n)_{n=1}^{\infty}$ is equivalent to the canonical basis of c_0 and therefore to $(\sum_{k \in F_n} e_k)_{n=1}^{\infty}$. We conclude that T cannot be strictly singular. □

From now on, whenever we say that a Banach space X *contains a copy of a Banach space Y* we mean that X contains a closed subspace E which is isomorphic to Y. Using Theorem 2.4.10 we obtain a very nice characterization of spaces that contain a copy of c_0.

Theorem 2.4.11. *In order that every WUC series in a Banach space X be unconditionally convergent it is necessary and sufficient that X contains no copy of c_0.*

Proof. Suppose that X contains no copy of c_0 and that $\sum_{n=1}^{\infty} x_n$ is a WUC series in X. By Proposition 2.4.7 there exists a bounded operator $T : c_0 \to X$ such that $Te_n = x_n$ for all n. T must be strictly singular since every infinite-dimensional subspace of c_0 contains a copy of c_0 (Proposition 2.2.1) so T is compact by Theorem 2.4.10. Hence the series $\sum_{n=1}^{\infty} x_n$ converges unconditionally by Proposition 2.4.8. The converse follows trivially from Example 2.4.5. □

Remark 2.4.12. This theorem of Bessaga and Pełczyński is a prototype for exclusion theorems which say that if we can exclude a certain subspace from a Banach space then it will have a particular property. It had considerable influence in suggesting that such theorems might be true. In Chapter 10 we will see a similar and much more difficult result for Banach spaces not containing ℓ_1 (due to Rosenthal [197]) which when combined with the Bessaga-Pełczyński theorem gives a very elegant pair of bookends in Banach space theory. It is also worth noting that the hypothesis that a Banach space fails to contain c_0 becomes ubiquitous in the theory precisely because of Theorem 2.4.11.

We have seen that a series $\sum_{n=1}^{\infty} x_n$ in a Banach space X converges unconditionally in norm if and only if each subseries $\sum_{k=1}^{\infty} x_{n_k}$ does. In particular every subseries of an unconditionally convergent series is weakly convergent. The Orlicz-Pettis theorem establishes that the converse is true as well. First we see an auxiliary result.

Lemma 2.4.13. *Let m_0 be the set of all sequences of scalars assuming only finitely many different values. Then m_0 is dense in ℓ_∞.*

Proof. Let $a = (a_n)_{n=1}^{\infty}$ be a sequence of scalars with $\|a\|_\infty \leq 1$. For any $\epsilon > 0$ pick $N \in \mathbb{N}$ such that $\frac{1}{N} < \epsilon$. Then the sequence $b = (b_n)_{n=1}^{\infty} \in m_0$ given by

$$b_n = (\operatorname{sgn} a_n)\frac{j}{N} \quad \text{if} \quad \frac{j}{N} \leq |a_n| \leq \frac{j+1}{N}, \quad j = 1, \ldots, N$$

satisfies $\|a - b\|_\infty \leq \frac{1}{N} < \epsilon$.

$\square$

Theorem 2.4.14 (The Orlicz-Pettis Theorem). *Suppose $\sum_{n=1}^{\infty} x_n$ is a series in a Banach space X for which every subseries $\sum_{k=1}^{\infty} x_{n_k}$ converges weakly. Then $\sum_{n=1}^{\infty} x_n$ converges unconditionally in norm.*

Proof. The hypothesis easily yields that $\sum_{n=1}^{\infty} x_n$ is a WUC series so, by Proposition 2.4.7, there exists a bounded operator $T : c_0 \to X$ with $Te_n = x_n$ for all n. We will show that T is actually compact.

Let us look at $T^{**} : \ell_\infty \to X^{**}$. For every $A \subset \mathbb{N}$ let us denote by $\chi_A = (\chi_A(k))_{k=1}^{\infty}$ the element of ℓ_∞ such that $\chi_A(k) = 1$ if $k \in A$ and 0 otherwise. By hypothesis $\sum_{n \in A} x_n$ converges weakly in X and it follows that $T^{**}(\chi_A) \in X$. The linear span of all such χ_A consists of the space of scalar sequences taking only finitely many different values, m_0, which by Lemma 2.4.13 is dense in ℓ_∞. Hence T^{**} maps ℓ_∞ into X. This means that T is a weakly compact operator. Now Theorem 2.4.10 implies that T is a compact operator and Proposition 2.4.8 completes the proof.

$\square$

Now, as a corollary, we can give a reciprocal of Proposition 2.4.4 *(iii)*.

Corollary 2.4.15. *If a Banach space X is weakly sequentially complete then every WUC series in X is unconditionally convergent.*

Proof. If $\sum_{n=1}^{\infty} x_n$ is WUC then $\sum_{n=1}^{\infty} x^*(x_n)$ is absolutely convergent for every $x^* \in X^*$, which is equivalent to saying that $\sum_{k=1}^{\infty} x^*(x_{n_k})$ converges for each subseries $\sum_{k=1}^{\infty} x_{n_k}$ and each $x^* \in X^*$. Hence $\sum_{k=1}^{\infty} x_{n_k}$ is weakly Cauchy and therefore weakly convergent by hypothesis. We deduce that $\sum_{n=1}^{\infty} x_n$ converges unconditionally in norm by the Orlicz-Pettis theorem.

$\square$

The Orlicz-Pettis theorem predates basic sequence techniques. It was first proved by Orlicz in 1929 [162] and referenced in Banach's book [8]. He attributes the result to Orlicz in the special case when X is weakly sequentially complete so that every WUC series has the property of the theorem. However, it seems that Orlicz did know the more general statement. Independently, Pettis published a proof in 1938 [178]. Pettis was interested in such a result as a by-product of the study of vector measures. If Σ is a σ-algebra of sets and $\mu : \Sigma \to X$ is a map such that for every $x^* \in X^*$ the set function $x^* \circ \mu$ is a (countably additive) measure then the Orlicz-Pettis theorem implies that μ is countably additive in the norm topology. Thus weakly countably additive set functions are norm countably additive.

This is an attractive theorem and as a result it has been proved, reproved, and generalized many times since then. It is not clear that there is much left to say on this subject! We will suggest some generalizations in the Problems.

2.5 Complementability of c_0

Let us discuss the following extension problem. Suppose that X and Y are Banach spaces and that E is a subspace of X. Let $T : E \to Y$ be a bounded operator. Can we extend T to a bounded operator $\tilde{T} : X \to Y$? If we consider the special case when $Y = E$ and T is the identity map on E, we are asking simply if E is the range of a projection on X, i.e., if E is complemented in X.

The Hahn-Banach theorem asserts that if Y has dimension one then such an extension is possible with preservation of norm. However, in general such an extension is not possible and we have discussed the fact that there are noncomplemented subspaces in almost all Banach spaces. For instance we have seen that ℓ_1 must have an uncomplemented subspace, but the construction of this subspace as the kernel of a certain quotient map means that it is rather difficult to see exactly what it is. In this section we will study a very natural example. Let us formalize the notion of an injective Banach space.

Definition 2.5.1. A Banach space Y is called *injective* if whenever X is a Banach space, E is a closed subspace of X, and $T : E \to Y$ is a bounded operator then there is a bounded linear operator $\tilde{T} : X \to Y$ which is an extension of T. Y is called *isometrically injective* if $\tilde{T}$ can be additionally chosen to have $\|\tilde{T}\| = \|T\|$.

We will defer our discussion of injective spaces to later and restrict ourselves to one almost trivial observation:

Proposition 2.5.2. *The space ℓ_∞ is an isometrically injective space. Hence, if a Banach space X has a subspace E isomorphic to ℓ_∞, then E is necessarily complemented in X.*

Proof. Suppose E is a subspace of X and $T : E \to \ell_\infty$ is bounded. Then $Te = (e_n^*(e))_{n=1}^\infty$ for some sequence $(e_n^*)_{n=1}^\infty$ in E^*; clearly $\|T\| = \sup_n \|e_n^*\|$. By the Hahn-Banach theorem we choose extensions $x_n^* \in X^*$ with $\|x_n^*\| = \|e_n^*\|$ for each n. By letting $\tilde{T}x = (x_n^*(x))_{n=1}^\infty$ we are done.

$\square$

c_0 is a subspace of ℓ_∞ (its bidual) and it is easy to see that c_0 will be injective if and only if it is complemented in ℓ_∞. Must a Banach space be complemented in its bidual? Certainly this is true for any space which is the dual of another space since for any Banach space X the space X^* is always complemented in its bidual, X^{***}. To see this consider the natural embedding $j : X \to X^{**}$. Then $j^* : X^{***} \to X^*$ is a norm-one operator. Denote by J the canonical injection of X^* into X^{***}. We claim that j^*J is the identity I_{X^*} on X^*. Indeed, suppose $x^* \in X^*$ and that $x \in X$. Then $\langle x, j^*J(x^*) \rangle = \langle jx, Jx^* \rangle = \langle x, x^* \rangle$. Thus j^* is a norm-one projection of X^{***} onto X^*. If X is isomorphic (but not necessarily isometric) to a dual space we leave for the reader the details to check that X will still be complemented in its bidual. So we may also ask if c_0 is isomorphic to a dual space.

As we will see next, c_0 is *not* complemented in ℓ_∞. This was proved essentially by Phillips [180] in 1940 although first formally observed by Sobczyk [208] the following year. Phillips in fact proved the result for the subspace c of convergent sequences. The proof we give is due to Whitley [220] and requires a simple lemma:

Lemma 2.5.3. *Every countably infinite set $\mathbb{S}$ has an uncountable family of infinite subsets $\{\mathbb{A}_i\}_{i \in \mathcal{I}}$ such that any two members of the family have finite intersection.*

Proof. The proof is very simple but rather difficult to spot! Without loss of generality we can identify $\mathbb{S}$ with the set of the rational numbers $\mathbb{Q}$. For each irrational number θ, take a sequence of rational numbers $(q_n)_{n=1}^\infty$ converging to θ. Then the sets of the form $\mathbb{A}_\theta = \{(q_n)_{n=1}^\infty : q_n \to \theta\}$ verify the lemma.

$\square$

If $\mathbb{A}$ is any subset of $\mathbb{N}$ we denote by $\ell_\infty(\mathbb{A})$ the subspace of ℓ_∞ given by

$$\ell_\infty(\mathbb{A}) = \left\{ \xi = (\xi(k))_{k=1}^\infty \in \ell_\infty : \xi(k) = 0 \text{ if } k \notin \mathbb{A} \right\}.$$

Theorem 2.5.4. *Let $T : \ell_\infty \to \ell_\infty$ be a bounded operator such that $T\xi = 0$ for all $\xi \in c_0$. Then there is an infinite subset $\mathbb{A}$ of $\mathbb{N}$ so that $T\xi = 0$ for every $\xi \in \ell_\infty(\mathbb{A})$.*

Proof. We use the family $(\mathbb{A}_i)_{i \in \mathcal{I}}$ of infinite subsets of $\mathbb{N}$ given by Lemma 2.5.3. Suppose that for every such set we can find $\xi_i \in \ell_\infty(\mathbb{A}_i)$ with $T\xi_i \neq 0$. We can assume by normalization that $\|\xi_i\|_\infty = 1$ for every $i \in \mathcal{I}$. There must exist

$n \in \mathbb{N}$ so that the set $\mathcal{I}_n = \{i \in \mathcal{I} : \xi_i(n) \neq 0\}$ is uncountable. Similarly, there exists $k \in \mathbb{N}$ so that the set $\mathcal{I}_{n,k} = \{i : |\xi_i(n)| \geq k^{-1}\}$ is also uncountable. For each $i \in \mathcal{I}_{n,k}$ choose α_i with $|\alpha_i| = 1$ and $\alpha_i \xi_i(n) = |\xi_i(n)|$.

Let $\mathbb{F}$ be a finite subset of $\mathcal{I}_{n,k}$. Consider $y = \sum_{i \in \mathbb{F}} \alpha_i \xi_i$. Since the intersection of the supports of any two distinct ξ_i is finite we can write $y = u + v$ where $\|u\|_\infty \leq 1$ and v has finite support. Thus

$$\|Ty\|_\infty = \|Tu\|_\infty \leq \|T\|,$$

and so

$$e_n^*(Ty) = \sum_{i \in \mathbb{F}} |\xi_i(n)| \leq \|T\|.$$

It follows that if $|\mathbb{F}| = m$ we have $mk^{-1} \leq \|T\|$, i.e., $m \leq k\|T\|$. Since this holds for every finite subset of $\mathcal{I}_{n,k}$ we have shown that $\mathcal{I}_{n,k}$ is in fact finite, which is a contradiction.

$\square$

Theorem 2.5.5 (Phillips-Sobczyk, 1940-1). *There is no bounded projection from ℓ_∞ onto c_0.*

Proof. If P is such a projection we can apply Theorem 2.5.4 to $T = I - P$, with I the identity operator on ℓ_∞, and then it is clear that $P\xi = \xi$ for all $\xi \in \ell_\infty(\mathbb{A})$ for some infinite set $\mathbb{A}$, which gives a contradiction.

$\square$

Corollary 2.5.6. c_0 *is not isomorphic to a dual space.*

Proof. If c_0 were isomorphic to a dual space then, by the comments that follow the proof of Proposition 2.5.2, c_0 should be complemented in c_0^{**}, which would lead to contradiction with Theorem 2.5.5.

$\square$

Several comments are in order here. Theorem 2.5.4 proves more than is needed for Phillips-Sobczyk's theorem. It shows that there is no bounded, one-to-one operator from the quotient space ℓ_∞/c_0 into ℓ_∞; in other words the points of ℓ_∞/c_0 cannot be separated by countably many bounded linear functionals. (Of course, if E is a complemented subspace of a Banach space X, then X/E must be isomorphic to a subspace of X which is complementary to E.)

Now we are also in position to note that c_0 is not an injective space. Actually there are no separable injective spaces, but we will see this later, when we discuss the structure of ℓ_∞ in more detail. For the moment let us notice the dual statement of Theorem 2.3.1.

Theorem 2.5.7. *If X is a separable Banach space then X embeds isometrically into ℓ_∞.*

Proof. Let $(x_n)_{n=1}^{\infty}$ be a dense sequence in X. For each integer n pick $x_n^* \in X^*$ so that $\|x_n^*\| = 1$ and $x_n^*(x_n) = \|x_n\|$. The sequence $(x_n^*)_{n=1}^{\infty} \subset X^*$ is norming in X. Therefore the operator $T : X \to \ell_{\infty}$ defined for each x in X by $T(x) = (x_n^*(x))_{n=1}^{\infty}$ provides the desired embedding.

$\square$

Thus X separable can only be injective if it is isomorphic to a complemented subspace of ℓ_{∞}. Therefore classifying the complemented subspaces of ℓ_{∞} becomes important; we will see in Chapter 5 the (already mentioned) theorem of Lindenstrauss [129] that ℓ_{∞} is a prime space and this will answer our question.

In the meantime we turn to Sobczyk's main result in his 1941 paper, which gives some partial answers to these questions. The proof we present here is due to Veech [219].

Theorem 2.5.8 (Sobczyk, 1941). *Let X be a separable Banach space. If E is a closed subspace of X and $T : E \longrightarrow c_0$ is a bounded operator then there exists an operator $\tilde{T} : X \longrightarrow c_0$ such that $\tilde{T}|_E = T$ and $\|\tilde{T}\| \leq 2\|T\|$.*

Proof. Without loss of generality we can assume that $\|T\| = 1$. It is immediate to realize that the operator T must be of the form

$$Tx = (f_n^*(x))_{n=1}^{\infty}, \qquad x \in E$$

for some $(f_n^*) \subset E^*$. Moreover $\|f_n^*\| \leq 1$ for all n and (f_n^*) converges to 0 in the weak* topology of E^*. By the Hahn-Banach theorem, for each $n \in \mathbb{N}$ there exists $\varphi_n^* \in X^*$, $\|\varphi_n^*\| \leq 1$, such that $\varphi_n^*|_E = f_n^*$.

X separable implies that (B_{X^*}, w^*) is metrizable (Lemma 1.4.1). Let ρ be the metric on B_{X^*} that induces the weak* topology on B_{X^*}. We claim that $\lim_{n \to \infty} \rho(\varphi_n^*, B_{X^*} \cap E^{\perp}) = 0$. If this is not the case, there would be some $\epsilon > 0$ and a subsequence $(\varphi_{n_k}^*)$ of (φ_n^*) such that $\rho(\varphi_{n_k}^*, B_{X^*} \cap E^{\perp}) \geq \epsilon$ for every k. Let $(\varphi_{n_{k_j}}^*)$ be a subsequence of $(\varphi_{n_k}^*)$ such that $\varphi_{n_{k_j}}^* \xrightarrow{w^*} \varphi^*$. Then $\varphi^* \in E^{\perp} \cap B_{X^*}$ since for each $e \in E$ we have

$$\varphi^*(e) = \lim_j \varphi_{n_{k_j}}^*(e) = \lim_j f_{n_{k_j}}^*(e) = 0.$$

Hence

$$\rho(\varphi_{n_{k_j}}^*, \varphi^*) \geq \epsilon \text{ for all } j. \tag{2.1}$$

On the other hand

$$\lim_{j \to \infty} \rho(\varphi_{n_{k_j}}^*, B_{X^*} \cap E^{\perp}) = \rho(\varphi^*, B_{X^*} \cap E^{\perp}) = 0 \tag{2.2}$$

since the function $\rho(\,\cdot\,, B_{X^*} \cap E^{\perp})$ is weak* continuous on B_{X^*}. Clearly we cannot have (2.1) and (2.2) at the same time, so our claim holds.

Recall that $E^{\perp}$ is weak* closed, hence $B_{X^*} \cap E^{\perp}$ is weak* compact. Therefore for each n we can pick $v_n^* \in B_{X^*} \cap E^{\perp}$ such that

$$\rho(\varphi_n^*, v_n^*) = \rho(\varphi_n^*, B_{X^*} \cap E^\perp).$$

Let $x_n^* = \varphi_n^* - v_n^*$ and define the operator $\tilde{T}$ on X by $\tilde{T}(x) = (x_n^*(x))$. Notice that $\tilde{T}(x) \in c_0$ because $x_n^* \xrightarrow{w^*} 0$. Moreover, for each $x \in X$ we have

$$\|\tilde{T}(x)\| = \sup_n |x_n^*(x)| = \sup_n(|\varphi_n^*(x) - v_n^*(x)|) \le \sup_n(\|\varphi_n^*\| + \|v_n^*\|) \|x\| \le 2 \|x\|,$$

so $\|\tilde{T}\| \le 2$.

$\square$

Corollary 2.5.9. *If E is a closed subspace of a separable Banach space X and E is isomorphic to c_0, then there is a projection P from X onto E.*

Proof. Suppose that $T : E \to c_0$ is an isomorphism and let $\tilde{T} : X \to c_0$ be the extension of T given by the preceding theorem. Then $P = T^{-1}\tilde{T}$ is a projection from X onto E. (Note that since $\|\tilde{T}\| \le 2\|T\|$, if E is *isometric* to c_0 then $\|P\| \le 2$.)

$\square$

Remark 2.5.10. It follows that if a separable Banach space X contains a copy of c_0 then X is not injective.

We finish this chapter by observing that in light of Theorem 2.5.8 it is natural to define a Banach space Y to be *separably injective* if whenever X is a separable Banach space, E is a closed subspace of X and $T : E \to Y$ is a bounded operator then T can be extended to an operator $\tilde{T} : X \to Y$. It was for a long time conjectured that c_0 is the only separable and separably injective space. This was solved by Zippin in 1977 [225], who showed that, indeed, c_0 is, up to isomorphism, the only separable space which is separably injective.

We also note that the constant 2 in Theorem 2.5.8 is the best possible (see Problem 2.7).

Problems

2.1. Let $T : X \to Y$ be an operator between the Banach spaces X, Y.

(a) Show that if T is strictly singular then in every infinite-dimensional subspace E of X there is a normalized basic sequence (x_n) with $\|Tx_n\| < 2^{-n}\|x_n\|$ for all n.

(b) Deduce that T is strictly singular if and only if every infinite-dimensional closed subspace E contains a further infinite-dimensional closed subspace F so that the restriction of T to F is compact.

2.2. Show that the sum of two strictly singular operators is strictly singular. Show also that if $T_n : X \to Y$ are strictly singular and $\|T_n - T\| \to 0$ then T is strictly singular.

2.3. Show that the set of all strictly singular operators on a Banach space X forms a closed two-sided ideal in the algebra $\mathcal{L}(X)$ of all bounded linear operators from X to X.

2.4. Show that if $1 < p < \infty$ and $T : \ell_p \to \ell_p$ is not compact then there is a complemented subspace E of ℓ_p so that T is an isomorphism of E onto a complemented subspace $T(E)$. Deduce that the Banach algebra $\mathcal{L}(\ell_p)$ contains exactly one proper closed two-sided ideal (the ideal of compact operators). Note that every strictly singular operator is compact in these spaces.

2.5. Show that $\mathcal{L}(\ell_p \oplus \ell_r)$ for $p \neq r$ contains at least two nontrivial closed two-sided ideals.

2.6. Suppose X is a Banach space whose dual is separable. Suppose that $\sum x_n^*$ is a series in X^* which has the property that every subseries $\sum x_{n_k}^*$ converges weak*. Show that $\sum x_n$ converges in norm. [*Hint*: Every $x^{**} \in X^{**}$ is the limit of a weak* converging sequence from X.]

2.7. Let c be the subspace of ℓ_∞ of converging sequences. Show that for any bounded projection P of c onto c_0 we have $\|P\| \geq 2$. This proves that 2 is the best possible constant in Sobczyk's theorem (Theorem 2.5.8).

2.8. In this exercise we will focus on the special properties of ℓ_1 as a target space for operators and show its *projectivity*.

(a) Suppose $T : X \to \ell_1$ is an operator from a Banach space X *onto* ℓ_1. Show that then X contains a complemented subspace isomorphic to ℓ_1.

(b) Prove that if Y is a separable infinite-dimensional Banach space with the property that whenever $T : X \to Y$ is a bounded surjective operator then Y is isomorphic to a complemented subspace of X, then Y is isomorphic to ℓ_1.

2.9. Let X be a Banach space.

(a) Show that for any $x^{**} \in X^{**}$ and any finite-dimensional subspace E of X^* there exists $x \in X$ such that

$$\|x\| < (1 + \epsilon)\|x^{**}\|,$$

and
$$x^*(x) = x^{**}(x^*), \qquad x^* \in E.$$

(b) Use part (a) to deduce the following result of Bessaga and Pełczyński ([12]): If X^* contains a subspace isomorphic to c_0 then X contains a complemented subspace isomorphic to ℓ_1, and hence X^* contains a subspace isomorphic to ℓ_∞. In particular, no separable dual space can contain an isomorphic copy of c_0. [This may also be used in Problem 2.6.]

2.10. For an arbitrary set Γ we define $c_0(\Gamma)$ as the space of functions $\xi : \Gamma \to \mathbb{R}$ such that for each $\epsilon > 0$ the set $\{\gamma : |\xi(\gamma)| > \epsilon\}$ is finite. When normed by $\|\xi\| = \max_{\gamma \in \Gamma} |\xi(\gamma)|$, the space $c_0(\Gamma)$ becomes a Banach space.

(a) Show that $c_0(\Gamma)^*$ can be identified with $\ell_1(\Gamma)$ the space of functions $\eta : \Gamma \to \mathbb{R}$ such that $\eta \in c_0(\Gamma)$ and $\|\eta\| = \sum_{\gamma \in \Gamma} |\eta(\gamma)| < \infty$.

(b) Show that $\ell_1(\Gamma)^* = \ell_\infty(\Gamma)$.

(c) Show, using the methods of Lemma 2.5.3 and Theorem 2.5.4, that $c_0(\mathbb{R})$ is isomorphic to a subspace of ℓ_∞/c_0.

2.11. Let Γ be an infinite set and let $\mathcal{P}\Gamma$ denote its power set $\mathcal{P}\Gamma = \{A : A \subset \Gamma\}$.

(a) Show that $\ell_1(\mathcal{P}\Gamma)$ is isometric to a subspace of $\ell_\infty(\Gamma)$. [*Hint*: For each $\gamma \in \Gamma$ define $\varphi_\gamma \in \ell_\infty(\mathcal{P}\Gamma)$ by $\varphi_\gamma = 1$ when $\gamma \in A$ and -1 when $\gamma \notin A$.]

(b) Show that if $\ell_1(\Gamma)$ is a quotient of a subspace of X then $\ell_1(\Gamma)$ embeds into X (compare with Problem 2.8).

(c) Deduce that if $\ell_1(\Gamma)$ embeds into X then $\ell_1(\mathcal{P}\Gamma)$ embeds into X^*.

(d) Deduce that ℓ_1^{**} contains an isometric copy of $\ell_1(\mathcal{P}\mathbb{R})$.

Special Types of Bases

We are next going to look a bit more carefully at special classes of bases. In particular we will consider the notion of an unconditional basis already hinted at in the previous chapter. Much of this chapter is based on classical work of James in the early 1950s.

3.1 Unconditional bases

Definition 3.1.1. A basis $(e_n)_{n=1}^\infty$ of a Banach space X is called *unconditional* if for each $x \in X$ the series $\sum_{n=1}^\infty e_n^*(x)e_n$ converges unconditionally.

Obviously, $(e_n)_{n=1}^\infty$ is an unconditional basis of X if and only if $(e_{\pi(n)})_{n=1}^\infty$ is a basis of X for all permutations $\pi : \mathbb{N} \to \mathbb{N}$.

Example 3.1.2. The standard unit vector basis is an unconditional basis of c_0 and ℓ_p for $1 \le p < \infty$. An example of a basis which is *conditional* (i.e., not unconditional) is the *summing basis* of c_0, $(f_n)_{n=1}^\infty$, defined as

$$ f_n = e_1 + \cdots + e_n, \quad n \in \mathbb{N}. $$

To see that (f_n) is a basis for c_0 we prove that for each $\xi = (\xi(n))_{n=1}^\infty \in c_0$ we have $\xi = \sum_{n=1}^\infty f_n^*(\xi)f_n$, where $f_n^* = e_n^* - e_{n+1}^*$ are the biorthogonal functionals of (f_n). Given $N \in \mathbb{N}$,

$$ \sum_{n=1}^N f_n^*(\xi)f_n = \sum_{n=1}^N \left(e_n^*(\xi) - e_{n+1}^*(\xi)\right)f_n $$

$$ = \sum_{n=1}^N (\xi(n) - \xi(n+1))f_n $$

$$ = \sum_{n=1}^N \xi(n)f_n - \sum_{n=2}^{N+1} \xi(n)f_{n-1} $$

$$= \sum_{n=1}^{N} \xi(n)(f_n - f_{n-1}) - \xi(N+1)f_N$$

$$= \left(\sum_{n=1}^{N} \xi(n)e_n \right) - \xi(N+1)f_N.$$

Therefore,

$$\left\| \xi - \sum_{n=1}^{N} f_n^*(\xi)f_n \right\|_\infty = \left\| \sum_{N+1}^{\infty} \xi(n)e_n + \xi(N+1)f_N \right\|_\infty$$

$$\leq \left\| \sum_{N+1}^{\infty} \xi_n e_n \right\|_\infty + |\xi(N+1)| \, \|f_N\|_\infty \overset{N \to \infty}{\to} 0,$$

and $(f_n)_{n=1}^{\infty}$ is a basis.

Now we will identify the set, S, of coefficients $(\alpha_n)_{n=1}^{\infty}$ such that the series $\sum_{n=1}^{\infty} \alpha_n f_n$ converges. In fact we have that $(\alpha_n) \in S$ if and only if there exists $\xi = (\xi(n)) \in c_0$ so that $\alpha_n = \xi(n) - \xi(n+1)$ for all n. Then, clearly, unless the series $\sum_{n=1}^{\infty} \alpha_n$ converges absolutely, the convergence of $\sum_{n=1}^{\infty} \alpha_n f_n$ in c_0 is not equivalent to the convergence of $\sum_{n=1}^{\infty} \epsilon_n \alpha_n f_n$ for any choice of signs $(\epsilon_n)_{n=1}^{\infty}$. Hence (f_n) cannot be unconditional.

Proposition 3.1.3. *A basis $(e_n)_{n=1}^{\infty}$ of a Banach space X is unconditional if and only if there is a constant $K \geq 1$ such that for all $N \in \mathbb{N}$, whenever $a_1, \ldots, a_N, b_1, \ldots, b_N$ are scalars satisfying $|a_n| \leq |b_n|$ for $n = 1, \ldots, N$, then the following inequality holds:*

$$\left\| \sum_{n=1}^{N} a_n e_n \right\| \leq K \left\| \sum_{n=1}^{N} b_n e_n \right\|. \tag{3.1}$$

Proof. Assume $(e_n)_{n=1}^{\infty}$ is unconditional. If $\sum_{n=1}^{\infty} a_n e_n$ is convergent then $\sum_{n=1}^{\infty} t_n a_n e_n$ converges for all $(t_n) \in \ell_\infty$ by Proposition 2.4.9. By the Banach-Steinhaus theorem, the linear map

$$T_{(t_n)} : X \to X, \quad \sum_{n=1}^{\infty} a_n e_n \to \sum_{n=1}^{\infty} t_n a_n e_n$$

is continuous. Now the Uniform Boundedness principle yields K so that equation (3.1) holds.

Conversely, let us take a convergent series $\sum_{n=1}^{\infty} a_n e_n$ in X. We are going to prove that the subseries $\sum_{k=1}^{\infty} a_{n_k} e_{n_k}$ is convergent for any increasing sequence of integers $(n_k)_{k=1}^{\infty}$. By Lemma 2.4.2, given $\epsilon > 0$ there is $N = N(\epsilon) \in \mathbb{N}$ such that if $m_2 > m_1 \geq N$ then

$$\left\| \sum_{n=m_1+1}^{m_2} a_n e_n \right\| < \frac{\epsilon}{K}.$$

By hypothesis, if $N \leq n_k < \cdots < n_{k+l}$ we have

$$\left\| \sum_{j=k+1}^{k+l} a_{n_j} e_{n_j} \right\| \leq K \left\| \sum_{j=n_k+1}^{n_{k+l}} a_j e_j \right\| < \epsilon,$$

and so $\sum_{k=1}^{\infty} a_{n_k} e_{n_k}$ is Cauchy.

□

Definition 3.1.4. Let (e_n) be an unconditional basis of a Banach space X. The *unconditional basis constant*, K_u, of (e_n) is the least constant K so that equation (3.1) holds. We then say that (e_n) is K-*unconditional* whenever $K \geq K_u$.

Remark 3.1.5. Suppose $(e_n)_{n=1}^{\infty}$ is an unconditional basis for a Banach space X. For each sequence of scalars (α_n) with $|\alpha_n| = 1$, let $T_{(\alpha_n)} : X \to X$ be the isomorphism defined by $T_{(\alpha_n)}(\sum_{n=1}^{\infty} a_n e_n) = \sum_{n=1}^{\infty} \alpha_n a_n e_n$. Then

$$K_u = \sup \left\{ \|T_{(\alpha_n)}\| : (\alpha_n) \text{ scalars, } |\alpha_n| = 1 \text{ for all } n \right\}.$$

If $(e_n)_{n=1}^{\infty}$ is an unconditional basis of X and A is any subset of the integers then there is a linear projection P_A from X onto $[e_k : k \in A]$ defined for each $x = \sum_{k=1}^{\infty} e_k^*(x) e_k$ by

$$P_A(x) = \sum_{k \in A} e_k^*(x) e_k.$$

P_A is bounded by the same argument used in the proof of Proposition 3.1.3. $\{P_A : A \subset \mathbb{N}\}$ are the natural projections associated to the unconditional basis (e_n) and the number

$$K_s = \sup_A \|P_A\|$$

(which is finite by the Uniform Boundedness principle) is called the *suppression constant* of the basis. Let us observe that in general we have

$$1 \leq K_s \leq K_u \leq 2K_s.$$

In the older literature the term *absolute basis* is often used in place of unconditional basis, but this usage has largely disappeared. Unconditional bases seem to have first appeared in work of Karlin in 1948 [107]. In particular Karlin proved that $\mathcal{C}[0,1]$ fails to have an unconditional basis. We will prove this later in this chapter.

3.2 Boundedly-complete and shrinking bases

Suppose $(e_n)_{n=1}^{\infty}$ is a basis for a Banach space X with biorthogonal functionals $(e_n^*)_{n=1}^{\infty} \subset X^*$. One of our goals in this section is to establish necessary and sufficient conditions for $(e_n^*)_{n=1}^\infty$ to be a basis for X^*. This is not always the case. For example, the coordinate functionals of the standard basis of ℓ_1 cannot be a basis for ℓ_1^* since ℓ_1^* is not separable. We will first prove that, at least, $(e_n^*)_{n=1}^\infty$ is a basic sequence in X^*.

Proposition 3.2.1. *Suppose that $(e_n^*)_{n=1}^\infty$ is the sequence of biorthogonal functionals associated to a basis $(e_n)_{n=1}^\infty$ of a Banach space X. Then $(e_n^*)_{n=1}^\infty$ is a basic sequence in X^* with basis constant no bigger than that of $(e_n)_{n=1}^\infty$.*

Proof. Given $(e_n^*)_{n=1}^\infty$, consider the subspace H of X^* given by

$$H = \left\{ x^* \in X^* \ : \ \|S_N^*(x^*) - x^*\| \to 0 \right\}, \tag{3.2}$$

where $(S_N^*)_{N=1}^\infty$ is the sequence of adjoint operators of the partial sum projections associated to $(e_n)_{n=1}^\infty$:

$$S_N^* : X^* \to X^*, \quad S_N^*(x^*) = \sum_{k=1}^N x^*(e_k)e_k^*.$$

Clearly $(e_n^*)_{n=1}^\infty$ is a basis for H, hence $(e_n^*)_{n=1}^\infty$ is basic. Notice that

$$\sup_N \|S_N^*|_H\|_{H \to H} \le \sup_N \|S_N^*\|_{X^* \to X^*} = \sup_N \|S_N\|,$$

which gives the latter statement in the proposition.

$\square$

Definition 3.2.2. Suppose that X is a normed space and that Y is a subspace of X^*. Let us consider a new norm on X defined by

$$\|x\|_Y = \sup \left\{ |y^*(x)| : y^* \in Y, \|y^*\| = 1 \right\}.$$

If there is a constant $c \le 1$ such that for all $x \in X$ we have

$$c\|x\| \le \|x\|_Y \le \|x\|,$$

then Y is said to be a *c-norming subspace for X in X^**.

The next result shows that if $(e_n)_{n=1}^\infty$ is a basis for a Banach space X with basis constant K then the subspace $[e_n^*] = H$ of X^* is reasonably big, in the sense that it is $1/K$-norming for X.

Lemma 3.2.3. *Let $(e_n)_{n=1}^\infty$ be a basis for a Banach space X with basis constant K and biorthogonal functionals $(e_n^*)_{n=1}^\infty$. Then $H = [e_n^*]$ is a K^{-1}-norming subspace for X in X^*. Thus the norm on X defined by*

$$\|x\|_H = \sup \left\{ |h(x)| \ : \ h \in H, \|h\| \le 1 \right\},$$

satisfies

$$\frac{\|x\|}{K} \le \|x\|_H \le \|x\| \tag{3.3}$$

for all $x \in X$.

Proof. Let $x \in X$. Since $H \subset X^*$, it follows immediately that $\|x\|_H \leq \sup\{|x^*(x)| : x^* \in X^*, \|x^*\| \leq 1\} = \|x\|$. For the other inequality, pick $x^* \in S_{X^*}$ so that $x^*(x) = \|x\|$. Then for each N,

$$\frac{|(S_N^* x^*)x|}{K} \leq \frac{|(S_N^* x^*)x|}{\|S_N^* x^*\|} \leq \sup\{|h(x)| : h \in H, \|h\| \leq 1\} = \|x\|_H.$$

Now we let N tend to infinity and use that if $\|S_N x - x\| \to 0$ then $|S_N^* x^*(x)| = |x^*(S_N x)| \to \|x\|$.

□

Remark 3.2.4. The previous result can be interpreted as saying that X embeds isomorphically in H^* via the map $x \mapsto j(x)|_H$, where j is the natural embedding of X in its second dual X^{**}. In the case that the basis $(e_n)_{n=1}^\infty$ is monotone, equation (3.3) implies that X embeds isometrically in H^*.

Definition 3.2.5. A basis $(e_n)_{n=1}^\infty$ of a Banach space X is *shrinking* if the sequence of its biorthogonal functionals $(e_n^*)_{n=1}^\infty$ is a basis for X^*, i.e., if $[e_n^*] = X^*$.

Proposition 3.2.6. *A basis $(e_n)_{n=1}^\infty$ of a Banach space X is shrinking if and only if whenever $x^* \in X^*$,*

$$\lim_{N \to \infty} \left\|x^*|_{[e_n]_{n>N}}\right\| = 0, \tag{3.4}$$

where

$$\left\|x^*|_{[e_n]_{n>N}}\right\| = \sup\left\{|x^*(y)| : y \in [e_n]_{n>N}\right\}.$$

Proof. Suppose that $(e_n^*)_{n=1}^\infty$ is a basis for X^*. Every $x^* \in X^*$ can be decomposed as $(x^* - S_N^* x^*) + S_N^* x^*$ for each N. Then the claim follows because

$$\left\|x^*|_{[e_n]_{n>N}}\right\| \leq \left\|\left(x^* - S_N^* x^*\right)|_{[e_n]_{n>N}}\right\| + \underbrace{\left\|S_N^* x^*|_{[e_n]_{n>N}}\right\|}_{\text{this term is 0}} \leq \|x^* - S_N^* x^*\|$$

and we know that $\lim_{N \to \infty} \|x^* - S_N^* x^*\| = 0$.

For the converse, assume that (3.4) holds. Let K be the basis constant of $(e_n)_{n=1}^\infty$ and x^* be an element in X^*. Since for any $x \in X$, $(I_X - S_N)(x)$ is in the subspace $[e_n]_{n>N}$, we have

$$|(x^* - S_N^* x^*)(x)| = |x^*(I_X - S_N)(x)|$$
$$\leq \left\|x^*|_{[e_n]_{n \geq N+1}}\right\| \|I_X - S_N\| \|x\|$$
$$\leq (K+1) \left\|x^*|_{[e_n]_{n \geq N+1}}\right\| \|x\|.$$

Hence $\|x^* - S_N^* x^*\| \leq (K+1) \left\|x^*|_{[e_n]_{n \geq N+1}}\right\|$ and so $\lim_{N \to \infty} \|x^* - S_N^* x^*\| = 0$. Thus $X^* = [e_n^*]$ and we are done.

□

Proposition 3.2.7. *A basis $(e_n)_{n=1}^\infty$ of a Banach space X is shrinking if and only if every bounded block basic sequence of $(e_n)_{n=1}^\infty$ is weakly null.*

Proof. Assume $(e_n)_{n=1}^\infty$ is not shrinking. Then $H \neq X^*$, hence there is x^* in $X^* \setminus [e_n^*]$, $\|x^*\| = 1$, such that the series $\sum_{n=1}^\infty x^*(e_n)e_n^*$ converges to x^* in the weak* topology of X^* but it does not converge in the norm topology of X^*. Using the Cauchy condition we can find two sequences of positive integers (p_n), (q_n) and $\delta > 0$ such that $p_1 \leq q_1 < p_2 \leq q_2 < p_3 \leq q_3 < \cdots$ and $\|\sum_{n=p_k}^{q_k} x^*(e_n)e_n^*\| > \delta$ for all $k \in \mathbb{N}$. Thus for each k there exists $x_k \in X$, $\|x_k\| = 1$, for which $\sum_{n=p_k}^{q_k} x^*(e_n)e_n^*(x_k) > \delta$. Put

$$y_k = \sum_{n=p_k}^{q_k} e_n^*(x_k)e_n, \quad k = 1, 2, \ldots$$

$(y_k)_{k=1}^\infty$ is a block basis of $(e_n)_{n=1}^\infty$ which is not weakly null since $x^*(y_k) > \delta$ for all k.

The converse implication follows readily from Proposition 3.2.6. □

Definition 3.2.8. Let X be a Banach space. A basis $(e_n)_{n=1}^\infty$ for X is *boundedly-complete* if whenever $(a_n)_{n=1}^\infty$ is a sequence of scalars such that

$$\sup_N \left\| \sum_{n=1}^N a_n e_n \right\| < \infty,$$

then the series $\sum_{n=1}^\infty a_n e_n$ converges.

Example 3.2.9. (a) The canonical basis of ℓ_p for $1 < p < \infty$ is both shrinking and boundedly-complete. In ℓ_1 the canonical basis is obviously boundedly-complete, but ℓ_1 cannot have a shrinking basis because its dual, ℓ_∞, is not separable.

(b) As for c_0, its natural basis is shrinking but not boundedly complete: the series $\sum_{n=1}^\infty e_n$ is not convergent in c_0 despite the fact that

$$\sup_N \left\| \sum_{n=1}^N e_n \right\|_\infty = \sup_N \left\| (\underbrace{1, 1, \ldots, 1}_{N}, 0, 0, \ldots) \right\|_\infty = 1.$$

On the other hand, the summing basis of c_0, $(f_n)_{n=1}^\infty$, is not shrinking because the linear functional e_1^* satisfies $e_1^*(f_n) = 1$ for all n, so equation (3.4) cannot hold. $(f_n)_{n=1}^\infty$ is not boundedly-complete either:

$$\sup_N \left\| \sum_{n=1}^N (-1)^n f_n \right\|_\infty = 1,$$

but the series $\sum_{n=1}^\infty (-1)^n f_n$ is not convergent.

Theorem 3.2.10. *Let* $(e_n)_{n=1}^{\infty}$ *be a basis for a Banach space* X *with biorthogonal functionals* $(e_n^*)_{n=1}^{\infty}$. *The following are equivalent:*

(i) $(e_n)_{n=1}^{\infty}$ *is a boundedly-complete basis for* X,
(ii) $(e_n^*)_{n=1}^{\infty}$ *is a shrinking basis for* H,
(iii) The canonical map $j : X \to H^*$ *defined by* $j(x)(h) = h(x)$, *for all* $x \in X$ *and* $h \in H$, *is an isomorphism.*

Proof. $(i) \Rightarrow (iii)$ Using Remark 3.2.4 we need only show that j is onto. For each $h^* \in H^*$ there exists $x^{**} \in X^{**}$ so that $x^{**}|_H = h^*$. Let us consider the formal series $\sum_{n=1}^{\infty} x^{**}(e_n^*)e_n$ in X. For each $N \in \mathbb{N}$,

$$\sum_{n=1}^{N} x^{**}(e_n^*)e_n = S_N^{**}x^{**},$$

where S_N^{**} is the double adjoint of S_N. Hence

$$\left\| \sum_{n=1}^{N} x^{**}(e_n^*)e_n \right\| = \|S_N^{**}x^{**}\| \leq \sup_N \|S_N^{**}\| \, \|x^{**}\| = K \, \|x^{**}\|.$$

$(e_n)_{n=1}^{\infty}$ boundedly-complete implies that $\sum_{n=1}^{\infty} x^{**}(e_n^*)e_n$ converges to some $x \in X$. Now $j(x) = h^*$ since for each $k \in \mathbb{N}$ we have

$$j(x)(e_k^*) = e_k^*(x) = x^{**}(e_k^*) = h^*(e_k^*).$$

$(iii) \Rightarrow (ii)$ Assume that $j : X \to H^*$ is an isomorphism onto. Then $(j(e_n))_{n=1}^{\infty}$ is a basis for H^* and it is also the sequence of coordinate functionals for $(e_n^*)_{n=1}^{\infty}$. That means $(e_n^*)_{n=1}^{\infty}$ is a shrinking basis for H.

$(ii) \Rightarrow (i)$ Let (a_n) be a sequence of scalars for which

$$\sup_N \left\| \sum_{n=1}^{N} a_n e_n \right\| < \infty. \tag{3.5}$$

For each N the norm of $j(\sum_{n=1}^{N} a_n e_n)$ as a linear functional on H is equivalent to the norm of $\sum_{n=1}^{N} a_n e_n$ in X. Therefore, by (3.5), $(\sum_{n=1}^{N} a_n j(e_n))_{N=1}^{\infty}$ is a bounded sequence in X^{**}. The Banach-Alaoglu theorem yields the existence of a weak* cluster point, $h^* \in X^{**}$, of that sequence. In particular we have $h^*(e_n^*) = a_n$ for each n. Using the hypothesis we can write

$$h^* = \sum_{n=1}^{\infty} h^*(e_n^*)j(e_n) = \sum_{n=1}^{\infty} a_n j(e_n),$$

where the series converges in the norm topology of H^*. Since j is an isomorphism, the series $\sum_{n=1}^{\infty} a_n e_n$ converges in the norm topology of X.

$\square$

Corollary 3.2.11. c_0 *has no boundedly-complete basis.*

Proof. It follows from Theorem 3.2.10, taking into account that c_0 is not isomorphic to a dual space (Corollary 2.5.6). $\qquad\qquad\square$

Theorem 3.2.12. *Let* $(e_n)_{n=1}^\infty$ *be a basis for a Banach space* X *with biorthogonal functionals* $(e_n^*)_{n=1}^\infty$. *The following are equivalent:*

(i) $(e_n)_{n=1}^\infty$ *is a shrinking basis for* X,
(ii) $(e_n^*)_{n=1}^\infty$ *is a boundedly-complete basis for* H,
(iii) $H = X^*$.

Proof. $(i) \Rightarrow (ii)$ Suppose that $(a_n)_{n=1}^\infty$ is a sequence of scalars such that the sequence $(\sum_{n=1}^N a_n e_n^*)_{N=1}^\infty$ is bounded in X^* and let $x^* \in X^*$ be a weak* cluster point of this sequence. Since $\lim_{N\to\infty}(\sum_{n=1}^N a_n e_n^*)(e_k) = a_k$, it follows that $x^*(e_k) = a_k$ for each k. Thus the series $\sum_{n=1}^\infty a_n e_n^*$ converges to x^*.

$\quad (ii) \Rightarrow (i)$ Suppose now that $(e_n^*)_{n=1}^\infty$ is boundedly-complete. For any x^* in X^* we know that the series $\sum_{n=1}^\infty x^*(e_n)e_n^*$ converges in the weak* topology of X^* to x^*. In particular, the sequence $(\sum_{n=1}^N x^*(e_n)e_n^*)_{N=1}^\infty$ is norm-bounded in X^*. Hence, by the bounded-completeness of $(e_n^*)_{n=1}^\infty$, the series $\sum_{n=1}^\infty x^*(e_n)e_n^*$ must converge to x^* in norm, so $(e_n^*)_{n=1}^\infty$ is a basis for X^*.

$\quad (i) \Leftrightarrow (iii)$ is obvious. $\qquad\qquad\square$

Now we come to the main result of the section, which is due to James [80].

Theorem 3.2.13 (James, 1951). *Let* X *be a Banach space. If* X *has a basis* $(e_n)_{n=1}^\infty$ *then* X *is reflexive if and only if* $(e_n)_{n=1}^\infty$ *is both boundedly-complete and shrinking.*

Proof. Assume that X is reflexive and that $(e_n)_{n=1}^\infty$ is a basis for X. Then $X^* = H$. If not, using the Hahn-Banach theorem, one could find $0 \neq x^{**} \in X^{**}$ such that $x^{**}(h) = 0$ for all $h \in H$. By reflexivity there is $0 \neq x = \sum_{n=1}^\infty e_n^*(x)e_n \in X$ such that $x = x^{**}$. In particular we would have $0 = x^{**}(e_n^*) = e_n^*(x)$ for all n, which would imply $x = 0$. Thus $(e_n)_{n=1}^\infty$ is shrinking. Notice that $(e_n)_{n=1}^\infty$ is a basis for X^{**} and is also the sequence of biorthogonal functionals associated to $(e_n^*)_{n=1}^\infty$. That implies that $(e_n^*)_{n=1}^\infty$ is a shrinking basis of $X^* = H$, hence by Theorem 3.2.10, $(e_n)_{n=1}^\infty$ is boundedly-complete.

$\quad$ Conversely, $(e_n)_{n=1}^\infty$ shrinking implies $H = X^*$, and since $(e_n)_{n=1}^\infty$ is boundedly-complete as well, the canonical map $j : X \to H^*$ in Theorem 3.2.10 (iii) is now the canonical embedding of X *onto* X^{**}. $\qquad\qquad\square$

This theorem gives a criterion for reflexivity which is enormously useful, particularly in the construction of examples. Notice that the facts that the canonical basis of ℓ_1 fails to be shrinking and that the canonical basis of c_0 fails to be boundedly-complete are explained now in the nonreflexivity of these spaces.

During the 1960s it was very fashionable to study the structure of Banach spaces by understanding the properties of their bases. Of course, this viewpoint was somewhat undermined when Enflo showed that not every separable Banach space has a basis [54]. One of the high points of this theory was the theorem of Zippin [224] that a Banach space with a basis is reflexive if and only if *every* basis is boundedly complete or if and only if *every* basis is shrinking. Thus, any nonreflexive Banach space which has a basis must have at least one non-boundedly-complete basis and at least one nonshrinking basis.

3.3 Nonreflexive spaces with unconditional bases

Now let us consider the boundedly-complete and shrinking unconditional bases. Again we follow the classic paper of James [80].

Theorem 3.3.1. *Let X be a Banach space with unconditional basis $(u_n)_{n=1}^{\infty}$. The following are equivalent:*

(i) $(u_n)_{n=1}^{\infty}$ *fails to be shrinking,*
(ii) X *contains a complemented subspace isomorphic to* ℓ_1,
(iii) *There exists a complemented block basic sequence* $(y_n)_{n=1}^{\infty}$ *with respect to* $(u_n)_{n=1}^{\infty}$ *which is equivalent to the canonical basis of* ℓ_1,
(iv) X *contains a subspace isomorphic to* ℓ_1.

Proof. The implications $(iii) \Rightarrow (ii) \Rightarrow (iv)$ are obvious.

$(iv) \Rightarrow (i)$ is also immediate because if X contains ℓ_1 then X^* cannot be separable and so $(u_n)_{n=1}^{\infty}$ is not shrinking.

$(i) \Rightarrow (iii)$ If $(u_n)_{n=1}^{\infty}$ is not shrinking, by Proposition 3.2.7 we can find a bounded block basic sequence $(y_k)_{k=1}^{\infty}$ of $(u_n)_{n=1}^{\infty}$, $\delta > 0$, and $x^* \in X^*$ with $\|x^*\| = 1$, such that $x^*(y_k) > \delta$ for all k. Then for any scalars $(a_k) \in c_{00}$ we have

$$\left\| \sum_{k=1}^{\infty} a_k y_k \right\| \geq \left| \sum_{k=1}^{\infty} x^*(y_k) a_k \right|.$$

By picking $\epsilon_k = \text{sgn } a_k$ for each k we obtain

$$\left\| \sum_{k=1}^{\infty} \epsilon_k a_k y_k \right\| \geq \sum_{k=1}^{\infty} |x^*(y_k) a_k| \geq \delta \sum_{k=1}^{\infty} |a_k|.$$

Being a block basis of $(u_n)_{n=1}^{\infty}$, $(y_k)_{k=1}^{\infty}$ is an unconditional basic sequence with unconditional basis constant $\leq K$. Therefore,

$$\left\| \sum_{k=1}^{\infty} a_k y_k \right\| \geq \delta K^{-1} \sum_{k=1}^{\infty} |a_k|.$$

On the other hand, since (y_k) is bounded, the triangle law yields an upper ℓ_1-estimate for $\| \sum_{k=1}^{\infty} a_k y_k \|$ and hence (y_k) is equivalent to the standard ℓ_1-basis. It remains to define a linear projection from X onto $[y_k]$.

For each k put

$$y_k^* = \frac{1}{x^*(y_k)} \sum_{n=p_k}^{q_k} x^*(u_n)u_n^*.$$

Clearly, the sequence (y_k^*) is orthogonal to (y_k) and $\|y_k^*\| \leq \delta^{-1}K$. For every $N \in \mathbb{N}$ let us consider the projection from X onto $[y_k]_{1 \leq k \leq N}$ defined as

$$P_N(x) = \sum_{k=1}^{N} y_k^*(x)y_k.$$

(P_N) is a bounded sequence: given any $x \in X$ if we pick $\epsilon_k = \operatorname{sgn} y_k^*(x)$ we have

$$\|P_N(x)\| \leq K \sum_{k=1}^{N} |y_k^*(x)|$$

$$= K \sum_{k=1}^{N} \epsilon_k y_k^*(x)$$

$$= K \sum_{k=1}^{N} \sum_{n=p_k}^{q_k} \frac{\epsilon_k}{x^*(y_k)} x^*(u_n)u_n^*(x)$$

$$= K x^* \Big(\sum_{k=1}^{N} \sum_{n=p_k}^{q_k} \frac{\epsilon_k}{x^*(y_k)} u_n^*(x)u_n \Big)$$

$$\leq K^2 \max_k \Big| \frac{1}{x^*(y_k)} \Big| \|x\|$$

$$\leq K^2 \delta^{-1} \|x\|.$$

Since $\lim_{N \to \infty} P_N(x)$ exists for each x, by the Banach-Steinhaus theorem, the operator

$$P : X \to [y_k], \quad x \mapsto P(x) = \sum_{k=1}^{\infty} y_k^*(x)y_k$$

is bounded by $K^2 \delta^{-1}$ and is obviously the desired projection. $\qquad \square$

Theorem 3.3.2. *Let X be a Banach space with unconditional basis $(u_n)_{n=1}^{\infty}$. The following are equivalent:*

(i) $(u_n)_{n=1}^{\infty}$ fails to be boundedly-complete,
(ii) X contains a complemented subspace isomorphic to c_0,
(iii) There exists a complemented block basic sequence $(y_n)_{n=1}^{\infty}$ with respect to $(u_n)_{n=1}^{\infty}$ equivalent to the canonical basis of c_0,
(iv) X contains a subspace isomorphic to c_0.

Proof. Note that (ii) and (iv) are equivalent since c_0 is separably injective (Sobczyk's theorem, Theorem 2.5.8).

$(i) \Rightarrow (iii)$ If $(u_n)_{n=1}^\infty$ is not boundedly-complete there exists a sequence of scalars (a_n) such that $\sup_N \| \sum_{n=1}^N a_n u_n \| < \infty$ but the series $\sum_{n=1}^\infty a_n u_n$ does not converge in X.

Given any $x^* \in X^*$, pick $\epsilon_n = \mathrm{sgn}\, x^*(u_n)$. By the unconditionality of the basis there exists K so that

$$\sum_{n=1}^N |a_n||x^*(u_n)| = \sum_{n=1}^N \epsilon_n a_n x^*(u_n) \leq K \|x\| \left\| \sum_{n=1}^N a_n u_n \right\|.$$

So the series of scalars $\sum_{n=1}^\infty |x^*(a_n u_n)|$ converges for all $x^* \in X^*$. That is, $\sum_{n=1}^\infty a_n u_n$ is a WUC series in X that is not unconditionally convergent. Proposition 2.4.7 yields a bounded operator $T : c_0 \to X$ such that $T(e_n) = a_n u_n$ for all n, where (e_n) denotes the standard unit vector basis of c_0. Furthermore, by Proposition 2.4.8, T cannot be compact. Using Theorem 2.4.10 we can extract a block basic sequence (x_k) with respect to the canonical basis of c_0 such that $T|_{[x_k]}$ is an isomorphism onto its range. Then $y_k = Tx_k$ defines a block basic sequence in X with respect to the basis $(u_n)_{n=1}^\infty$ such that $[y_k]$ is isomorphic to c_0. Corollary 2.5.9 implies that $[y_k]$ is complemented in X.

$(iii) \Rightarrow (ii)$ is obvious.

$(ii) \Rightarrow (i)$ Suppose that (ii) holds and that $(u_n)_{n=1}^\infty$ is boundedly-complete. Then, by Theorem 3.2.10, X is a dual space and so there is a bounded projection of X^{**} onto X (see the discussion after Proposition 2.5.2). Hence there is a projection of X^{**} onto a subspace E of X isomorphic to c_0. However, if E is a subspace of X then E^{**} embeds as a subspace of X^{**} (it can be identified with $E^{\perp\perp}$ which is also the weak* closure of E). Hence there is a projection of E^{**} onto E. This contradicts Theorem 2.5.5. $\qquad\square$

The following theorem is again due to James [80] except that the last statement was proved earlier, using different techniques, by Karlin [107].

Theorem 3.3.3. *Suppose that X is a Banach space with an unconditional basis. If X is not reflexive then either c_0 is complemented in X, or ℓ_1 is complemented in X (or both). In either case X^{**} is nonseparable.*

Proof. The first statement of the theorem follows immediately from Theorem 3.2.13, Theorem 3.3.1, and Theorem 3.3.2. Now, for the latter statement, if c_0 were complemented in X then X^{**} would contain a (complemented) copy ℓ_∞. If ℓ_1 were complemented in X then X^* would be nonseparable since it would contain a (complemented) copy of ℓ_∞. In either case, X^{**} is nonseparable. $\qquad\square$

3.4 The James space $\mathcal{J}$

Continuing with the classic paper of James [80] we come to his construction
of one of the most important examples in Banach space theory. This space,
nowadays known as the James space, is, in fact, quite a natural space consisting of sequences of bounded 2-variation. The James space will provide an
example of a Banach space with a basis but with no unconditional basis; it
also answered several other open questions at the time. For example, it was
not known if a Banach space X was necessarily reflexive if its bidual was
separable. The James space $\mathcal{J}$ is separable and has codimension one in $\mathcal{J}^{**}$,
and so gives a counterexample. Later, James [81] went further and modified
the definition of the norm to make $\mathcal{J}$ isometric to $\mathcal{J}^{**}$, thus showing that
a Banach space can be isometrically isomorphic to its bidual yet fail to be
reflexive!

Let us define $\tilde{\mathcal{J}}$ to be the space of all sequences $\xi = (\xi(n))_{n=1}^{\infty}$ of real
numbers with finite square variation; that is, $\xi \in \tilde{\mathcal{J}}$ if and only if there is a
constant M so that for every choice of integers $(p_j)_{j=0}^{n}$ with $1 \leq p_0 < p_1 < \cdots < p_n$ we have

$$\sum_{j=1}^{n} (\xi(p_j) - \xi(p_{j-1}))^2 \leq M^2.$$

It is easy to verify that if $\xi \in \tilde{\mathcal{J}}$ then $\lim_{n \to \infty} \xi(n)$ exists. We then define
$\mathcal{J}$ as the subspace of $\tilde{\mathcal{J}}$ of all ξ so that $\lim_{n \to \infty} \xi(n) = 0$.

Definition 3.4.1. The *James space $\mathcal{J}$* is the (real) Banach space of all se-
quences $\xi = (\xi(n))_{n=1}^{\infty} \in \tilde{\mathcal{J}}$ such that $\lim_{n \to \infty} \xi(n) = 0$, endowed with the
norm

$$\|\xi\|_{\mathcal{J}} = \frac{1}{\sqrt{2}} \sup \left\{ \left((\xi(p_n) - \xi(p_0))^2 + \sum_{k=1}^{n} (\xi(p_k) - \xi(p_{k-1}))^2 \right)^{1/2} \right\},$$

where the supremum is taken over all $n \in \mathbb{N}$, and all choices of integers $(p_j)_{j=0}^{n}$
with $1 \leq p_0 < p_1 < \cdots < p_n$.

The definition of the norm in the James space is not quite natural; clearly,
the norm is equivalent to the alternative norm given by the formula

$$\|\xi\|_0 = \sup \left\{ \left(\sum_{k=1}^{n} (\xi(p_k) - \xi(p_{k-1}))^2 \right)^{1/2} \right\},$$

where, again, the supremum is taken over all sequences of integers $(p_j)_{j=0}^{n}$
with $1 \leq p_0 < p_1 < \cdots < p_n$. In fact,

$$\frac{1}{\sqrt{2}} \|\xi\|_0 \leq \|\xi\|_{\mathcal{J}} \leq \sqrt{2} \|\xi\|_0, \qquad \xi \in \mathcal{J}.$$

Notice that $\|e_n\|_{\mathcal{J}} = 1$ for all n, but $\|e_n\|_0 = \sqrt{2}$ for $n \geq 2$.

We also note that $\| \cdot \|_{\mathcal{J}}$ can be canonically extended to $\tilde{\mathcal{J}}$ by

$$\|\xi\|_{\mathcal{J}} = \frac{1}{\sqrt{2}} \sup \left\{ \left((\xi(p_n) - \xi(p_0))^2 + \sum_{k=1}^{n} (\xi(p_k) - \xi(p_{k-1}))^2 \right)^{1/2} \right\},$$

but this defines only a seminorm on $\tilde{\mathcal{J}}$ vanishing on all constant sequences.

Proposition 3.4.2. *The sequence $(e_n)_{n=1}^{\infty}$ of standard unit vectors is a monotone basis for $\mathcal{J}$ in both norms $\| \cdot \|_{\mathcal{J}}$ and $\| \cdot \|_0$.*

Proof. We will leave for the reader the verification that $(e_n)_{n=1}^{\infty}$ is a monotone basic sequence in both norms. To prove it is a basis we need only consider the norm $\| \cdot \|_0$.

Suppose $\xi \in \mathcal{J}$. For each N let

$$\xi_N = \xi - \sum_{j=1}^{N} \xi(j) e_j.$$

Given $\epsilon > 0$, pick $1 \le p_0 < p_1 < \cdots < p_n$ for which

$$\sum_{j=1}^{n} (\xi(p_j) - \xi(p_{j-1}))^2 > \|\xi\|_0^2 - \epsilon^2.$$

In order to estimate the norm of ξ_N when $N > p_n$ it is enough to consider positive integers $q_0 \le q_1 < q_2 < \cdots < q_m$, where $N \le q_0$. Then for the partition $1 \le p_0 < p_1 < \cdots < p_n < q_0 < q_2 < \cdots < q_m$ we have

$$\|\xi\|_0^2 \ge \sum_{j=1}^{n} (\xi(p_j) - \xi(p_{j-1}))^2 + (\xi(q_0) - \xi(p_n))^2 + \sum_{j=1}^{m} (\xi(q_j) - \xi(q_{j-1}))^2$$

$$\ge \sum_{j=1}^{n} (\xi(p_j) - \xi(p_{j-1}))^2 + \sum_{j=1}^{m} (\xi(q_j) - \xi(q_{j-1}))^2.$$

Hence

$$\sum_{j=1}^{m} (\xi(q_j) - \xi(q_{j-1}))^2 \le \epsilon^2.$$

Thus, $\|\xi_N\|_0 < \epsilon$ for $N > p_n$.

Proposition 3.4.3. *Let $(\eta_k)_{k=1}^{\infty}$ be a normalized block basic sequence with respect to $(e_n)_{n=1}^{\infty}$ in $(\mathcal{J}, \| \cdot \|_0)$. Then, for any sequence of scalars $(\lambda_k)_{k=1}^{n}$ the following estimate holds:*

$$\left\| \sum_{k=1}^{n} \lambda_k \eta_k \right\|_0 \le \sqrt{5} \left(\sum_{k=1}^{n} \lambda_k^2 \right)^{1/2}.$$

Proof. For each k let

$$\eta_k = \sum_{j=q_{k-1}+1}^{q_k} \eta_k(j)e_j$$

where $0 = q_0 < q_1 < \ldots$, and put

$$\xi = \sum_{k=1}^{n} \lambda_k \eta_k.$$

Suppose $1 \le p_0 < p_1 < \cdots < p_m$. Fix $i \le n$. Let A_i be the set of k so that $q_{i-1} < p_{k-1} < p_k \le q_i$. If $k \in A_i$,

$$\xi(p_k) - \xi(p_{k-1}) = \lambda_i(\eta_i(p_k) - \eta_i(p_{k-1})).$$

Hence

$$\sum_{k \in A_i} (\xi(p_k) - \xi(p_{k-1}))^2 \le \lambda_i^2.$$

If $A = \cup_i A_i$ we thus have

$$\sum_{k \in A} (\xi(p_k) - \xi(p_{k-1}))^2 \le \sum_{i=1}^{n} \lambda_i^2.$$

Let B be the set of $1 \le k \le m$ with $k \notin A$. For each such k there exist $i = i(k), j = j(k)$ so that $q_{i-1} < p_{k-1} \le q_i$ and $q_{j-1} < p_k \le q_j$. Then,

$$\begin{aligned}(\xi(p_k) - \xi(p_{k-1}))^2 &= (\lambda_j \eta_j(p_k) - \lambda_i \eta_i(p_{k-1}))^2 \\ &\le 2(\lambda_j^2 \eta_j(p_k)^2 + \lambda_i^2 \eta_j(p_{k-1})^2) \\ &\le 2(\lambda_j^2 + \lambda_i^2).\end{aligned}$$

Thus,

$$\sum_{k=1}^{m} (\xi(p_k) - \xi(p_{k-1}))^2 \le \sum_{i=1}^{n} \lambda_i^2 + 2\sum_{k \in B} \lambda_{i(k)}^2 + 2\sum_{k \in B} \lambda_{j(k)}^2.$$

Since the $i(k)$'s and similarly the $j(k)$'s are distinct for $k \in B$, it follows that

$$\sum_{k=1}^{m} (\xi(p_k) - \xi(p_{k-1}))^2 \le 5\sum_{i=1}^{n} \lambda_i^2,$$

and this completes the proof. □

Proposition 3.4.4. *The sequence $(e_n)_{n=1}^{\infty}$ is a shrinking basis for $\mathcal{J}$ (for both norms $\|\cdot\|_{\mathcal{J}}$ and $\|\cdot\|_0$).*

Proof. We will prove that every bounded block basic sequence of (e_n) is weakly null and then we will appeal to Proposition 3.2.7. Let $(\eta_k)_{k=1}^{\infty}$ be a normalized block basic sequence in $(\mathcal{J}, \|\cdot\|_0)$. Using Proposition 3.4.3, the operator S : $\ell_2 \to [\eta_k] \subset \mathcal{J}$ defined for each $\lambda = (\lambda_k) \in \ell_2$ by

$$S(\lambda) = \sum_{k=1}^{\infty} \lambda_k \eta_k$$

is bounded. The norm-continuity of S implies that S is weak-to-weak continuous. Since the sequence of the unit vector basis of ℓ_2 is weakly null, it follows that their images, the block basic sequence $(\eta_k)_{k=1}^{\infty}$, must converge to 0 weakly as well.

$\square$

Remark 3.4.5. Notice that the standard unit vector basis of $\mathcal{J}$ is not boundedly-complete since

$$\left\| \sum_{n=1}^{N} e_n \right\|_{\mathcal{J}} = \|(1, 1, \ldots, 1, 0, \ldots)\|_0 = 1$$

for all N, but the series $\sum_{n=1}^{\infty} e_n$ does not converge in $\mathcal{J}$.

Since $(e_n)_{n=1}^{\infty}$ is shrinking we can identify each $x^{**} \in \mathcal{J}^{**}$ with the sequence $\xi(n) = x^{**}(e_n^*)$. Under this identification $\mathcal{J}^{**}$ becomes the space of sequences ξ such that

$$\|\xi\|_{\mathcal{J}^{**}} = \sup_n \|(\xi(1), \ldots, \xi(n), 0, \ldots)\|_{\mathcal{J}} < \infty.$$

Note that we now specialize to the use of the norm $\|\cdot\|_{\mathcal{J}}$ on $\mathcal{J}$. That $\|\cdot\|_{\mathcal{J}^{**}}$ is the bidual norm on $\mathcal{J}^{**}$ follows easily from the fact that the basis $(e_n)_{n=1}^{\infty}$ is monotone. It is clear from the definition that $\mathcal{J}^{**}$ coincides with $\tilde{\mathcal{J}}$, i.e., the space of sequences of bounded square variation.

We have already noticed that the canonical extension of $\|\cdot\|_{\mathcal{J}}$ to $\tilde{\mathcal{J}} = \mathcal{J}^{**}$ is only a seminorm. In fact the relationship between $\|\cdot\|_{\mathcal{J}^{**}}$ and $\|\cdot\|_{\mathcal{J}}$ is

$$\|\xi\|_{\mathcal{J}^{**}} = \max(\|\xi\|_{\mathcal{J}}, \|\xi\|_1),$$

where

$$\|\xi\|_1 = \frac{1}{\sqrt{2}} \sup \left\{ \left(\xi(p_n)^2 + \xi(p_0)^2 + \sum_{k=1}^{n} (\xi(p_k) - \xi(p_{k-1}))^2 \right)^{1/2} \right\},$$

and, as usual, the supremum is taken over all $n \in \mathbb{N}$, and all choices of integers $(p_j)_{j=0}^{n}$ with $1 \le p_0 < p_1 < \cdots < p_n$.

Theorem 3.4.6. $\mathcal{J}$ *is a subspace of codimension 1 in* $\mathcal{J}^{**}$ *and* $\mathcal{J}^{**}$ *is isometric to* $\mathcal{J}$.

Proof. Clearly, $\mathcal{J} = \{\xi \in \mathcal{J}^{**} : \lim_{n\to\infty} \xi(n) = 0\}$ has codimension one in its bidual. To prove the fact that it is isometric to its bidual we observe that

$$\|\xi\|_{\mathcal{J}^{**}} = \|(0, \xi(1), \xi(2), \ldots)\|_{\mathcal{J}}, \qquad \xi \in \mathcal{J}^{**}.$$

Let

$$L(\xi) = \lim_{n\to\infty} \xi(n), \qquad \xi \in \mathcal{J}^{**}.$$

We define

$$S(\xi) = (-L(\xi), \xi(1) - L(\xi), \xi(2) - L(\xi), \ldots).$$

S maps $\mathcal{J}^{**}$ onto $\mathcal{J}$ and is one-to-one. Since $\|\cdot\|_{\mathcal{J}}$ is a seminorm on $\mathcal{J}^{**}$ vanishing on constants,

$$\|S(\xi)\|_{\mathcal{J}} = \|(0, \xi(1), \ldots)\|_{\mathcal{J}} = \|\xi\|_{\mathcal{J}^{**}}.$$

Thus S is an isometry.

$\square$

Corollary 3.4.7. *$\mathcal{J}$ does not have an unconditional basis.*

Proof. It follows immediately from the separability of $\mathcal{J}^{**}$, Theorem 3.3.3, and Theorem 3.4.6.

$\square$

After the appearance of James's example the term *quasi-reflexive* was often used for Banach spaces X so that X^{**}/X is finite-dimensional.

The ideas of the James construction have been repeatedly revisited to produce more sophisticated examples of similar type. For example, Lindenstrauss [130] showed that for any separable Banach space X there is a Banach space $\mathcal{Z}$ with a shrinking basis such that $\mathcal{Z}^{**}/\mathcal{Z}$ is isomorphic to X (see Section 13.1).

3.5 A litmus test for unconditional bases

We now want to go a little further and show that $\mathcal{J}$ cannot even be isomorphic to a subspace of a Banach space with an unconditional basis. We therefore need to identify a property of subspaces of spaces with unconditional bases which we can test. For this we use Pełczyński's property (u) introduced in 1958 [168].

Definition 3.5.1. A Banach space X *has property (u)* if whenever $(x_n)_{n=1}^{\infty}$ is a weakly Cauchy sequence in X, there is a WUC series $\sum_{k=1}^{\infty} u_k$ in X so that

$$x_n - \sum_{k=1}^{n} u_k \to 0 \text{ weakly.}$$

Proposition 3.5.2. *If a Banach space X has property (u) then every closed subspace Y of X has property (u).*

Proof. Let (y_s) be a weakly Cauchy sequence in a closed subspace Y of X. Since X has property (u), there is a WUC series $\sum_{i=1}^{\infty} u_i$ in X so that the sequence $(y_s - \sum_{i=1}^{s} u_i)$ converges to 0 weakly. By Mazur's theorem there is a sequence of convex combinations of members of $(y_s - \sum_{i=1}^{s} u_i)$ that converges to 0 in norm. Using the Cauchy condition we find integers (p_k), $0 = p_0 < p_1 < p_2 < \ldots$, and convex combinations $(\sum_{j=p_{k-1}+1}^{p_k} \lambda_j(y_j - \sum_{i=1}^{j} u_i))_{k=1}^{\infty}$ such that

$$\left\| \sum_{j=p_{k-1}+1}^{p_k} \lambda_j(y_j - \sum_{i=1}^{j} u_i) \right\| \le 2^{-k} \quad \text{for all } k.$$

Put $z_0 = 0$, and for each integer $k \ge 1$ let

$$z_k = \sum_{j=p_{k-1}+1}^{p_k} \lambda_j y_j \in Y.$$

Then for any $x^* \in X^*$, $\|x^*\| = 1$, we have

$$|x^*(z_k - z_{k-1})| \le 2^{-k} + 2^{1-k}$$
$$+ \left| x^* \left(\sum_{j=p_{k-1}+1}^{p_k} \lambda_j \sum_{i=p_{k-2}+1}^{j} u_i - \sum_{j=p_{k-2}+1}^{p_{k-1}} \lambda_j \sum_{i=p_{k-2}+1}^{j} u_i \right) \right|.$$

Thus,

$$|x^*(z_k - z_{k-1})| \le 3 \cdot 2^{-k} + 2 \sum_{j=p_{k-2}+1}^{p_k} |x^*(u_j)|,$$

which implies

$$\sum_{k=1}^{\infty} |x^*(z_k - z_{k-1})| \le \frac{3}{2} + 4 \sum_{j=1}^{\infty} |x^*(u_j)| < \infty.$$

Therefore, $\sum_{k=1}^{\infty}(z_k - z_{k-1})$ is a WUC series in Y. Now one easily checks that the sequence

$$\left(y_n - \sum_{k=1}^{n}(z_k - z_{k-1}) \right)_{n=1}^{\infty} = (y_n - z_n)_{n=1}^{\infty}$$

converges weakly to 0.

$\square$

Proposition 3.5.3 (Pełczyński [168]). *If a Banach space X has an unconditional basis then X has property (u).*

Proof. Let $(u_n)_{n=1}^{\infty}$ be a K-unconditional basis of X with biorthogonal functionals $(u_n^*)_{n=1}^{\infty}$. If (x_n) is a weakly Cauchy sequence in X then for each k the scalar sequence $(u_k^*(x_n))_{n=1}^{\infty}$ converges, say, to α_k. Hence the sequence

$(\sum_{k=1}^{N} t_k u_k^*(x_n) u_k)_{n=1}^{\infty}$ converges weakly to $\sum_{k=1}^{N} t_k \alpha_k u_k$ for each N and any scalars (t_k). Therefore,

$$\left\| \sum_{k=1}^{N} \epsilon_k \alpha_k u_k \right\| \leq K \sup_n \|x_n\|$$

for all N and any sequence of signs (ϵ_k). Being weakly Cauchy, (x_n) is norm-bounded thus $\sum_{k=1}^{\infty} \alpha_k u_k$ is a WUC series. Put

$$y_n = x_n - \sum_{k=1}^{n} \alpha_k u_k.$$

(y_n) is weakly Cauchy. Also, $\lim_{n \to \infty} u_s^*(y_n) = 0$ for all $s \in \mathbb{N}$. We claim that (y_n) converges weakly to 0. If not, there is $x^* \in X^*$ so that $\lim_{n \to \infty} x^*(y_n) = 1$. Using the Bessaga-Pełczyński selection principle (Proposition 1.3.10) we can extract a subsequence (y_{n_j}) of (y_n) and find a block basic sequence (z_j) of (u_n) such that (z_j) is equivalent to (y_{n_j}) and $\|y_{n_j} - z_j\| \to 0$. We deduce that $x^*(z_j) \to 1$ since

$$|x^*(z_j) - 1| \leq |x^*(z_j - y_{n_j})| + |x^*(y_{n_j}) - 1| \leq \|x^*\| \underbrace{\|z_j - y_{n_j}\|}_{\text{this tends to } 0} + \underbrace{|x^*(y_{n_j}) - 1|}_{\text{this tends to } 0}.$$

Without loss of generality we can assume that $|x^*(z_j)| > 1/2$ for all j. Given $(a_j) \in c_{00}$, by letting $\epsilon_j = \operatorname{sgn} a_j x^*(z_j)$ we have

$$\sum_{j=1}^{\infty} |a_j| |x^*(z_j)| = \left| \sum_{j=1}^{\infty} \epsilon_j a_j x^*(z_j) \right|$$

$$= \left| x^* \left(\sum_{j=1}^{\infty} \epsilon_j a_j z_j \right) \right|$$

$$\leq \|x^*\| K \left\| \sum_{j=1}^{\infty} a_j z_j \right\|.$$

Hence

$$\left\| \sum_{j=1}^{\infty} a_j z_j \right\| \geq \frac{1}{2K \|x^*\|} \sum_{j=1}^{\infty} |a_j|.$$

On the other hand we obtain an upper ℓ_1-estimate for $\| \sum_{j=1}^{\infty} a_j z_j \|$ using the boundedness of the sequence (z_j) and the triangle law. We conclude that (z_j) is equivalent to the standard ℓ_1-basis. This is a contradiction because (z_j) is weakly Cauchy whereas the canonical basis of ℓ_1 is not. Therefore our claim holds and this finishes the proof.

$\square$

Proposition 3.5.4. *(i)* $\mathcal{J}$ *does not have property (u) and so cannot be embedded in any Banach space with an unconditional basis.*

(ii) **(Karlin [107])** $\mathcal{C}[0,1]$ *does not have an unconditional basis, and cannot be embedded in a space with unconditional basis.*

Proof. (*i*) Assume that $\mathcal{J}$ has property (u). Since the sequence defined for each n by $s_n = \sum_{k=1}^{n} e_k$ is weakly Cauchy in $\mathcal{J}$, there exists a WUC series in $\mathcal{J}$, $\sum_{k=1}^{\infty} u_k$, so that the sequence $(\sum_{k=1}^{n} e_k - \sum_{k=1}^{n} u_k)_{n=1}^{\infty}$ converges weakly to 0. One easily notices that the series $\sum_{k=1}^{\infty} u_k$ cannot be unconditionally convergent in $\mathcal{J}$ because that would force the sequence (s_n) to converge weakly to the same limit, when (s_n) is not weakly convergent in $\mathcal{J}$ (it does converge weakly, though, to $(1,1,1,\ldots,1,\ldots) \in \tilde{\mathcal{J}}$). Therefore using Theorem 2.4.11, c_0 embeds in $\mathcal{J}$, which implies that ℓ_∞ embeds in $\mathcal{J}^{**}$, contradicting the separability of $\mathcal{J}^{**}$.

That $\mathcal{J}$ does not embed into any space with unconditional basis follows immediately from Proposition 3.5.2 and Proposition 3.5.3.

(*ii*) This follows from (*i*) because $\mathcal{J}$ embeds isometrically into $\mathcal{C}[0,1]$ by the Banach-Mazur theorem (Theorem 1.4.3). □

Thus we have seen that having an unconditional basis is very special and one cannot rely on the existence of such bases in most spaces. It is, however, true that most of the spaces which are useful in harmonic analysis or partial differential equations such as the spaces L_p for $1 < p < \infty$ do have unconditional bases (which we will see in Chapter 6). We will see also that L_1 fails to have an unconditional basis. It is perhaps reasonable to argue that the reason the spaces L_p for $1 < p < \infty$ seem to be more useful for applications in these areas is precisely because they admit unconditional bases!

From the point of view of abstract Banach space theory, in this context it was natural to ask:

The unconditional basic sequence problem. *Does every Banach space contain at least an unconditional basic sequence?*

This problem was regarded as perhaps the single most important problem in the area after the solution of the approximation problem by Enflo in 1973. Eventually a counterexample was found by Gowers and Maurey in 1993 [71]. The construction is extremely involved but has led to a variety of other applications, some of which we have already met (see e.g. [115], [70], and [72]).

Problems

3.1. Let (u_n) be a K_u-unconditional basis in a Banach space X.

(a) Show that if (y_n) is a block basic sequence of (u_n) then (y_n) is an unconditional basic sequence in X with unconditional constant $\leq K_u$.

(b) Show that the sequence of biorthogonal functionals (u_n^*) of (u_n) is an unconditional basic sequence in X^* with unconditional constant $\leq K_u$.

3.2. Let (u_n) be an unconditional basis for a Banach space X with suppression constant K_s. Prove that for all N, whenever $a_1, \ldots, a_N, b_1, \ldots, b_N$ are scalars so that $|a_n| \le |b_n|$ for all $1 \le n \le N$ and $a_n b_n > 0$ we have

$$\left\| \sum_{n=1}^{N} a_n u_n \right\| \le K_s \left\| \sum_{n=1}^{N} b_n u_n \right\|.$$

That is, the suppression constant can replace the unconditional constant in equation (3.1) when the sign of the coefficients in the linear combinations of the basis coincide.

3.3. Show that the sequence $(e_n)_{n=1}^{\infty}$ of standard unit vectors is a monotone basic sequence for $\mathcal{J}$ in both norms $\| \cdot \|_{\mathcal{J}}$ and $\| \cdot \|_0$ (see Proposition 3.4.2).

3.4. Orlicz sequence spaces.
An *Orlicz function* is a continuous convex function $F : [0, \infty) \to [0, \infty)$ with $F(0) = 0$ and $F(x) > 0$ for $x > 0$. Let us assume that for suitable $1 < q < \infty$ we have that $F(x)/x^q$ is a decreasing function (caution: this is a mild additional assumption; see [138] for the full picture). The corresponding *Orlicz sequence space* ℓ_F is the space of (real) sequences $(\xi(n))_{n=1}^{\infty}$ such that

$$\sum_{n=1}^{\infty} F(|\xi(n)|) < \infty.$$

(a) Prove that ℓ_F is a linear space which becomes a Banach space under the norm

$$\|\xi\|_{\ell_F} = \inf\{\lambda > 0 : \sum_{n=1}^{\infty} F(\lambda^{-1}|\xi(n)|) \le 1\}.$$

(b) Show that the canonical basis $(e_n)_{n=1}^{\infty}$ is an unconditional basis for ℓ_F.
(c) Show the canonical bases of ℓ_F and ℓ_G are equivalent if and only if there is a constant C so that

$$F(x)/C \le G(x) \le CF(x), \qquad 0 \le x \le 1.$$

3.5. (Continuation of the previous problem)
(a) By considering the behavior of block basic sequences, show that ℓ_F contains no subspace isomorphic to c_0.
(b) Now assume additionally that there exists $1 < p < \infty$ so that $F(x)/x^p$ is an increasing function. Show that ℓ_F is reflexive.

3.6. Let X be a subspace of a space with unconditional basis. Show that if X contains no copy of c_0 or ℓ_1 then X is reflexive.

3.7. Let X be a Banach space with property (u) and separable dual. Suppose Y is a Banach space containing no copy of c_0. Show that every bounded operator $T : X \to Y$ is weakly compact.

3.8. Let X be a Banach space.

(a) Show that if X contains a non-boundedly-complete basic sequence then X contains a basic sequence $(x_n)_{n=1}^\infty$ with $\inf_n \|x_n\| > 0$ and $\sup_n \|\sum_{i=1}^n x_i\| < \infty$.

(b) (Continuation of (a)) Show that $y_n = \sum_{i=1}^n x_i$ is also a basic sequence.

(c) Show that if X contains a nonshrinking basic sequence then X contains a basic sequence $(x_n)_{n=1}^\infty$ such that $\sup_n \|x_n\| < \infty$ but for some $x^* \in X^*$ we have $x^*(x_n) = 1$ for all n.

(d) (Continuation of (c)) Show that if $y_1 = x_1$ and $y_n = x_n - x_{n-1}$ for $n \geq 2$ then $(y_n)_{n=1}^\infty$ is also a basic sequence. [We remind the reader of Problem 1.3.]

3.9. Let X be a Banach space. Show that the following conditions are equivalent:

(i) Every basic sequence in X is shrinking;
(ii) Every basic sequence in X is boundedly complete;
(iii) X is reflexive.

This result is due to Singer [206]; later Zippin [224] improved the result to replace *basic sequence* by *basis* when X is known to have a basis (see Problem 9.7).

3.10. Let $(e_n)_{n=1}^\infty$ be the canonical basis of the James space $\mathcal{J}$. Show that the sequence defined by $f_n = e_1 + \cdots + e_n$ is a boundedly-complete basis and that the regular norm on $\mathcal{J}$ is equivalent to the norm given by

$$\left\| \sum_{j=1}^\infty a_j f_j \right\| = \sup \left\{ \left(\sum_{j=1}^n \left(\sum_{i=p_{j-1}+1}^{p_j} a_i \right)^2 \right)^{1/2} \right\},$$

where the supremum is taken over all n and all integers $(p_j)_{j=0}^n$ with $0 = p_0 < p_1 < \cdots < p_n$.

4

Banach Spaces of Continuous Functions

We are now going to shift our attention from sequence spaces to spaces of functions, and we start in this chapter by considering spaces of type $\mathcal{C}(K)$. If K is a compact Hausdorff space, $\mathcal{C}(K)$ will denote the space of all real-valued, continuous functions on K. $\mathcal{C}(K)$ is a Banach space with the norm $\|f\|_\infty = \max_{s \in K} |f(s)|$.

It can be argued that the space $\mathcal{C}[0,1]$ was the first Banach space studied in Fredholm's 1903 paper [61]. Indeed, prior to the development of Lebesgue measure, the spaces of continuous functions were the only readily available Banach spaces!

We will begin by establishing some well-known classical facts. We include an optional section on characterization of real $\mathcal{C}(K)$-spaces. Then we turn to the classification of isometrically injective spaces. Continuing in the spirit of considering the isomorphic theory of Banach spaces, we will also be interested in classifying $\mathcal{C}(K)$-spaces at least for K metrizable. This will give us the opportunity to use some of the techniques we have already developed in Chapters 2 and 3.

The highlight of the chapter is a celebrated result of Miljutin from 1966 which states that if K and L are uncountable compact metric spaces then $\mathcal{C}(K)$ and $\mathcal{C}(L)$ are isomorphic as Banach spaces. This is a very elegant application of some of the ideas developed in the previous chapters. However, we will not use this result later, so the more impatient reader can safely skip it.

4.1 Basic properties

Most of the material in this section is classical. For convenience we will always consider spaces of real-valued functions, although the extension of the main results to complex-valued functions is not difficult.

Let us start by recalling some of the basic facts about spaces of continuous functions. The first is the classical Riesz Representation theorem.

Theorem 4.1.1 (Riesz Representation Theorem). *If K is a compact Hausdorff topological space, then $C(K)^*$ is isometrically isomorphic to the space $\mathcal{M}(K)$ of all finite regular signed Borel measures on K with the norm $\|\mu\| = |\mu|(K)$. The duality is given by*

$$\langle f, \mu \rangle = \int_K f \, d\mu.$$

If, in addition, K is metrizable then every Borel measure is regular and so $\mathcal{M}(K)$ coincides with the space of all finite Borel measures.

Theorem 4.1.2 (The Stone-Weierstrass Theorem). *Suppose that K is a compact Hausdorff topological space.*

(a) (Real case) Let $\mathcal{A}$ be a subalgebra of $C(K)$ (i.e., $\mathcal{A}$ is a linear subspace of $C(K)$ and sums, products, and scalar multiples of functions from $\mathcal{A}$ are in $\mathcal{A}$) containing constants. If $\mathcal{A}$ separates the points of K (i.e., for every $s_1, s_2 \in K$ with $s_1 \neq s_2$ there is some $f \in \mathcal{A}$ such that $f(s_1) \neq f(s_2)$), then $\overline{\mathcal{A}} = C(K)$.

(b) (Complex case) Let $\mathcal{A}$ be a subalgebra of $C_{\mathbb{C}}(K)$ containing constants. If $\mathcal{A}$ is self-adjoint (i.e., $f \in \mathcal{A}$ implies $\overline{f} \in \mathcal{A}$) then $\overline{\mathcal{A}} = C_{\mathbb{C}}(K)$.

Theorem 4.1.3. *If K is compact Hausdorff then the space $C(K)$ is separable if and only if K is metrizable.*

Proof. There is a natural embedding $s \to \delta_s$ (the point mass at s) of K into $\mathcal{M}(K)$. This is a homeomorphism for the the weak* topology of $\mathcal{M}(K)$. By Lemma 1.4.1 (i) this shows that K is metrizable if $C(K)$ is separable. For the converse, let us begin by observing that if K is a metrizable compact Hausdorff space then, in particular, it is separable. Let d be a metric inducing the topology and let $(s_n)_{n=1}^{\infty}$ be a dense countable subset of K. For $n = 1, 2, \ldots$, let $d_n : K \to \mathbb{R}$ be the (continuous) function defined for each $s \in K$ by $d_n(s) = d(s, s_n)$. The algebra A generated in $C(K)$ by the countable set $D = \{1, d_1, d_2, \ldots\}$ (here 1 denotes the constantly one function) is dense in $C(K)$ by the Stone-Weierstrass theorem. The set of all polynomials of several variables in the functions from D with rational coefficients is a countable dense set in A, hence it is dense in $C(K)$, so $C(K)$ is separable. $\qquad\square$

Let us recall that a *separation* of a topological space X is a pair U, V of disjoint open subsets of X whose union is X. Then, the space X is said to be *connected* if there does not exist a separation of X, i.e., if and only if the only subsets of X that are both open and closed in X (or *clopen*) are the empty set and X itself. On the other hand, a space is *totally disconnected* if its only connected subsets are one-point sets. This is equivalent to saying that each point in X has a base of neighborhoods consisting of sets which are both open and closed in X. The Cantor set $\Delta = \{0, 1\}^{\mathbb{N}}$ is an example of a totally disconnected compact metric space. We will need the following elementary fact:

Proposition 4.1.4. *If K is a totally disconnected compact Hausdorff space, then the collection of simple continuous functions (i.e., function f of the form $f = \sum_{j=1}^{n} a_j \chi_{U_j}$ where $U_1, \ldots, U_n$ are disjoint clopen sets) is dense in $\mathcal{C}(K)$.*

Proof. This is an easy deduction from the Stone-Weierstrass theorem as the simple functions form a subalgebra of $\mathcal{C}(K)$.

□

We conclude this section with another basic theorem from the classical theory, the Banach-Stone theorem, whose proof is proposed as an exercise (see Problem 4.2).

Theorem 4.1.5 (Banach-Stone). *Suppose K and L are two compact Hausdorff spaces such that $\mathcal{C}(K)$ and $\mathcal{C}(L)$ are isometrically isomorphic Banach spaces. Then K and L are homeomorphic.*

The Banach-Stone theorem appears for K, L metrizable in Banach's 1932 book [8]. In full generality it was proved by M. H. Stone in 1937. In fact, general topology was in its infancy in that period, and Banach was constrained by the imperfect state of development of nonmetrizable topology; thus, for example, Alaoglu's theorem on the weak* compactness of the dual unit ball was not obtained till 1941 because it required Tychonoff's theorem.

One needs to know that certain spaces such as ℓ_∞ and $L_\infty(0, 1)$ are $\mathcal{C}(K)$-spaces in disguise. The standard derivation of such facts requires considering the complex versions of these spaces as commutative C^*-algebras (or B^*-algebras) and invoking the standard representation of such algebras as $\mathcal{C}(K)$-spaces via the Gelfand transform ([32], pp. 242ff). Readers familiar with this approach can skip the next section, which is presented to remain within the category of real spaces.

4.2 A characterization of real $\mathcal{C}(K)$-spaces

The approach in this section allows us to avoid some relatively sophisticated ideas in Banach algebra theory and gives a direct proof that ℓ_∞ and $L_\infty[0, 1]$ are indeed $\mathcal{C}(K)$-spaces.

Definition 4.2.1. Suppose $\mathcal{A}$ is a commutative real Banach algebra with identity e such that $\|e\| = 1$. The *state space* of $\mathcal{A}$ is the set

$$\mathcal{S} = \{\varphi \in \mathcal{A}^* : \|\varphi\| = \varphi(e) = 1\}.$$

An element of $\mathcal{S}$ is called a *state*.

Remark 4.2.2. The set of states $\mathcal{S}$ of a commutative real Banach algebra $\mathcal{A}$ with identity is nonempty by the Hahn-Banach theorem, and $\mathcal{S}$ is obviously weak* compact.

$\mathcal{A}_+$ will denote the closure of the set of squares in $\mathcal{A}$, that is,

$$\mathcal{A}_+ = \overline{\{a^2 : a \in \mathcal{A}\}}.$$

The following lemma states two properties of $\mathcal{A}_+$ which are trivially verified, and therefore we omit its proof.

Lemma 4.2.3.

(i) If $x, y \in \mathcal{A}_+$ then $xy \in \mathcal{A}_+$.
(ii) If $x \in \mathcal{A}_+$ and $\lambda \geq 0$ then $\lambda x \in \mathcal{A}_+$.

Proposition 4.2.4.

(i) If $x \in \mathcal{A}$ is such that $\|x\| \leq 1$ then $e + x \in \mathcal{A}_+$.
(ii) $\mathcal{A} = \mathcal{A}_+ - \mathcal{A}_+$.

Proof. (i) Let $x \in \mathcal{A}$ such that $\|x\| < 1$. By writing $(1+t)^{1/2}$ in its binomial series $\sum_{n=1}^{\infty} c_n t^n$ (where, in fact, $c_n = \binom{1/2}{n}$), valid for scalars t with $|t| < 1$, we see that the series $\sum_{n=1}^{\infty} c_n t^n$ is absolutely convergent, therefore convergent to some $y \in \mathcal{A}$. By expanding out $(1+t)^{1/2}(1+t)^{1/2}$ for a real variable t when $|t| < 1$ it is clear that

$$\sum_{m+n=k} c_m c_n = \begin{cases} 1 & \text{if } k = 0, 1 \\ 0 & \text{if } k \geq 2. \end{cases}$$

We deduce that $y^2 = e + x$. Since $\mathcal{A}_+$ is closed we obtain that $e + x \in \mathcal{A}_+$ if $\|x\| \leq 1$.

(ii) follows immediately (using Lemma 4.2.3) since if $\|x\| \leq 1$ we can write

$$x = \tfrac{1}{2}(e + x) - \tfrac{1}{2}(e - x).$$

$\square$

We aim to show that a real Banach algebra $\mathcal{A}$ with identity is a $\mathcal{C}(K)$-space if it satisfies one additional condition, that is:

Theorem 4.2.5 ([1]). *Let $\mathcal{A}$ be a commutative real Banach algebra with an identity e such that $\|e\| = 1$. Then $\mathcal{A}$ is isometrically isomorphic to the algebra $\mathcal{C}(K)$ for some compact Hausdorff space K if and only if*

$$\|a^2 - b^2\| \leq \|a^2 + b^2\|, \quad a, b \in \mathcal{A}. \tag{4.1}$$

In our way to the proof of Theorem 4.2.5 we will need two preparatory Lemmas which rely on the following simple deductions from the hypothesis. Equation (4.1) gives

$$\|x - y\| \leq \|x + y\|, \quad x, y \in \mathcal{A}_+. \tag{4.2}$$

So, if $x, y \in \mathcal{A}_+$ we also have

$$\|x\| \leq \tfrac{1}{2}(\|x - y\| + \|x + y\|) \leq \|x + y\|. \tag{4.3}$$

Lemma 4.2.6. *Suppose $\mathcal{A}$ satisfies the condition (4.1). Then $\varphi(x) \geq 0$ whenever $\varphi \in \mathcal{S}$ and $x \in \mathcal{A}_+$.*

Proof. Take $x \in \mathcal{A}_+$ with $\|x\| = 1$. By Proposition 4.2.4, $e - x \in \mathcal{A}_+$ and, by (4.3),

$$\|e - x\| \leq \|(e - x) + x\| = 1.$$

Hence for $\varphi \in \mathcal{S}$ we have

$$1 = \|\varphi\| \geq \varphi(e - x) = 1 - \varphi(x),$$

and thus $\varphi(x) \geq 0$.

$\square$

Lemma 4.2.7. *Suppose $\mathcal{A}$ satisfies (4.1). Let K be the set of all multiplicative states of $\mathcal{A}$, i.e.,*

$$K = \{\varphi \in \mathcal{S} : \varphi(xy) = \varphi(x)\varphi(y) \text{ for all } x, y \in \mathcal{A}\}.$$

Then K is a compact Hausdorff space in the weak* topology of $\mathcal{A}^*$ which contains the set $\partial_e \mathcal{S}$ of extreme points of $\mathcal{S}$ (and in particular is nonempty).

Proof. It is trivial to show that K is a closed subset of the closed unit ball of $\mathcal{A}^*$ and so is compact for the weak* topology. Suppose $\varphi \in \partial_e \mathcal{S}$. Since $\mathcal{A} = \mathcal{A}_+ - \mathcal{A}_+$ it suffices to show that $\varphi(xy) = \varphi(x)\varphi(y)$ whenever $x \in \mathcal{A}_+$ and $y \in \mathcal{A}$.

Let $x \in \mathcal{A}_+$ such that $\|x\| \leq 1$ and $y \in \mathcal{A}$ with $\|y\| \leq 1$. By Proposition 4.2.4, $e \pm y \in \mathcal{A}_+$. Therefore, by Lemma 4.2.6

$$\varphi(x(e \pm y)) \geq 0,$$

which implies

$$|\varphi(xy)| \leq \varphi(x).$$

Similarly, $e - x \in \mathcal{A}_+$ by Proposition 4.2.4 and so

$$|\varphi((e - x)y)| \leq 1 - \varphi(x).$$

If $\varphi(x) = 0$ or $\varphi(x) = 1$, using the previous inequalities it is immediate that $\varphi(xy) = \varphi(x)\varphi(y)$.

If $0 < \varphi(x) < 1$, we can define states on $\mathcal{A}$ by $\psi_1(y) = \varphi(x)^{-1}\varphi(xy)$ and $\psi_2(y) = (1 - \varphi(x))^{-1}\varphi((e - x)y)$ and then write

$$\varphi = \varphi(x)\psi_1 + (1 - \varphi(x))\psi_2.$$

By the fact that φ is an extreme point of $\mathcal{S}$ we must have $\psi_1 = \varphi$ and, therefore,

$$\varphi(xy) = \varphi(x)\varphi(y), \qquad x \in \mathcal{A}_+, y \in \mathcal{A}.$$

□

Proof of Theorem 4.2.5. Suppose $\mathcal{A}$ satisfies the condition (4.1). Let $J : \mathcal{A} \to \mathcal{C}(K)$ be the natural map, given by

$$Jx(\varphi) = \varphi(x).$$

Clearly, J is an algebra homomorphism, $J(e) = 1$ and $\|J\| = 1$. In order to prove that J is an isometry we need the following:

Claim. *Suppose $x \in \mathcal{A}$ is such that $\|Jx\|_{\mathcal{C}(K)} \le 1$. Then for any $\epsilon > 0$ there exists $t_\varepsilon > 0$ so that*

$$\|e - t_\varepsilon(1 + \epsilon)e - t_\varepsilon x\| < 1.$$

If the Claim fails, there is $x \in \mathcal{A}$ with $\|Jx\|_{\mathcal{C}(K)} \le 1$ so that for some $\epsilon > 0$ we have

$$\|e - t(1 + \epsilon)e - tx\| \ge 1, \qquad t \ge 0.$$

By the Hahn-Banach theorem (separating the set $\{e - t(1 + \epsilon)e - tx : t \ge 0\}$ from the open unit ball) we can find a linear functional φ with $\|\varphi\| = 1$ and

$$\varphi(e - t(1 + \epsilon)e - tx) \ge 1, \qquad t \ge 0.$$

In particular $\varphi \in \mathcal{S}$ and $\varphi((1+\epsilon)e + x) \le 0$. Hence $|\varphi(x)| \ge 1 + \epsilon$. But now by the Krein-Milman theorem and Lemma 4.2.7, we deduce that $\|Jx\|_{\mathcal{C}(K)} > 1$, a contradiction.

Thus, combining the Claim with Proposition 4.2.4 *(i)*, we have that $\|Jx\|_{\mathcal{C}(K)} \le 1$ implies $(1 + \epsilon)e + x \in \mathcal{A}_+$ for all $\epsilon > 0$, so $e + x \in \mathcal{A}_+$.

Applying the same reasoning to $-x$ we have $e - x \in \mathcal{A}_+$. Hence, by (4.2), we obtain

$$\|x\| = \tfrac{1}{2}\|(e + x) - (e - x)\| \le \tfrac{1}{2}\|(e + x) + (e - x)\| = 1.$$

Thus J is an isometry.

Finally J is onto $\mathcal{C}(K)$ by the Stone-Weierstrass theorem.

□

Remark 4.2.8. We only needed the full hypothesis (4.1) at the very last step. Prior to that we only use the weaker hypothesis

$$\|a^2\| \le \|a^2 + b^2\|, \qquad a, b \in \mathcal{A}. \tag{4.4}$$

The condition (4.4) implies (4.3), which was used in Lemmas 4.2.6 and 4.2.7. However, this hypothesis only allows one to deduce that $\|Jx\|_{\mathcal{C}(K)} \ge \tfrac{1}{2}\|x\|$ and so $\mathcal{A}$ is only 2-isomorphic to $\mathcal{C}(K)$. That this is best possible is clear from the norm on $\mathcal{C}(K)$ given by

$$|||f||| = \|f_+\|_{\mathcal{C}(K)} + \|f_-\|_{\mathcal{C}(K)}$$

where $f_+ = \max(f, 0)$ and $f_- = \max(-f, 0)$. Under this norm $\mathcal{C}(K)$ is a commutative real Banach algebra satisfying equation (4.4) but not equation (4.1).

Let us observe that if we consider $\mathcal{A} = \ell_\infty$ (with the multiplication of two sequences defined coordinate-wise), Theorem 4.2.5 yields that $\mathcal{A} = \mathcal{C}(K)$ (isometrically) for some compact Hausdorff space K. This set K is usually denoted by $\beta\mathbb{N}$. We also note that if (Ω, Σ, μ) is any σ-finite measure space then $L_\infty(\Omega, \mu)$ is again a $\mathcal{C}(K)$-space. In each case the isomorphism *preserves order* (i.e., nonnegative functions are mapped to nonnegative functions) since squares are mapped to squares.

4.3 Isometrically injective spaces

We now turn to the problem of classifying isometrically injective spaces, originally introduced in Chapter 2 (Section 2.5). There we saw that ℓ_∞, which we identify with $\mathcal{C}(\beta\mathbb{N})$, is isometrically injective but that c_0 is not an (isomorphically) injective space (although it is separably injective). Let us recall that $\beta\mathbb{N}$ is the Stone-Čech compactification of $\mathbb{N}$ endowed with the discrete topology, i.e., $\beta\mathbb{N}$ is the unique compact Hausdorff space containing $\mathbb{N}$ as a dense subspace so that every bounded continuous function on $\mathbb{N}$ extends to a continuous function on $\beta\mathbb{N}$.

The complete classification of isometrically injective spaces was achieved in the early 1950s by the combined efforts of Nachbin [155], Goodner [68], and Kelley [109]. The basic approach developed by Nachbin and Goodner was to abstract the essential ingredient of the Hahn-Banach theorem, which is the order-completeness (i.e., the least upper bound axiom) of the real numbers.

Definition 4.3.1. We say that the space $\mathcal{C}(K)$ is *order-complete* if whenever A, B are nonempty subsets of $\mathcal{C}(K)$ with $f \leq g$ for all $f \in A$ and $g \in B$, then there exists $h \in \mathcal{C}(K)$ such that $f \leq h \leq g$ whenever $f \in A$ and $g \in B$.

Remark 4.3.2. (*a*) If $\mathcal{C}(K)$ is order-complete then any subset A of $\mathcal{C}(K)$ which has an upper bound has also a least upper bound, which we denote $\sup A$. Indeed, let B be the set of all upper bounds of A and apply the preceding definition. The (uniquely determined) function h must be the least upper bound. It is important to stress that h is a continuous function and may not coincide with the pointwise supremum $\tilde{h}(s) = \sup_{f \in A} f(s)$, which need not be a continuous function. Similar statements may be made about greatest lower bounds (i.e., infima).

(*b*) The previous definition can easily be extended to any space with a suitable order structure such as ℓ_∞ or L_∞. It is clear that ℓ_∞ is order-complete for its natural order and therefore $\mathcal{C}(\beta\mathbb{N})$ is also order-complete. To compute the supremum of A in ℓ_∞ one does indeed take the pointwise supremum, but the corresponding supremum in $\mathcal{C}(\beta\mathbb{N})$ is not necessarily a pointwise supremum.

We will say that a map $V : F \to \mathcal{C}(K)$, where F is a linear subspace of a Banach space X, is *sublinear* if

(i) $V(\alpha x) = \alpha V(x)$ for all $\alpha \geq 0$ and $x \in F$, and

(ii) $V(x + y) \leq V(x) + V(y)$ for all $x, y \in F$.

A sublinear map $V : X \to \mathcal{C}(K)$ is *minimal* provided there is no sublinear map $U : X \to \mathcal{C}(K)$ such that $U(x) \leq V(x)$ for all $x \in X$ and $U \neq V$.

Lemma 4.3.3. *Let X be a Banach space and F a linear subspace of X. Suppose $V : X \to \mathcal{C}(K)$ and $W : F \to \mathcal{C}(K)$ are sublinear maps such that $W(y) + V(-y) \geq 0$ for all $y \in F$. If $\mathcal{C}(K)$ is order-complete then the map $V \wedge W : X \to \mathcal{C}(K)$ given by*

$$V \wedge W(x) = \inf\{V(x - y) + W(y) : y \in F\},$$

is well defined and sublinear.

Proof. For each fixed $x \in X$ we have

$$V(x - y) + W(y) \geq V(-y) - V(-x) + W(y) \geq -V(-x)$$

for all $y \in F$. That is, $-V(-x)$ is a lower bound of the set $\{V(x - y) + W(y) : y \in F\}$. Thus, by the order-completeness of $\mathcal{C}(K)$, we can define a map $V \wedge W : F \to \mathcal{C}(K)$ by

$$V \wedge W(x) = \inf\{V(x - y) + W(y) : y \in F\}.$$

It is a straightforward verification to check that $V \wedge W$ is sublinear.   □

Lemma 4.3.4. *Let $V : X \to \mathcal{C}(K)$ be a sublinear map. If $\mathcal{C}(K)$ is order-complete then there is a minimal sublinear map $W : X \to \mathcal{C}(K)$ with $W(x) \leq V(x)$ for all $x \in X$.*

Proof. Put

$$\mathcal{S} = \{U : X \to \mathcal{C}(K) : U \text{ is sublinear and } U(x) \leq V(x) \text{ for all } x \in X\}.$$

$\mathcal{S}$ is nonempty ($V \in \mathcal{S}$) and partially ordered. Let $\Psi = (U_i)_{i \in I}$ be a chain (i.e., a totally ordered subset) in $\mathcal{S}$. Note that for each $i \in I$ we have $0 = U_i(x + (-x)) \leq U_i(x) + U_i(-x)$ for all $x \in X$, hence

$$U_i(x) \geq -U_i(-x) \geq -V(-x).$$

Thus, for each $x \in X$, the set $\{U_i(x) : i \in I\} \subset \mathcal{C}(K)$ has a lower bound. By the order-completeness of $\mathcal{C}(K)$, the map

$$U_\Psi(x) = \inf_{i \in I} U_i(x)$$

is well defined on X and sublinear. To see this, since Ψ is a totally ordered set, given $i \neq j \in I$, without loss of generality we can assume that $U_i \leq U_j$. Then, for any $x, y \in X$ we have

$$U_\Psi(x+y) \le U_i(x+y) \le U_j(x) + U_i(y),$$

therefore $U_\Psi(x+y) - U_j(x) \le U_\Psi(y)$, which yields $U_\Psi(x+y) - U_\Psi(y) \le U_\Psi(x)$. Moreover, $U_\Psi(x) \le V(x)$ for all $x \in X$. That is, $U_\Psi \in \mathcal{S}$ is a lower bound for the chain $(U_i)_{i \in I}$. Using Zorn's lemma we deduce the existence of a minimal element W in $\mathcal{S}$.

$\square$

Lemma 4.3.5. *Suppose that $\mathcal{C}(K)$ is order-complete and let $V : X \to \mathcal{C}(K)$ be a sublinear map. If V is minimal then V is linear.*

Proof. Given an element $x \in X$, let us call F its linear span, $F = \langle x \rangle$. Then, $W(\lambda x) = -\lambda V(-x)$ defines a linear map from F to $\mathcal{C}(K)$. Clearly, $W(\lambda x) \ge -V(-\lambda x)$ for every real λ. Using Lemma 4.3.3 we can define on X the sublinear map

$$V \wedge W(x) = \inf_{\lambda \in \mathbb{R}} \{V(x - \lambda x) + W(\lambda x)\}.$$

By the minimality of V, $V \wedge W = V$ on X. Therefore $V \le W$ on F, which implies that $V(x) \le -V(-x)$. On the other hand, $V(x) \ge -V(-x)$ by the sublinearity of V, so $V(-x) = -V(x)$. Since this holds for all $x \in X$, it is clear that V is linear.

$\square$

Theorem 4.3.6 (Goodner, Nachbin, 1949-1950). *Let K be a compact Hausdorff space. Then $\mathcal{C}(K)$ is isometrically injective if and only if $\mathcal{C}(K)$ is order-complete.*

Proof. Assume, first, that $\mathcal{C}(K)$ is order-complete. Let E be a subspace of a Banach space X and let $S : E \to \mathcal{C}(K)$ be a linear operator with $\|S\| = 1$. That is, for each $x \in E$ we have

$$-\|x\| \le (Sx)(k) \le \|x\| \quad \text{for all } k \in K,$$

which, if we let 1 denote the constant function 1 on K, is equivalent to writing

$$-\|x\| \cdot 1 \le S(x) \le \|x\| \cdot 1. \tag{4.5}$$

Thus, if we consider the sublinear map from X to $\mathcal{C}(K)$ given by $V_0(x) = \|x\| \cdot 1$, equation (4.5) tells us that $S(x) \ge -V_0(-x)$ for all $x \in E$ and so we can define on X the sublinear map $V = V_0 \wedge S$ as in Lemma 4.3.3:

$$V(x) = \inf \{V_0(x - y) + S(y) : y \in E\}.$$

By Lemma 4.3.4 there exists $T : X \to \mathcal{C}(K)$, a minimal sublinear map satisfying $T \le V$. Lemma 4.3.5 yields that T is linear.

On E, we have $T(x) \le S(x)$ and $T(-x) \le S(-x)$. Therefore, $T|_E = S$. Finally, $T(x) \le \|x\| \cdot 1$ and $T(-x) \le \|x\| \cdot 1$ for all $x \in X$, which implies that $\|T\| \le 1$. Thus, we have successfully extended S from E to X.

Suppose, conversely, that $\mathcal{C}(K)$ is isometrically injective. Then there is a norm-one projection P from $\ell_\infty(K)$ onto $\mathcal{C}(K)$, where $\ell_\infty(K)$ denotes the space of all bounded functions on K. Suppose that A, B are two nonempty subsets of $\mathcal{C}(K)$ such that $f \in A$ and $g \in B$ implies $f \leq g$. For each $s \in K$, put $a(s) = \sup_{f \in A} f(s)$. Obviously, $a \in \ell_\infty(K)$. Let $h = P(a)$. We will prove that $f \leq h \leq g$ for all $f \in A$ and all $g \in B$.

Since $P(1) = 1$ and P has norm one, it follows that for each $b \in \ell_\infty(K)$ with $b > 0$ we have

$$\|P(1 - \lambda b)\| \leq 1 \text{ for } 0 \leq \lambda \leq 2/\|b\|.$$

We deduce that P is a positive map, that is, $Pb \geq 0$ whenever $b \in \ell_\infty(K)$ and $b \geq 0$. Thus, if $f \in A$ then $f \leq a$ and, therefore, $f \leq h$. Analogously, if $g \in B$ we have $g \geq a$ and so $g \geq h$. Hence, $\mathcal{C}(K)$ is order-complete.

$\square$

The spaces K so that $\mathcal{C}(K)$ is order-complete are characterized by the property that the closure of any open set remains open; such spaces are called *extremally disconnected*. We refer the reader to the Problems for more information.

The natural question arises as to whether only $\mathcal{C}(K)$-spaces can be isometrically injective. Both Nachbin and Goodner showed that an isometrically injective Banach space X is (isometrically isomorphic to) a $\mathcal{C}(K)$-space provided the unit ball of X has at least one extreme point. The key here is that the constant function 1 is always an extreme point on the unit ball in $\mathcal{C}(K)$ and they needed to find an element in the space X to play this role. However, two years later, in 1952, Kelley completed the argument and proved the definitive result:

Theorem 4.3.7 (Kelley, 1952). *A Banach space X is isometrically injective if and only if it is isometrically isomorphic to an order-complete $\mathcal{C}(K)$-space.*

Proof. We need only show the forward implication. For that, we are going to identify X (via an isometric isomorphism) with a suitable $\mathcal{C}(K)$-space which, by the isometric injectivity of X, will be order-continuous appealing to Theorem 4.3.6.

The trick is to "find" K as a subset of the dual unit ball B_{X^*}. Consider the set $\partial_e B_{X^*}$ of extreme points of B_{X^*} with the weak* topology. There is a maximal open subset, U, of $\partial_e B_{X^*}$ subject to the property that $U \cap (-U) = \emptyset$. This is an easy consequence of Zorn's lemma again, as any chain of such open sets has an upper bound, namely, their union. Let K be the weak* closure of U in B_{X^*}. K is, of course, compact and Hausdorff for the weak* topology.

Let us observe that $K \cap \partial_e B_{X^*}$ cannot meet $-U$ since $\partial_e B_{X^*} \setminus (-U)$ is relatively weak* closed in $\partial_e B_{X^*}$. Then, $K \cap (-U) = \emptyset$.

We claim that $\partial_e B_{X^*} \subset (K \cup (-K))$. Indeed, suppose that there exists $x^* \in \partial_e B_{X^*} \setminus (K \cup (-K))$. Then there is an absolutely convex weak* open neighborhood, V, of 0 such that $x^* \notin V$ and $(x^* + V) \cap (K \cup (-K)) = \emptyset$. Let $U_1 = U \cup \left((x^* + V) \cap \partial_e B_{X^*} \right)$. Then U_1 strictly contains U since $x^* \in U_1$. Suppose $y^* \in U_1 \cap (-U_1)$. Then either $y^* \notin U$ or $-y^* \notin U$; thus replacing y^* by $-y^*$ if necessary we can assume $y^* \notin U$. Then $y^* \in x^* + V$; this implies that $y^* \notin K \cup (-K)$ and so $y^* \notin -U$. Hence $y^* \in -x^* - V$ and so $0 \in 2x^* + 2V$ or $x^* \in V$ yielding a contradiction. Thus $U_1 \cap (-U_1) = \emptyset$, which contradicts the maximality of U.

By the Krein-Milman theorem, B_{X^*} must be the weak* closed convex hull of $K \cup (-K)$ and, in particular, if $x \in X$ we have

$$\|x\| = \sup_{x^* \in B_{X^*}} |x^*(x)| = \max_{x^* \in K} |x^*(x)|.$$

Thus, the map J that assigns to each $x \in X$ the function $\hat{x} \in \mathcal{C}(K)$ given by $\hat{x}(x^*) = x^*(x)$, $x^* \in K$, is an isometry. We can therefore use the isometric injectivity of X (extending the map $J^{-1} : J(X) \to X$) to define an operator $T : \mathcal{C}(K) \to X$ such that $T(\hat{x}) = x$ for all $x \in X$ with $\|T\| = 1$.

Let us consider the adjoint map $T^* : X^* \to \mathcal{M}(K)$. If $u^* \in U$, then $T^* u^* = \mu \in \mathcal{M}(K)$ with $\|\mu\| \leq 1$. Let V be any weak* open neighborhood of u^* relative to K and put $K_0 = K \setminus V$. We can define $v^* \in X^*$ by

$$v^*(x) = \int_V x^*(x)\, d\mu(x^*), \qquad x \in X,$$

and $w^* \in X^*$ by

$$w^*(x) = \int_{K_0} x^*(x)\, d\mu(x^*), \qquad x \in X.$$

Then $\|v^*\| \leq |\mu|(V)$ and $\|w^*\| \leq |\mu|(K_0)$. But,

$$\int_K x^*(x)\, d\mu = \langle \hat{x}, T^*(u^*) \rangle = \langle x, u^* \rangle,$$

hence $v^* + w^* = u^*$. Since $\|u^*\| = 1 \geq \|\mu\|$, we must have $|\mu|(V) + |\mu|(K_0) = 1$. Thus, $\|v^*\| + \|w^*\| = 1$ and so the fact that u^* is an extreme point implies that $v^* = \|v^*\| u^*$ and $w^* = \|w^*\| u^*$.

Suppose $|\mu|(K_0) = \|w^*\| = \alpha > 0$. Then,

$$u^*(x) = \alpha^{-1} \int_{K_0} x^*(x)\, d\mu(x^*), \qquad x \in X,$$

and, in particular,

$$|u^*(x)| \leq \max_{x^* \in K_0} |x^*(x)|, \qquad x \in X.$$

This implies that u^* is in the weak* closed convex hull, C, of $K_0 \cup (-K_0)$. But u^* must be an extreme point in C also, so by Milman's theorem it must belong to the weak* closed set $K_0 \cup (-K_0)$. Since $u^* \notin K_0$ we have that $u^* \in (-K_0)$, i.e., $-u^* \in K_0$. Thus, K_0 meets $-U$, so K meets $-U$, which is a contradiction to our previous remarks.

Hence $|\mu|(K_0) = \|w^*\| = 0$ and so $|\mu(V)| = 1$ for every weak* open neighborhood V of u^*. By the regularity of μ we must have that $\mu = \pm\delta_{u^*}$ (δ_{u^*} is the point mass at u^*). Thus $\mu = \delta_{u^*}$ for $u^* \in U$. Since T^* is weak* continuous we infer that $T^*(x^*) = \delta_{x^*}$ for all $x^* \in K$. We are done because if $f \in \mathcal{C}(K)$, then

$$\langle Tf, x^* \rangle = f(x^*),$$

so J is onto $\mathcal{C}(K)$. This shows that X is a $\mathcal{C}(K)$-space.

$\square$

At this point we have only one example where $\mathcal{C}(K)$ is order-complete, namely, ℓ_∞ (although, of course, $\ell_\infty(\mathcal{I})$ for any index set $\mathcal{I}$ will also work). There are, however, less trivial examples as the next proposition shows.

Proposition 4.3.8.

(i) If $\mathcal{C}(K)$ is (isometrically isomorphic to) a dual space, then $\mathcal{C}(K)$ is isometrically injective.

(ii) If (Ω, Σ, μ) is any σ-finite measure space, then $L_\infty(\Omega, \Sigma, \mu)$ is isometrically injective.

(iii) For any compact Hausdorff space K the space $\mathcal{C}(K)^{**}$ is isometrically injective.

Proof. For (i) we will first show that $P = \{f \in \mathcal{C}(K) : f \geq 0\}$, the positive cone of $\mathcal{C}(K)$, is closed for the weak* topology of $\mathcal{C}(K)$ (regarded now as a dual Banach space by hypothesis). By the Banach-Dieudonné theorem it suffices to show that $P \cap \lambda B_{\mathcal{C}(K)}$ is weak* closed for each $\lambda > 0$. But $P \cap \lambda B_{\mathcal{C}(K)} = \{f : \|f - \frac{1}{2}\lambda \cdot 1\| \leq \frac{1}{2}\lambda\}$ is simply a closed ball, which must be weak* closed.

Let us see that $\mathcal{C}(K)$ is order-complete and then we will invoke Theorem 4.3.6 to deduce that $\mathcal{C}(K)$ is isometrically injective. Suppose A, B are nonempty subsets of $\mathcal{C}(K)$ such that $f \in A, g \in B$ imply $f \leq g$. For each $f \in A$ and $g \in B$, put

$$C_{f,g} = \{h \in \mathcal{C}(K) : f \leq h \leq g\}.$$

Every $C_{f,g}$ is a (nonempty) bounded and weak* closed set. If $f_1, \ldots, f_n \in A$ and $g_1, \ldots, g_n \in B$ then $\cap_{k=1}^n C_{f_k, g_k}$ is nonempty because it contains for example $\max(f_1, \ldots, f_n)$. Hence, by weak* compactness, the intersection $\cap_{\{f \in A, g \in B\}} C_{f,g}$ is nonempty. If we pick h in the intersection we are done.

(ii) follows directly from (i) since $L_\infty(\mu) = L_1(\mu)^*$.

(iii) Here we observe that $\mathcal{M}(K)$ is actually a vast ℓ_1-sum of $L_1(\mu)$-spaces. Precisely, using Zorn's lemma one can produce a maximal collection $(\mu_i)_{i \in \mathcal{I}}$ of probability measures on K with the property that any two members of the collection are mutually singular.

If $\nu \in \mathcal{M}(K)$, for each $i \in \mathcal{I}$ we define $f_i \in L_1(K, \mu_i)$ to be the Radon-Nikodym derivative $d\nu/d\mu_i$. Thus, $d\nu = f_i d\mu_i + \gamma$, where γ is singular with respect to μ_i. Then it is easy to show (we leave the details to the reader) that for any finite set $\mathbb{A} \subset \mathcal{I}$ we have

$$\sum_{i \in \mathbb{A}} \|f_i\|_{L_1(\mu_i)} \leq \|\nu\|.$$

Hence,

$$\sum_{i \in \mathcal{I}} \|f_i\|_{L_1(\mu_i)} \leq \|\nu\|.$$

Notice that the last statement implies that only countably many terms in the sum are nonzero. Put

$$\nu_0 = \sum_{i \in \mathcal{I}} f_i d\mu_i,$$

where the series converges in $\mathcal{M}(K)$. It is clear that the measure $\nu - \nu_0$ is singular with respect to every μ_i and, as a consequence, it must vanish on K. It follows that the map $\nu \mapsto (f_i)_{i \in \mathcal{I}}$ defines an isometric isomorphism between $\mathcal{M}(K)$ and the ℓ_1-sum of the spaces $L_1(\mu_i)$ for $i \in \mathcal{I}$.

This yields that $\mathcal{C}(K)^{**}$ can be identified with the ℓ_∞-sum of the spaces $L_\infty(\mu_i)$. Using (ii) we deduce that $\mathcal{C}(K)^{**}$ is isometrically injective.   □

Remark 4.3.9. We should note here that there are order-complete $\mathcal{C}(K)$-spaces which are not isometric to dual spaces. The first example was given in 1951 (in a slightly different context) by Dixmier [43] and we refer to Problem 4.8 and Problem 4.9 for details.

There is an easy but surprising application of the preceding proposition to the isomorphic theory [167]:

Theorem 4.3.10. $L_\infty[0, 1]$ is isomorphic to ℓ_∞.

Proof. First, observe that ℓ_∞ embeds isometrically into $L_\infty[0, 1]$ via the map

$$(\xi(n))_{n=1}^\infty \mapsto \sum_{n=1}^\infty \xi(n) \chi_{A_n}(t),$$

where $(A_n)_{n=1}^\infty$ is a partition of $[0, 1]$ into sets of positive measure. Since ℓ_∞ is an injective space, it follows that ℓ_∞ is complemented in $L_\infty[0, 1]$.

On the other hand, $L_\infty[0, 1]$ also embeds isometrically into ℓ_∞. To see this, pick $(\varphi_n)_{n=1}^\infty$, a dense sequence in the unit ball of L_1, and map $f \in L_\infty[0, 1]$ to $(\int_0^1 \varphi_n f \, dt)_{n=1}^\infty$. Therefore, being an injective space, $L_\infty[0, 1]$ is complemented in ℓ_∞.

Furthermore, $\ell_\infty \approx \ell_\infty \oplus \ell_\infty$ and

$$L_\infty[0,1] \approx L_\infty[0,1/2] \oplus L_\infty[1/2,1] \approx L_\infty[0,1] \oplus L_\infty[0,1].$$

Using Theorem 2.2.3 (a) (the Pełczyński decomposition technique) we deduce that $L_\infty[0,1]$ is isomorphic to ℓ_∞.

□

We conclude this section by showing that a separable isometrically injective space is necessarily finite-dimensional.

Proposition 4.3.11. *For any infinite compact Hausdorff space K, $\mathcal{C}(K)$ contains a subspace isometric to c_0. If K is metrizable this subspace is complemented.*

Proof. Let (U_n) be a sequence of nonempty, disjoint, open subsets of K. Such a sequence can be found by induction: simply pick U_1 so that $K_1 = K \setminus \overline{U_1}$ is infinite and then take $U_2 \subset K_1$ such that $K_2 = K_1 \setminus \overline{U_2}$ is infinite and so on. Next, pick a sequence $(\varphi_n)_{n=1}^\infty$ of continuous functions on K so that $0 \le \varphi_n \le 1$, $\max_{s \in K} \varphi_n(s) = 1$ and $\{s \in K : \varphi_n(s) > 0\} \subset U_n$, for all $n \in \mathbb{N}$. Then for any $(a_n) \in c_{00}$ we have

$$\left\| \sum_{n=1}^\infty a_n \varphi_n \right\| = \max_n |a_n|.$$

Thus $(\varphi_n)_{n=1}^\infty$ is a basic sequence isometrically equivalent to the unit vector basis of c_0.

If K is metrizable, Theorem 4.1.3 implies that $\mathcal{C}(K)$ is separable and we can apply Sobczyk's theorem (Theorem 2.5.8) to deduce that the space $[\varphi_n]_{n=1}^\infty$ is complemented by a projection of norm at most two.

□

Proposition 4.3.12. *If $\mathcal{C}(K)$ is order-complete and K is metrizable then K is finite.*

Proof. If K is infinite, $\mathcal{C}(K)$ contains a complemented copy of c_0 by Proposition 4.3.11. But if, moreover, $\mathcal{C}(K)$ is isometrically injective this would make c_0 injective, which is false because c_0 is uncomplemented in ℓ_∞ as we saw in Theorem 2.5.5.

□

Corollary 4.3.13. *The only isometrically injective separable Banach spaces are finite-dimensional and isometric to ℓ_∞^n for some $n \in \mathbb{N}$.*

Proof. If X is an isometrically injective Banach space, by Theorem 4.3.7, X can be identified with an order-complete $\mathcal{C}(K)$-space for some compact Hausdorff K. Since X is separable, Theorem 4.1.3 yields that K is metrizable and, by Proposition 4.3.12, K must be finite. Therefore $\mathcal{C}(K)$ is (isometrically isomorphic to) $\ell_\infty^{|K|}$.

□

In fact, there are no infinite-dimensional injective separable Banach spaces (even dropping isometrically) but this is substantially harder and we will see it in the next chapter.

4.4 Spaces of continuous functions on uncountable compact metric spaces

We now turn to the problem of isomorphic classification of $\mathcal{C}(K)$-spaces. The Banach-Stone theorem (Theorem 4.1.5) asserts that if K and L are non-homeomorphic compact Hausdorff spaces then the corresponding spaces of continuous functions $\mathcal{C}(K)$ and $\mathcal{C}(L)$ cannot be linearly isometric.

However, it is quite a different question to ask if they can be linearly *isomorphic*. In the 1950s and 1960s a complete classification of the isomorphism classes of $\mathcal{C}(K)$ for K metrizable (i.e., for $\mathcal{C}(K)$ separable) was found through the work of Bessaga, Pełczyński, and Miljutin. We will describe some of this work in this section and the next.

Let us note before we start that it is quite possible for $\mathcal{C}(K)$ and $\mathcal{C}(L)$ to be linearly isomorphic when K and L are not homeomorphic. We shall need the following:

Proposition 4.4.1. *If K is an infinite compact metric space then $\mathcal{C}(K) \approx \mathcal{C}(K) \oplus \mathbb{R}$. Hence $\mathcal{C}(K)$ is isomorphic to its hyperplanes.*

Proof. By Proposition 4.3.11, $\mathcal{C}(K) \approx E \oplus c_0 \approx E \oplus c_0 \oplus \mathbb{R}$ for some subspace E. Hence $\mathcal{C}(K) \approx \mathcal{C}(K) \oplus \mathbb{R}$.

The latter statement of the proposition follows from the fact that any two hyperplanes in a Banach space are isomorphic to each other and that, obviously, $\mathcal{C}(K)$ is a hyperplane of $\mathcal{C}(K) \oplus \mathbb{R}$. $\qquad\square$

Remark 4.4.2. This proposition really does need metrizability of $\mathcal{C}(K)$! Indeed, a remarkable and very recent result of Plebanek [190] is that there exists a compact Hausdorff space K so that $\mathcal{C}(K)$ fails to be isomorphic to its hyperplanes.

Given Proposition 4.4.1, note that if $K = [0,1] \cup \{2\}$ then $\mathcal{C}(K) \approx \mathcal{C}[0,1] \oplus \mathbb{R} \approx \mathcal{C}[0,1]$ but K and $[0,1]$ are not homeomorphic. Similarly $\mathcal{C}[0,1]$ is isomorphic to its (hyperplane) subspace $\{f : f(0) = f(1)\}$, which is trivially isometric to $\mathcal{C}(\mathbb{T})$. But it is more difficult to make general statements. In Banach's 1932 book [8] he raised the question whether $\mathcal{C}[0,1]$ and $\mathcal{C}[0,1]^2$ are linearly isomorphic. We will see that they are, but at this stage it is far from obvious.

To study $\mathcal{C}(K)$-spaces with K infinite and compact metric, we must consider two cases, namely, when K is countable and when K is uncountable. K must be separable, of course, but it could actually be already countable. Indeed, the simplest infinite K is the one-point compactification of $\mathbb{N}$, $\gamma\mathbb{N}$, which consists of the terms of a convergent sequence and its limit; e.g., we can take $K = \{1, \frac{1}{2}, \frac{1}{3}, \dots\} \cup \{0\}$. Then $\mathcal{C}(K)$ can be identified with the space c of convergent sequences. This is linearly isomorphic to c_0 since $c \approx c_0 \oplus \mathbb{R}$. If K is countable then $\mathcal{M}(K)$ consists only of purely atomic measures and is immediately seen to be isometric to ℓ_1. Thus $\mathcal{C}(K)^*$ is separable. However, $\mathcal{C}[0,1]^*$ is nonseparable (as $\mathcal{C}[0,1]$ contains a copy of ℓ_1 by the Banach-Mazur theorem (Theorem 1.4.3)).

In this section we will restrict to the case of uncountable K. The main result is the remarkable theorem of Miljutin [150], which asserts that for any uncountable compact metric space K, the space $\mathcal{C}(K)$ is isomorphic to $\mathcal{C}[0,1]$. This result was obtained by Miljutin in his thesis in 1952, but was not published until 1966. Miljutin's mathematical interests changed after his thesis and he apparently did not regard the result as important enough to merit publication. In fact, the result was discovered in Miljutin's thesis by Pełczyński on a visit to Moscow in the 1960s and it was only at his urging that a paper finally appeared in 1966.

The key players in the proof will be the Cantor set $\Delta = \{0,1\}^{\mathbb{N}}$, the unit interval $[0,1]$, and the Hilbert cube $[0,1]^{\mathbb{N}}$. We will need the following basic topological facts:

Proposition 4.4.3.

(i) If K is a compact metric space then K is homeomorphic to a closed subset of the Hilbert cube $[0,1]^{\mathbb{N}}$.

(ii) If K is an uncountable compact metric space then Δ is homeomorphic to a closed subset of K.

Proof. We have already showed (i) in the proof of Theorem 1.4.3. Just take $(f_n)_{n=1}^{\infty}$ a dense sequence in $\{f \in \mathcal{C}(K) : 0 \le f \le 1\}$ and define the map $\sigma : K \to [0,1]^{\mathbb{N}}$ by $\sigma(s) = (f_n(s))_{n=1}^{\infty}$. Then σ is continuous and one-to-one, hence a homeomorphism onto $\sigma(K)$. (We repeatedly use the standard fact that a one-to-one continuous map from a compact space to a Hausdorff topological space is a homeomorphism onto its range since closed sets must be mapped to compact, therefore closed, sets.)

To show part (ii) we first note that since K is uncountable, given any $\epsilon > 0$ we can find two disjoint uncountable closed subsets K_0, K_1 each with diameter at most ϵ. In fact the set E of all $s \in K$ with a countable neighborhood is necessarily countable by an application of Lindelöf's theorem (every open covering of a separable metric space has a countable subcover). If we take two distinct points s_0, s_1 outside E we can then choose K_0 and K_1 as suitable neighborhoods of s_0, s_1.

Now we proceed by induction: for $n \in \mathbb{N}$ and $t = (t_1, \dots, t_n) \in \{0,1\}^n$ define $K_{t_1,t_2,\dots,t_n}$ to be an uncountable compact subset of K of diameter at most 2^{-n} such that for each $n \in \mathbb{N}$ the sets $K_{t_1,\dots,t_n,0}$ and $K_{t_1,\dots,t_n,1}$ are disjoint subsets of $K_{t_1,\dots,t_n}$. For each $t = (t_k)_{k=1}^{\infty} \in \Delta$ define $\sigma(t)$ to be the unique point in $\cap_{n=1}^{\infty} K_{t_1,\dots,t_n}$. It is simple to see that σ is one-to-one and continuous and thus is an embedding.

$\square$

Let us use this proposition. Suppose that K is a compact, metric Hausdorff space and let E be a closed subset of K. We can naturally identify $\mathcal{C}(E)$ as a *quotient* of $\mathcal{C}(K)$ by considering the restriction operator

$$R : \mathcal{C}(K) \to \mathcal{C}(E), \qquad Rf = f|_E.$$

This is a genuine quotient map by the Tietze Extension theorem[1]. Let us suppose that we can find a bounded linear operator $T : \mathcal{C}(E) \to \mathcal{C}(K)$ which selects an element of each coset. Then T is a *linear* extension operator which defines an extension of each $f \in \mathcal{C}(E)$ to a member of $\mathcal{C}(K)$; note that RT is nothing other than the identity map I on $\mathcal{C}(E)$. T is an isomorphism of $\mathcal{C}(E)$ onto a subspace of $\mathcal{C}(K)$ and the subspace is complemented by the projection TR. Thus we could conclude that $\mathcal{C}(E)$ is isomorphic to a complemented subspace of $\mathcal{C}(K)$. Note that the kernel of the projection is $\{f \in \mathcal{C}(K) : f|_E = 0\}$ and this must also be a complemented subspace via $I - TR$.

We have met this problem in two special cases already. In the proof of the Banach-Mazur theorem we considered the case $K = [0, 1]$ and E a closed subset, and defined an extension operator by linear interpolation on the intervals of $K \backslash E$. Now, if we regard ℓ_∞ as $\mathcal{C}(\beta\mathbb{N})$, then the subspace c_0 is identified with $\{f : f_{\beta\mathbb{N}\backslash\mathbb{N}} = 0\}$ (here $\mathbb{N}$ is an open subset of $\beta\mathbb{N}$ since each point is isolated). This is uncomplemented (Theorem 2.5.5) so no linear extension operator can exist from $\beta\mathbb{N} \backslash \mathbb{N}$.

On the other hand, recall Sobczyk's theorem (Theorem 2.5.8). If we consider a separable closed subalgebra of ℓ_∞ containing c_0 (which corresponds to a metrizable compactification) then we have no problem with the extension. This suggests that metrizability of K is important here and leads us to the following classical theorem which actually implies Sobczyk's theorem. It was proved in 1933 by Borsuk [14].

Theorem 4.4.4 (Borsuk). *Let K be a compact metric space and suppose that E is a closed subset of K. Then there is a linear operator $T : \mathcal{C}(E) \to \mathcal{C}(K)$ such that $(Tf)|_E = f$, $\|T\| = 1$ and $T1 = 1$. In particular $\mathcal{C}(E)$ is isometric to a norm-one complemented subspace of $\mathcal{C}(K)$.*

Let us remark that the projection onto the kernel of T has then norm at most 2, and this explains the constant in Sobczyk's theorem.

Proof. The key point in the argument is that $U = K \backslash E$ is metrizable and hence paracompact, i.e., every open covering of U has a locally finite refinement. Let us consider the covering of U by the sets $V_u = \{s \in U : d(s, u) < \frac{1}{2}d(u, E)\}$. There is a locally finite refinement of $(V_u)_{u \in U}$, which implies that we can find a partition of the unity subordinate to $(V_u)_{u \in U}$, that is, a family of continuous functions $(\phi_j)_{j \in J}$ on U such that

1. $0 \le \phi_j \le 1$,

---

[1] *The Tietze Extension theorem* states that given a normal topological space X (i.e., a topological space satisfying the T_4 separation axiom), a closed subspace E of X and a continuous real-valued function on E, there exists a continuous real-valued function $\tilde{f}$ on X such that $\tilde{f}(x) = f(x)$ for all $x \in E$.

2. $\{\phi_j > 0\}$ is a locally finite covering of U,
3. $\sum_{j \in J} \phi_j(s) = 1$ for all $s \in U$,
4. For each $j \in J$ there exists $u_j \in U$ so that $\{\phi_j > 0\} \subset V_{u_j}$.

For each $j \in J$ pick $v_j \in E$ with $d(u_j, E) = d(u_j, v_j)$ (possible by compactness).

If $f \in \mathcal{C}(E)$ we define

$$Tf(s) = \begin{cases} f(s) & \text{if } s \in E \\ \sum_{j \in J} \phi_j(s) f(v_j) & \text{if } s \in U. \end{cases}$$

The theorem will be proved once we have shown that Tf is a continuous function on K, because T clearly is linear, $T1 = 1$ and $\|T\| = 1$. It is also clear that Tf is continuous on U.

Now suppose $t \in E$. If $\epsilon > 0$ fix $\delta > 0$ so that $d(s, t) < 4\delta$ implies that $|f(s) - f(t)| < \epsilon$. Assume $d(s, t) < \delta$. If $s \in E$ then $|Tf(s) - Tf(t)| < \epsilon$. If $s \in U$ then

$$|Tf(s) - Tf(t)| = \sum_{\phi_j(s) > 0} \phi_j(s) |f(v_j) - f(t)| \leq \max_{\phi_j(s) > 0} |f(v_j) - f(t)|.$$

If $\phi_j(s) > 0$ then

$$d(s, u_j) < \frac{1}{2} d(u_j, E) \leq \frac{1}{2} (d(s, u_j) + d(s, t)),$$

so $d(s, u_j) < d(s, t) < \delta$ and $d(u_j, E) = d(u_j, v_j) < 2\delta$. Thus,

$$d(t, v_j) \leq d(s, t) + d(s, u_j) + d(u_j, v_j) < 4\delta.$$

Therefore, $|Tf(s) - Tf(t)| < \epsilon$, and the proof is completed.

$\square$

If we combine Borsuk's theorem with Proposition 4.4.3 we see that an arbitrary $\mathcal{C}(K)$ with K an uncountable compact metric space (a) is isomorphic to a complemented subspace of $\mathcal{C}([0, 1]^{\mathbb{N}})$ and (b) contains a complemented subspace isomorphic to $\mathcal{C}(\Delta)$ where $\Delta = \{0, 1\}^{\mathbb{N}}$. To complete the proof of Miljutin's theorem we need to set up the conditions for the Pełczyński decomposition technique (Theorem 2.2.3). The first step is easy:

Proposition 4.4.5. $\mathcal{C}(\Delta) \approx c_0(\mathcal{C}(\Delta))$.

Proof. Since $\mathcal{C}(\Delta)$ is isomorphic to its hyperplanes (Proposition 4.4.1), it is isomorphic to the subspace $Z = \{f \in \mathcal{C}(\Delta) : f(0, 0, \dots) = 0\}$.

For each $n \in \mathbb{N}$ let $\Delta_n = \{(s_k)_{k=1}^\infty \in \Delta : s_k = 0 \text{ if } k < n \text{ and } s_n = 1\}$. Each Δ_n is homeomorphic to Δ and is a clopen subset of Δ.

If we define the map $S : Z \to \ell_\infty(\mathcal{C}(\Delta_n))$ by $Sf = (f|_{\Delta_n})_{n=1}^\infty$ then it is clear from continuity at $(0, 0, \dots)$ that S maps into $c_0(\mathcal{C}(\Delta_n))$ and, in fact, defines an isometric isomorphism between Z and this space.

□

At this point we need only one more ingredient, but it is the crux of the argument. We must show that $\mathcal{C}([0,1]^{\mathbb{N}})$ can be embedded complementably into $\mathcal{C}(\Delta)$. In order to understand the difficulty we will first look at the problem of embedding $\mathcal{C}[0,1]$ complementably into $\mathcal{C}(\Delta)$.

It is easy to embed $\mathcal{C}[0,1]$ into $\mathcal{C}(\Delta)$. Indeed, we saw in the proof of the Banach-Mazur theorem that there is a continuous surjection $\varphi : \Delta \to [0,1]$ defined by

$$\varphi((s_n)_{n=1}^{\infty}) = \sum_{n=1}^{\infty} \frac{s_n}{2^n}.$$

This induces an isometric embedding,

$$\mathcal{C}[0,1] \to \mathcal{C}(\Delta), \qquad f \to f \circ \varphi.$$

Unfortunately the image of this embedding is *not* complemented in $\mathcal{C}(\Delta)$. We will detour from the proof of Miljutin's theorem to explain this.

Let $\mathcal{B}[0,1]$ be the space of bounded Borel functions on $[0,1]$ with the usual supremum norm,

$$\|f\| = \sup_{0 \leq t \leq 1} |f(t)|.$$

Let $\mathcal{D}$ be the set of dyadic rationals in $(0,1)$, i.e., $q \in \mathcal{D}$ if and only if $q = k/2^n$ where $1 \leq k \leq 2^n - 1$. We will consider the subspace E of $\mathcal{B}[0,1]$ of all functions f which are right-continuous everywhere, continuous at all points $t \notin \mathcal{D}$, and have left-hand limits at each $t \in \mathcal{D}$. E consists of exactly those functions $f \in \mathcal{B}[0,1]$ such that

- $f(t) = \lim_{s \to t+} f(s)$ for all $0 \leq t < 1$,
- $f(t-) = \lim_{s \to t-} f(s)$ exists for all $0 < t \leq 1$, and
- $f(t-) = f(t)$ if $t \notin \mathcal{D}$.

Then E can be identified with $\mathcal{C}(\Delta)$. We utilize the fact that φ is quite close to a homeomorphism. In fact $\varphi^{-1}(t)$ consists of at most two points and is unique for $t \notin \mathcal{D}$. Let $\rho : [0,1] \to \Delta$ be the map defined by taking $\rho(t) = \varphi^{-1}(t)$ for $t \notin \mathcal{D}$ then extending it to be right-continuous. Thus $\varphi \circ \rho$ is the identity map on $[0,1]$ and ρ is right-continuous. We can define an isometry of $\mathcal{C}(\Delta)$ onto E by $Tf(t) = f(\rho(t))$.

For $s_1, s_2, \ldots, s_n \in \{0,1\}$ let

$$\Delta_{s_1,\ldots,s_n} = \{t = (t_k)_{k=1}^{\infty} \in \Delta : t_k = s_k \text{ for } 1 \leq k \leq n\}.$$

$\Delta_{s_1,\ldots,s_n}$ is a clopen subset of Δ. Let

$$q(s_1,\ldots,s_n) = \varphi(s_1,\ldots,s_n,0,\ldots) = \sum_{k=1}^{n} \frac{s_k}{2^k}.$$

Then for $n \in \mathbb{N}$ and q of the form $k/2^n$ with $0 \leq k \leq 2^n - 1$ let $I_{n,q}$ be the half open interval $[q, q + 2^{-n})$ when $q + 2^{-n} < 1$ and the closed interval $[q, 1]$ when $q + 2^{-n} = 1$. In this language we have

$$T\chi_{\Delta_{s_1,\ldots,s_n}} = \chi_{I_{n,q(s_1,\ldots,s_n)}}.$$

Now, the embedding of $\mathcal{C}[0,1]$ into $\mathcal{C}(\Delta)$ using φ is isometrically equivalent to the embedding of $\mathcal{C}[0,1]$ into E in the sense that there is an isometry of $\mathcal{C}(\Delta)$ onto E which sends $\mathcal{C}[0,1]$ to $\mathcal{C}[0,1]$.

Proposition 4.4.6. *There is no bounded projection from E onto $\mathcal{C}[0,1]$.*

Proof. We start by identifying the quotient space $E/\mathcal{C}[0,1]$. Define the map $S : E \to \ell_\infty(\mathcal{D})$ by

$$Sf(q) = \frac{1}{2}(f(q) - f(q-)).$$

If we consider a function in E of the form

$$f = \sum_{k=0}^{2^n-1} a_k \chi_{I_{n,k}}, \qquad n \in \mathbb{N}, \ a_0, \ldots, a_{2^n-1} \in \mathbb{R},$$

it is clear that $\|Sf\| = d(f, \mathcal{C}[0,1])$ and that S maps this space onto the subspace of all finitely nonzero functions on $\mathcal{D}$. Thus it follows that S maps onto $c_0(\mathcal{D})$ and the quotient may be identified isometrically with $c_0(\mathcal{D})$.

If $\mathcal{C}[0,1]$ is complemented in E then there is a *lifting* of S, i.e., a bounded linear map $R : c_0(\mathcal{D}) \to E$ so that $SR = I_{c_0(\mathcal{D})}$. Let e_d denote a canonical basis element in $c_0(\mathcal{D})$ and let $f_d = Re_d$. We will inductively select $(d_n)_{n=1}^\infty$ in $\mathcal{D}$, open intervals $(J_n)_{n=1}^\infty$ in $(0,1)$, and signs $(\epsilon_n)_{n=1}^\infty$ so that

$$\sum_{k=1}^{n} \epsilon_k f_{d_k}(t) \geq \frac{n}{2}, \qquad n \in \mathbb{N}, t \in J_n.$$

To start the induction pick $d_1 = \frac{1}{2}$ and then either $|f_{d_1}(d_1)|$ or $|f_{d_1}(d_1-)|$ is at least one. Hence we may pick a sign ϵ_1 and an open interval J_1 (with d_1 as an endpoint) so that $\epsilon_1 f_{d_1}(t) > \frac{1}{2}$ for $t \in J_1$.

If $d_1, \ldots, d_{n-1}$, $\epsilon_1, \ldots, \epsilon_{n-1}$ and $J_1, \ldots, J_{n-1}$ have been chosen we pick $d_n \in J_{n-1}$, and then ϵ_n so that either

$$\sum_{k=1}^{n} \epsilon_k f_{d_k}(d_n) \geq \frac{n-1}{2} + 1$$

or

$$\sum_{k=1}^{n} \epsilon_k f_{d_k}(d_n-) \geq \frac{n-1}{2} + 1.$$

Thus we can find an open interval J_n with d_n as an endpoint so that

$$\sum_{k=1}^{n} \epsilon_k f_{d_k}(t) \geq \frac{n}{2} \qquad t \in J_n.$$

This completes the induction.

It follows that

$$\frac{n}{2} \leq \|R(\epsilon_1 e_{d_1} + \cdots + \epsilon_n e_{d_n})\| \leq \|R\|, \quad n \in \mathbb{N},$$

which is clearly absurd.

$\square$

The next result, known as Miljutin's lemma, is the key step in the argument. Miljutin was able to show that $\mathcal{C}[0,1]$ can be embedded as a complemented subspace of $\mathcal{C}(\Delta)$. Indeed, we can construct an alternative continuous surjection $\psi : \Delta \to [0,1]$ so that there is a norm-one linear operator $R : \mathcal{C}(\Delta) \to \mathcal{C}[0,1]$ with $R(f \circ \psi) = f$.

Lemma 4.4.7 (Miljutin's Lemma). *There exist a continuous surjection* $\phi : \Delta \times \Delta \to [0,1]$ *and a norm-one operator* $S : \mathcal{C}(\Delta \times \Delta) \to \mathcal{C}[0,1]$ *such that* $S(f \circ \phi) = f$ *for all* $f \in \mathcal{C}[0,1]$.

Proof. We start using a very similar approach as in the previous case. This time we consider an isometric embedding T of $\mathcal{C}(\Delta \times \Delta)$ into $\mathcal{B}[0,1]^2$ induced by the formula

$$Tf(s,t) = f(\rho(s), \rho(t)), \qquad 0 \leq s, t \leq 1,$$

where ρ is the right-continuous left-inverse of the function φ that we considered above. Thus,

$$T(\chi_{\Delta(r_1,\ldots,r_m) \times \Delta(s_1,\ldots,s_n)}) = \chi_{I_{m,q(r_1,\ldots,r_m)} \times I_{n,q(s_1,\ldots,s_n)}},$$

where $r_1, \ldots, r_m, s_1, \ldots s_n \in \{0,1\}$. T maps $\mathcal{C}(\Delta \times \Delta)$ isometrically onto a subspace F of $\mathcal{B}[0,1]^2$.

Let us define a homeomorphism θ of $[0,1]^2$ onto itself by the formula

$$\theta(t,u) = (t, u^2 t + (1-t)u), \qquad (t,u) \in [0,1]^2.$$

Notice that for each fixed choice of t the map $u \to u^2 t + u(1-t)$ is a monotone increasing homeomorphism of $[0,1]$ onto itself and that $(t,u) \to (t, u^2 t + u(1-t))$ is a homeomorphism of the square onto itself. Let the (continuous) inverse map be given by $(t,v) \to (t, \sigma(t,v))$, where for each fixed t the map $v \to \sigma(t,v)$ is an increasing homeomorphism of $[0,1]$ onto itself.

Let $\phi : \Delta \times \Delta \to [0,1]$ be given by $\phi(r,s) = \sigma(\varphi(r), \varphi(s))$.

Next define a norm-one operator $V : \mathcal{B}[0,1]^2 \to \mathcal{B}[0,1]$ via the formula

$$Vf(u) = \int_0^1 f \circ \theta(t,u) dt.$$

Notice that $VT(f \circ \phi) = f$ if $f \in \mathcal{C}[0,1]$. Indeed, if $g \in C(\Delta \times \Delta)$ and $(t, u) \in [0,1]^2$ then $Tg(t,u) = g(\rho(t), \rho(u))$ and hence $Tf \circ \phi(t,u) = f \circ \phi(\rho(t), \rho(u)) = f \circ \sigma(t, u)$ and thus $T(f \circ \phi)(\theta(t, u)) = f \circ \sigma \circ \theta(t, u) = f(u)$ for all $0 \le t \le 1$.

All that remains is to show that VT actually maps $\mathcal{C}(\Delta \times \Delta)$ into $\mathcal{C}[0,1]$. To this end we need to show that V maps F into $\mathcal{C}[0,1]$ and it is therefore more than enough to show that $g = V(\chi_{[0,a) \times [0,b)}) = V(\chi_{[0,a] \times [0,b]}) \in \mathcal{C}[0,1]$ for any $0 < a \le 1$ and $0 < b \le 1$.

Notice that $g(u)$ can be computed as the measure of the set of t so that $0 \le t \le a$ and $u^2 t + u(1-t) \le b$. The later inequality reduces to $t \ge (u - b)(u - u^2)^{-1}$. The single nonnegative solution of the quadratic equation $u - b = (u - u^2)a$ will be denoted by $h(a, b)$. Note that $h(a, b) > b$ unless $a = 0$. We thus have

$$g(u) = \begin{cases} a & \text{if } u \le b \\ a - \frac{u-b}{u-u^2} & \text{if } b < u \le h(a,b) \\ 0 & \text{if } h(a,b) < u < 1. \end{cases}$$

Since g is continuous this completes our proof.   □

We are now in position to complete Miljutin's theorem:

Theorem 4.4.8 (Miljutin's Theorem). *Suppose K is an uncountable compact metric space. Then $\mathcal{C}(K)$ is isomorphic to $\mathcal{C}[0,1]$.*

Proof. The first step is to show that $\mathcal{C}([0,1]^{\mathbb{N}})$ is isomorphic to a complemented subspace of $\mathcal{C}(\Delta)$. By Lemma 4.4.7 there is a continuous surjection $\psi : \Delta \to [0,1]$ so that we can find a norm one operator $R : \mathcal{C}(\Delta) \to \mathcal{C}[0,1]$ with $Rf \circ \psi = f$ for $f \in \mathcal{C}[0,1]$. Then $R(\chi_\Delta) = \chi_{[0,1]}$. For fixed $t \in [0,1]$ the linear functional $f \to Rf(t)$ is given by a probability measure μ_t so that

$$Rf(t) = \int_\Delta f \, d\mu_t.$$

The map $\tilde{\psi} : \Delta^{\mathbb{N}} \to [0,1]^{\mathbb{N}}$ given by

$$\tilde{\psi}(s_1, \dots, s_n, \dots) = (\psi(s_1), \dots, \psi(s_n), \dots)$$

is a continuous surjection. We will define $\tilde{R} : \mathcal{C}(\Delta^{\mathbb{N}}) \to \mathcal{C}([0,1]^{\mathbb{N}})$ in such a way that $\tilde{R}f \circ \tilde{\psi} = f$ for $f \in \mathcal{C}([0,1]^{\mathbb{N}})$. Indeed, the subalgebra $\mathcal{A}$ of $\mathcal{C}(\Delta^{\mathbb{N}})$ of all f which depend only on a finite number of coordinates is dense by the Stone-Weierstrass theorem. If $f \in \mathcal{A}$ depends only on $s_1, \dots, s_n$ we define

$$\tilde{R}f(t_1, \dots, t_n) = \int_\Delta \cdots \int_\Delta f(s_1, \dots, s_n) \, d\mu_{t_1}(s_1) \dots d\mu_{t_n}(s_n).$$

This map is clearly linear into $\ell_\infty[0,1]$ and has norm one. It therefore extends to a norm-one operator $\tilde{R} : \mathcal{C}(\Delta^{\mathbb{N}}) \to \ell_\infty[0,1]$. If $f \in \mathcal{C}(\Delta^{\mathbb{N}})$ is of the form $f_1(s_1) \dots f_n(s_n)$ then

$$\tilde{R}f(t) = Rf_1(t)\dots Rf_n(t),$$

so $\tilde{R}f \in \mathcal{C}[0,1]$. The linear span of such functions is again dense by the Stone-Weierstrass theorem so $\tilde{R}$ maps into $\mathcal{C}[0,1]$.

If $f \in \mathcal{C}([0,1]^{\mathbb{N}})$ is of the form $f_1(t_1)\dots f_n(t_n)$ then it is clear that $\tilde{R}f \circ \tilde{\psi} = f$. It follows that this equation holds for all $f \in \mathcal{C}([0,1]^{\mathbb{N}})$.

Thus $\mathcal{C}([0,1]^{\mathbb{N}})$ is isomorphic to a norm-one complemented subspace of $\mathcal{C}(\Delta^{\mathbb{N}})$ or $\mathcal{C}(\Delta)$ as Δ is homeomorphic to $\Delta^{\mathbb{N}}$.

Now, suppose K is an uncountable compact metric space. Then $\mathcal{C}(K)$ is isomorphic to a complemented subspace of $\mathcal{C}([0,1]^{\mathbb{N}})$ by combining Proposition 4.4.3 and Theorem 4.4.4. Hence, by the preceding argument, $\mathcal{C}(K)$ is isomorphic to a complemented subspace of $\mathcal{C}(\Delta)$. On the other hand $\mathcal{C}(\Delta)$ is isomorphic to a complemented subspace of $\mathcal{C}(K)$ again by Proposition 4.4.3 and Theorem 4.4.4. We also have Proposition 4.4.5 which gives $c_0(\mathcal{C}(\Delta)) \approx \mathcal{C}(\Delta)$. We can apply Theorem 2.2.3 to deduce that $\mathcal{C}(K) \approx \mathcal{C}(\Delta)$. Of course, the same reasoning gives $\mathcal{C}[0,1] \approx \mathcal{C}(\Delta)$.

□

4.5 Spaces of continuous functions on countable compact metric spaces

We will now briefly discuss the case when K is countable. The simplest such example as we saw in the previous section is when $K = \gamma\mathbb{N}$, the one-point compactification of the natural numbers $\mathbb{N}$, in which case $\mathcal{C}(\gamma\mathbb{N}) = c \approx c_0$.

In 1960, Bessaga and Pełczyński [13] gave a complete classification of all $\mathcal{C}(K)$-spaces when K is countable and compact. To fully describe this classification requires some knowledge of ordinals and ordinal spaces, and we prefer to simply discuss the case when K has the simplest structure.

If K is any countable compact metric space, the Baire Category theorem implies that the union of all its isolated points, U, is dense and open in K. The *Cantor-Bendixson derivative of K* is the set $K' = K \setminus U$ of accumulation points of K. Analogously, we can define $K'' = (K')'$ and, in general, for any natural number n, $K^{(n)} = (K^{(n-1)})'$.

K is said to have finite Cantor-Bendixson index if $K^{(n)}$ is finite for some n and, hence, $K^{(n+1)}$ is empty. When this happens, $\sigma(K)$ will denote the first n for which $K^{(n)}$ is finite.

Example 4.5.1. It is easy to make examples of spaces K without finite Cantor-Bendixson index. Let us note, first, that if E is any closed subset of K then $E' \subset K'$, therefore $\sigma(E) \le \sigma(K)$. If K is a countable compact metric space, then $K_1 = K \times \gamma\mathbb{N}$ has the property that $(K_1)'$ contains a subset homeomorphic to K, so $\sigma(K_1) > \sigma(K)$. In this way we can build a sequence $(K_r)_{r=1}^{\infty}$ with $\sigma(K_r) \to \infty$. If we let K_∞ be the one-point compactification of the disjoint union $\bigsqcup_{r=1}^{\infty} K_r$, then K_∞ does not have finite Cantor-Bendixson index.

If K does not have finite index then its index can be defined as a countable ordinal. This was used by Bessaga and Pełczyński to give a complete classification, up to linear isomorphism, of all $\mathcal{C}(K)$ for K countable. But we will not pursue this; instead we will give one result in the direction of classifying such $\mathcal{C}(K)$-spaces.

Theorem 4.5.2. *Let K be a compact metric space. The following conditions are equivalent:*

(i) K is countable and has finite Cantor-Bendixson index;
(ii) $\mathcal{C}(K) \approx c_0$;
(iii) $\mathcal{C}(K)$ embeds in a space with unconditional basis;
(iv) $\mathcal{C}(K)$ has property (u).

Let us point out that this theorem greatly extends Karlin's theorem (see Proposition 3.5.4 (ii)) that $\mathcal{C}[0,1]$ has no unconditional basis.

Proof. $(i) \Rightarrow (ii)$. Let us suppose, first, that $\sigma(K) = 1$. Then K' is a finite set, say $K' = \{s_1, \ldots, s_n\}$. Let $V_1, \ldots, V_n$ be disjoint open neighborhoods of $s_1, \ldots, s_n$, respectively. $V_1, V_2, \ldots, V_n$ must also be closed sets since, for each j, no sequence in V_j can converge to a point which does not belong to V_j. If we denote $V_{n+1} = K \setminus (V_1 \cup \cdots \cup V_n)$, V_{n+1} must be a finite set of isolated points and is also clopen; we therefore can absorb it into, say, V_1 without changing the conditions. Now, K splits into n-clopen sets $V_1, \ldots, V_n$ and each V_j is homeomorphic to $\gamma\mathbb{N}$. Hence $\mathcal{C}(K)$ is isometric to the ℓ_∞-product of n copies of c, thus it is isomorphic to c_0.

The proof of this implication is completed by induction. Assume we have shown that $\mathcal{C}(K) \approx c_0$ if $\sigma(K) < n$, $n \geq 2$, and suppose that $\sigma(K) = n$. Then $\mathcal{C}(K') \approx c_0$. Consider the restriction map $f \to f|_{K'}$. By Theorem 4.4.4, $\mathcal{C}(K)$ is isomorphic to $\mathcal{C}(K') \oplus E$, where E denotes the kernel of the restriction $f \to f|_{K'}$. If $U = K \setminus K'$ is the set of isolated points of K then E can be identified with $c_0(U)$, which is isometric to c_0. Hence $\mathcal{C}(K)$ is isomorphic to c_0.

$(ii) \Rightarrow (iii)$ is trivial, and $(iii) \Rightarrow (iv)$ is a consequence of Proposition 3.5.3.

$(iv) \Rightarrow (i)$ First observe that if $\mathcal{C}(K)$ has property (u), then it immediately follows that K is countable by combining Theorem 4.4.8 with the fact that the space $\mathcal{C}[0,1]$ fails to have property (u). This means that $\mathcal{M}(K)$ contains only purely atomic measures and that $\mathcal{C}(K)^* = \ell_1(K)$ is separable. Thus $\mathcal{C}(K)^{**} = \ell_\infty(K)$.

Suppose h is an arbitrary element in $\ell_\infty(K)$ with $\|h\| \leq 1$. Then, since $B_{\mathcal{C}(K)}$ is weak* dense in $B_{\ell_\infty(K)}$ by Goldstine's theorem, and $B_{\ell_\infty(K)}$ is weak* metrizable by Lemma 1.4.1, it follows that we can find a sequence $(g_n)_{n=1}^\infty$ in $\mathcal{C}(K)$ with $\|g_n\| \leq 1$ which converges weak* to h. $(g_n)_{n=1}^\infty$ is a weakly-Cauchy sequence in $\mathcal{C}(K)$, so by property (u) we can find a WUC series $\sum_{n=1}^\infty f_n$ such that $\left(g_n - \sum_{k=1}^n f_k\right)_n$ converges weakly to zero in $\mathcal{C}(K)$. This means that $\sum_{k=1}^\infty f_k = h$ for the weak* topology. In particular we have that

$$\sum_{k=1}^{\infty} f_k(s) = h(s), \qquad s \in K.$$

Since $\sum f_n$ is a WUC series, there is a constant M such that

$$\sup_{N} \sup_{\epsilon_j = \pm 1} \left| \sum_{k=1}^{N} \epsilon_k f_k(s) \right| = \sum_{k=1}^{\infty} |f_k(s)| \leq M$$

for every $s \in K$.

Put $\phi(s) = \sum_{k=1}^{\infty} |f_k(s)|$ and $\psi(s) = \sum_{k=1}^{\infty} \left(|f_k(s)| - f_k(s) \right) = \phi(s) - h(s)$. Both ϕ and ψ are lower semicontinuous functions on K, that is, for every $a \in \mathbb{R}$ the sets $\phi^{-1}(a, \infty)$ and $\psi^{-1}(a, \infty)$ are open. We also have $\|\phi\|, \|\psi\| \leq M$ and $h = \phi - \psi$.

Suppose that K fails to have finite Cantor-Bendixson index. Then each of the sets $E_n = K^{(n-1)} - K^{(n)}$ is nonempty for $n = 1, 2, \ldots$ (here, $K^{(0)} = K$). We pick a particular $h \in \ell_\infty(K)$ with $\|h\| \leq 1$ so that

$$h(s) = (-1)^n, \qquad s \in E_n.$$

Since K fails to have finite index, the set $K \setminus \cup_{n=1}^{\infty} E_n$ is nonempty and we can define h to be zero on this set. Thus, we can write $h = \phi - \psi$ as above. If we put

$$a_n = \sup_{s \in E_{2n}} \phi(s), \qquad n = 1, 2, \ldots$$

then $|a_n| \leq M$ for all n.

Suppose $\epsilon > 0$ and that $n \geq 1$. Then, there exists $s_0 \in E_{2n}$ so that $\phi(s_0) > a_n - \epsilon$. Thus by the lower semicontinuity of ϕ there is an open set U_0 containing s_0 so that $\phi(s) > a_n - \epsilon$ for every $s \in U_0$. In particular $U_0 \cap K^{(2n-2)}$ is relatively open in $K^{(2n-2)}$ and $U_0 \cap E_{2n-1} \neq \emptyset$. Hence there exists $s_1 \in U_0 \cap E_{2n-1}$ so that $\phi(s_1) > a_n - \epsilon$. Thus $\psi(s_1) > a_n + 1 - \epsilon$. Next we find an open set U_1 containing s_1 so that $\psi(s) > a_n + 1 - \epsilon$ for $s \in U_1$. Reasoning as above we can find $s_2 \in U_1 \cap E_{2n-2}$ with $\psi(s_2) > a_n + 1 - \epsilon$. But this implies $\phi(s_2) > a_n + 2 - \epsilon$ and so $a_{n-1} \geq a_n + 2 - \epsilon$. Since $\epsilon > 0$ is arbitrary we have:

$$a_n \leq a_{n-1} - 2, \qquad n = 1, 2, \ldots.$$

Clearly this contradicts the lower bound of $-M$ on the sequence $(a_n)_{n=1}^{\infty}$. The contradiction shows that K has finite Cantor-Bendixson index.

$\square$

If K and L are countable compact metric spaces with different but finite Cantor-Bendixson indices then K and L are not homeomorphic but the spaces $\mathcal{C}(K)$ and $\mathcal{C}(L)$ are both isomorphic to c_0. Later we will see that, up to equivalence, there is only one unconditional basis of c_0, in the sense that any normalized unconditional basis is equivalent to the canonical basis.

Remark 4.5.3. Notice that since $\mathcal{C}(K)^*$ is isometric to ℓ_1 for *every* countable compact metric space K, the Banach space ℓ_1 is isometric to the dual of many nonisomorphic Banach spaces.

Problems

4.1. Let K be any compact Hausdorff space. Show that any extreme point of $B_{\mathcal{C}(K)^*}$ is of the form $\pm \delta_s$ where δ_s is the probability measure defined on the Borel sets of K by $\delta_s(B) = 1$ if $s \in B$ and 0 otherwise.

4.2. The Banach-Stone Theorem. Suppose K and L are compact Hausdorff spaces such that $\mathcal{C}(K)$ and $\mathcal{C}(L)$ are isometric. Show that K and L are homeomorphic. [*Hint* : Argue that if $U : \mathcal{C}(K) \to \mathcal{C}(L)$ is any (onto) isometry, then U^* maps extreme points of the dual ball to extreme points.]

4.3. Ransford's proof of the Stone-Weierstrass Theorem [193].
(a) If E is a closed subset of K, let $\|f\|_E = \sup\{|f(t)| : t \in E\}$. Assume $A \neq \mathcal{C}(K)$; pick $f \in \mathcal{C}(K)$ with $d(f, A) = \inf\{\|f - a\| : a \in K\} = 1$. Show by a Zorn's lemma argument that there is a minimal compact subset E of K with $d_E(f, A) = \inf\{\|f - a\|_E : a \in A\} = 1$.

(b) Show that E cannot consist of one point and that there exists $h \in A$ with $\min_{s \in E} h(s) = 0$ and $\max_{s \in E} h(s) = 1$.

(c) Let $E_0 = \{s \in E : h(s) \leq 2/3\}$ and $E_1 = \{s \in E : h(s) \geq 1/3\}$. Show that there exist $a_0, a_1 \in E$ so that $\|f - a_0\|_{E_0} < 1$ and $\|f - a_1\|_{E_1} < 1$.

(d) Let $g_n = (1 - (1 - h)^n)^{2^n} \in A$. Show that for large enough n we have $\|(1 - g_n)a_0 + g_n a_1 - f\|_E < 1$. This contradiction proves the theorem.

4.4. De Branges's proof of the Stone-Weierstrass Theorem [37].
(a) Let μ be a regular probability measure on K and let E be the intersection of all compact sets $F \subset K$ with $\mu(F) = 1$. Show that $\mu(E) = 1$. (E is called the *support* of μ.)

(b) Suppose $A \neq \mathcal{C}(K)$. Let $V = B_{\mathcal{M}(K)} \cap A^\perp \subset \mathcal{C}(K)^*$. Show that A is weak* compact and convex and deduce that it has an extreme point ν with $\|\nu\| = 1$.

(c) If $a \in A$ with $0 \leq a \leq 1$, show that $\nu_a \in A^\perp$, where

$$\int h \, d\nu_a = \int ha \, d\nu.$$

Show that $\|\nu_a\| = \int a \, d|\nu|$. Deduce from the fact that ν is an extreme point that a is constant ν-a.e. on the support of $|\nu|$.

(d) Deduce that the support of $|\nu|$ is a single point and hence obtain a contradiction.

4.5. A compact Hausdorff space K is called *extremally disconnected* if the closure of every open set is again open (and hence clopen!). Prove that if $\mathcal{C}(K)$ is order-complete then K is extremally disconnected. [*Hint*: If U is open, apply order-completeness to the set of $f \in \mathcal{C}(K)$ with $f \geq \chi_U$.]

4.6. (a) If K is extremally disconnected, show that for every bounded lower semicontinuous function f, the upper semicontinuous regularization

$$\tilde{f}(s) = \inf\{g(s) : g \in C(K), g \geq f\}$$

is continuous.

(b) Deduce that if K is extremally disconnected then $C(K)$ is order-complete.

4.7. Let K be any topological space.
(a) Show that for every Borel set there is an open set U so that the symmetric difference $B \triangle U$ has first category. (Of course, this is vacuous unless K is of second category in itself!)
(b) Deduce that for every real Borel function f on K there is a lower semicontinuous function g such that $\{f \neq g\}$ has first category.
(c) Show that if K is compact and extremally disconnected then for every bounded Borel function there is a continuous function g so that $\{f \neq g\}$ has first Baire category.

4.8. Let K be a compact Hausdorff space and consider the space $\mathcal{B}(K)$ of all bounded Borel functions on K. Consider $\mathcal{B}(K)$ modulo the equivalence relation $f \sim g$ if and only if $\{s \in K : f(s) \neq g(s)\}$ has first category. Define a norm on the space $\mathcal{B}^\sim(K) = \mathcal{B}(K)/\sim$ by

$$\|f\| = \inf\{\lambda : \{|f| > \lambda\} \text{ is of first category}\}.$$

Show that $\mathcal{B}^\sim(K)$ is a Banach space which can be identified with a space $C(L)$ where L is compact Hausdorff. Show further that $C(L)$ is order-complete and hence L is extremally disconnected.

Note that if K is extremally disconnected then $\mathcal{B}^\sim(K) = C(K)$ (in the sense that there is a unique continuous function in each equivalence class).

4.9. (Continuation of 4.8.) (a) Now suppose $\mathcal{B}^\sim(K)$ is isometrically a dual space. Show that if φ belongs to the predual then there is a regular Borel measure μ on K so that $\mu(B) = \varphi(\chi_B)$ for every Borel set. Show that μ must vanish on every set of first category. [*Hint*: Use the fact that the positive cone must be closed for the weak* topology.]

(b) Deduce that if K is compact and metrizable and has no isolated points (e.g., $K = [0, 1]$) then $\mathcal{B}^\sim(K)$ cannot be a dual space.

4.10. Let K be metrizable and let E denote the smallest subspace of $C(K)^{**}$ containing $C(K)$ which is weak* sequentially closed (i.e., is closed under the weak* convergence of sequences). Show that $E = \mathcal{B}(K)$ where $\mathcal{B}(K)$ is considered as a subspace of $C(K)^{**}$ via the action

$$\langle f, \mu \rangle = \int f \, d\mu, \qquad \mu \in \mathcal{M}(K).$$

4.11. The Amir-Cambern Theorem [4], [22].

Let K and L be compact spaces and suppose $T : \mathcal{C}(K) \to \mathcal{C}(L)$ is an isomorphism such that $\|T\| = 1$ and $\|T^{-1}\| < c < 2$. For the proof of the theorem that we outline here we shall impose the additional assumption that K and L are metrizable.

(a) Show that T^{**} maps $\mathcal{B}(K)$ onto $\mathcal{B}(L)$.

(b) For $t \in K$ define $e_t \in \mathcal{B}(K)$ by $e_t(t) = 1$ and $e_t(s) = 0$ for $s \neq t$. Show that, for fixed $t \in K$,

$$|T^{**}e_t(x)| > \frac{1}{c}$$

for exactly one choice of $x \in L$. [*Hint:* If this holds for $x \neq y$ consider $T^*(a\delta_x + b\delta_y)$ where a, b are chosen suitably.]

Show also that, for fixed $x \in L$, $|T^{**}e_t(x)| > \frac{1}{2}$ for at most one of $t \in K$.

(c) Use (b) to define an injective map $\phi : K \to L$ such that

$$|\langle T^*\delta_{\phi(t)}, e_t \rangle| > \frac{1}{c}, \qquad t \in K.$$

Show that ϕ is continuous and that

$$\|Tf - f \circ \phi\| \leq 2(1 - c^{-1})\|f\|, \qquad f \in \mathcal{C}(K).$$

(d) Deduce that ϕ is onto and K and L are homeomorphic.

The Amir-Cambern theorem is an extension of the Banach-Stone theorem. Of course, Miljutin's theorem means that we must have some restriction on $\|T^{-1}\|$; in fact 2 is sharp in the sense that one can find nonhomeomorphic K and L and T with $\|T\| = 1$, $\|T^{-1}\| = 2$; this is due to Cohen [31].

5

$L_1(\mu)$-Spaces and $\mathcal{C}(K)$-Spaces

In this chapter we will prove some very classical results concerning weak compactness and weakly compact operators on $\mathcal{C}(K)$-spaces and $L_1(\mu)$-spaces, and exploit them to give further information about complemented subspaces of such spaces. We have proved forerunners of these results in Chapter 2 for the corresponding sequence spaces. If $T : c_0 \to X$ or $T : X \to \ell_1$ is weakly compact then T is in fact compact (Theorem 2.4.10 and Theorem 2.3.7). These results are essentially consequences of the fact that ℓ_1 is a Schur space.

We can regard c_0 as being a space of continuous functions (it is isomorphic to c which is isometrically a space of continuous functions) and ℓ_1 is a very special example of a space $L_1(\mu)$ where μ is counting measure on the natural numbers. It is therefore natural to consider to what extent we can find substitutes for more general $\mathcal{C}(K)$-spaces and $L_1(\mu)$-spaces.

Much of the material in this chapter dates back in some form or other to some remarkable and very early work of Dunford and Pettis [45] in 1940, later developed by Grothendieck [75]. However, we will take a modern approach based on the techniques we have built up in the preceding chapters; this approach to the study of function spaces may be said to date to the paper of Kadets and Pełczyński [98].

5.1 General remarks about $L_1(\mu)$-spaces

Let (Ω, Σ, μ) be a probability measure space, that is, μ is a measure on the σ-algebra Σ of sets of Ω of total mass $\mu(\Omega) = 1$. Although it might appear restrictive to consider probability spaces, this covers much more general situations. Indeed, if ν is merely assumed to be a σ-finite measure on Σ then we can always find a ν-integrable function φ so that $\varphi > 0$ everywhere and $\int \varphi \, d\nu = 1$. If we define $d\mu = \varphi \cdot d\nu$ then μ is a probability measure and $L_1(\Omega, \mu)$ is isometric to $L_1(\Omega, \nu)$ via the isometry $U : L_1(\nu) \to L_1(\mu)$ given by $Uf(\omega) = f(\omega)(\varphi(\omega))^{-1}$.

In most practical examples Ω is a complete separable metric space K (also called a *Polish space*), Σ coincides with the Borel sets $\mathcal{B}$ and μ is nonatomic. In this case it is important to note that there is only one such space $L_1(K, \mathcal{B}, \mu)$. More precisely, if μ is a nonatomic probability measure on K then there is a bijection $\sigma : [0, 1] \to K$ so that both σ and σ^{-1} are Borel maps and

$$\mu(B) = \lambda(\sigma^{-1}B), \qquad B \in \mathcal{B}(K),$$

where λ denotes Lebesgue measure on $[0, 1]$. Thus $f \to f \circ \sigma$ defines an isometry between $L_1(K, \mu)$ and $L_1 = L_1([0, 1], \lambda)$. See e.g. [166] or [200].

Let us first note that, unlike ℓ_1, L_1 is not a Schur space. To see this, take for example the sequence of functions $f_n(x) = \sqrt{2}\sin n\pi x$, $n \in \mathbb{N}$. $(f_n)_{n=1}^\infty$ is orthonormal in $L_2[0, 1]$ and by Bessel's inequality we have

$$\lim_{n \to \infty} \int_0^1 f_n(x)g(x)\, dx = 0$$

for all $g \in L_2[0, 1]$. In particular $(f_n)_{n=1}^\infty$ converges to 0 weakly in L_1 but not in norm.

On the other hand, since it is separable and its dual is nonseparable, L_1 is not reflexive. Therefore the relatively weakly compact sets of $L_1[0, 1]$ are not simply the bounded sets.

We start by trying to imitate the techniques which we developed to handle sequence spaces. First we give an analogue for Lemma 2.1.1:

Lemma 5.1.1. *Let $(f_n)_{n=1}^\infty$ be a sequence of norm-one, disjointly supported functions in $L_1(\mu)$. Then $(f_n)_{n=1}^\infty$ is a norm-one complemented basic sequence, isometrically equivalent to the canonical basis of ℓ_1.*

Proof. For any scalars $(\alpha_i)_{i=1}^n$ and any $n \in \mathbb{N}$,

$$\left\| \sum_{i=1}^n \alpha_i f_i \right\|_1 = \int_\Omega \left| \sum_{i=1}^n \alpha_i f_i \right| d\mu$$

$$= \int_\Omega \left(\sum_{i=1}^n |\alpha_i f_i| \right) d\mu$$

$$= \sum_{i=1}^n |\alpha_i| \int_\Omega |f_i|\, d\mu = \sum_{i=1}^n |\alpha_i|.$$

Let us consider the operator $P : L_1(\mu) \to L_1(\mu)$ given by

$$P(f) = \sum_{n=1}^\infty \left(\int_\Omega f h_n\, d\mu \right) f_n,$$

where, for each n,

$$h_n(\omega) = \begin{cases} \frac{\overline{f_n(\omega)}}{|f_n(\omega)|} & \text{if } |f_n(\omega)| > 0 \\ 0 & \text{if } f_n(\omega) = 0. \end{cases}$$

(This covers both the case of real and complex scalars.) P is a projection onto $[f_n]$. Furthermore,

$$\|Pf\|_1 = \sum_{n=1}^{\infty} \left| \int_{\Omega} f h_n \, d\mu \right|$$

$$= \sum_{n=1}^{\infty} \int_{\{|f_n|>0\}} |f| \, d\mu$$

$$= \int_{\cup_{n=1}^{\infty}\{|f_n|>0\}} |f| \, d\mu$$

$$\leq \int_{\Omega} |f| \, d\mu.$$

$\square$

5.2 Weakly compact subsets of $L_1(\mu)$

In this section we will consider the problem of identifying the weakly compact subsets of $L_1(\mu)$ when (Ω, Σ, μ) is a probability measure space. Our approach is through certain subsequence principles. In Chapters 1 and 2 we made heavy use of so-called *gliding hump techniques*. For example a sequence in ℓ_1 which converges coordinatewise to zero but not in norm has a subsequence which is basic and equivalent to the canonical basis of ℓ_1. The appropriate generalization to $L_1(\mu)$-spaces replaces coordinatewise convergence by almost everywhere convergence or convergence in measure.

Lemma 5.2.1. *Let $(h_n)_{n=1}^{\infty}$ be a bounded sequence in $L_1(\mu)$ that converges to 0 in measure. Then there is a subsequence $(h_{n_k})_{k=1}^{\infty}$ of $(h_n)_{n=1}^{\infty}$ and a sequence of disjoint measurable sets $(A_k)_{k=1}^{\infty}$ such that*

$$\|h_{n_k} - h_{n_k}\chi_{A_k}\|_1 \to 0.$$

Proof. We are going to extract such a subsequence by an inductive procedure based on a similar technique to the "gliding hump" argument for sequences.

Let us first note that $(h_n)_{n=1}^{\infty}$ has a subsequence which converges to 0 a.e. and so we may assume without loss of generality that $\lim_{n\to\infty} h_n(\omega) = 0$ μ-a.e.

Let $h_{n_1} = h_1$ and take $F_1 = \{\omega \in \Omega : |h_{n_1}(\omega)| > \frac{1}{2}\}$. The function h_{n_1} is integrable, therefore there exists $\delta_1 > 0$ such that $\mu(E) < \delta_1$ implies $\int_E |h_{n_1}| \, d\mu < \frac{1}{2}$. Next, pick $n_2 > n_1$ such that $\mu(|h_{n_2}| > \frac{1}{2^2}) < \delta_1$ and consider $F_2 = \{\omega \in \Omega : |h_{n_2}(\omega)| > \frac{1}{2^2}\}$.

Similarly there exists $\delta_2 > 0$ such that $\mu(E) < \delta_2$ implies $\int_E |h_{n_i}| \, d\mu < \frac{1}{2^2}$ for $i = 1, 2$. Pick $n_3 > n_2$ such that $\mu(|h_{n_3}| > \frac{1}{2^3}) < \delta_2$ and consider $F_3 = \{\omega \in \Omega : |h_{n_3}(\omega)| > \frac{1}{2^3}\}$.

Continuing by induction, we produce a subsequence (h_{n_k}) of (h_n) and a sequence of sets $(F_k)_{k=1}^\infty$ such that $\|h_{n_k} - h_{n_k}\chi_{F_k}\|_1 \le \frac{1}{2^k}$ for all k.

Now we take the sequence of disjoint subsets of Ω, (A_j), given by

$$A_1 = F_1 \setminus \bigcup_{k>1} F_k, \quad A_2 = F_2 \setminus \bigcup_{k>2} F_k, \quad \ldots \quad A_j = F_j \setminus \bigcup_{k>j} F_k, \quad \ldots.$$

Clearly, for each k we have

$$\int_{F_k} |h_{n_k}| \, d\mu - \int_{A_k} |h_{n_k}| \, d\mu \le \sum_{j>k} \int_{F_j} |h_{n_k}| \, d\mu \le \sum_{j>k} \frac{1}{2^{j-1}} = \frac{1}{2^{k-1}},$$

i.e.,

$$\|h_{n_k}\chi_{F_k} - h_{n_k}\chi_{A_k}\|_1 \le \frac{1}{2^{k-1}}.$$

Hence

$$\|h_{n_k} - h_{n_k}\chi_{A_k}\|_1 \le \|h_{n_k} - h_{n_k}\chi_{F_k}\|_1 + \|h_{n_k}\chi_{F_k} - h_{n_k}\chi_{A_k}\|_1 \le \frac{1}{2^k} + \frac{1}{2^{k-1}},$$

and so $\|h_{n_k} - h_{n_k}\chi_{A_k}\|_1 \to 0$.

$\square$

Definition 5.2.2. A bounded subset $\mathcal{F} \subset L_1(\mu)$ is called *equi-integrable* (or *uniformly integrable*) if given $\epsilon > 0$ there is $\delta = \delta(\epsilon) > 0$ so that for every set $E \subset \Omega$ with $\mu(E) < \delta$ we have $\sup_{f \in \mathcal{F}} \int_E |f| \, d\mu < \epsilon$, i.e.,

$$\lim_{\mu(E) \to 0} \sup_{f \in \mathcal{F}} \int_E |f| \, d\mu = 0.$$

Remark 5.2.3. In the previous definition we can omit the word "bounded" if μ is nonatomic, since then given any $\delta > 0$ it is possible to partition Ω into a finite number of sets of measure less than δ.

Example 5.2.4. (i) Given a nonnegative $h \in L_1(\mu)$, the set

$$\mathcal{F} = \{f \in L_1(\mu) \, ; \, |f| \le h\}$$

is equi-integrable.

(ii) The closed unit ball of $L_2(\mu)$ is an equi-integrable subset of $L_1(\mu)$. Indeed, for any $f \in B_{L_2(\mu)}$ and any measurable set E, by the Cauchy-Schwarz inequality,

$$\int_E |f| \, d\mu \le \left(\int_E 1 \, d\mu\right)^{1/2} \left(\int_E |f|^2 \, d\mu\right)^{1/2} \le (\mu(E))^{1/2}.$$

Then,

$$\lim_{\mu(E)\to 0} \sup_{f\in F} \int_E |f|d\mu = 0.$$

(iii) The closed unit ball of $L_1(\mu)$ is not equi-integrable as one can easily check by taking the subset $\mathcal{F} = \{\delta^{-1}\chi_{[0,\delta]} \; ; \; 0 < \delta < 1\}$.

Lemma 5.2.5. *Let $\mathcal{F}$ and $\mathcal{G}$ be bounded sets of equi-integrable functions in $L_1(\mu)$. Then the sets $\mathcal{F}\cup\mathcal{G}$ and $\mathcal{F}+\mathcal{G} = \{f+g\,;\, f\in\mathcal{F}, g\in\mathcal{G}\} \subset L_1(\mu)$ are (bounded and) equi-integrable.*

This is a very elementary deduction from the definition, and we leave the proof to the reader. Next we give an alternative formulation of equi-integrability.

Lemma 5.2.6. *Suppose $\mathcal{F}$ is a bounded subset of $L_1(\mu)$. Then the following are equivalent:*

(i) $\mathcal{F}$ *is equi-integrable;*

(ii) $\displaystyle\lim_{M\to\infty} \sup_{f\in\mathcal{F}} \int_{\{|f|>M\}} |f|\,d\mu = 0.$

Proof. $(i) \Rightarrow (ii)$ Since $\mathcal{F}$ is bounded, there is a constant $A > 0$ such that $\sup_{f\in\mathcal{F}} \|f\|_1 \le A$. Given $f \in \mathcal{F}$, by Chebyshev's inequality

$$\mu(\{|f| > M\}) \le \frac{\|f\|_1}{M} \le \frac{A}{M}.$$

Therefore, $\lim_{M\to\infty} \mu(\{|f| > M\}) = 0$. Using the equi-integrability of $\mathcal{F}$, we conclude that

$$\lim_{M\to\infty} \sup_{f\in\mathcal{F}} \int_{\{|f|>M\}} |f|\,d\mu = 0.$$

$(ii) \Rightarrow (i)$ Given $f \in \mathcal{F}$ and $E \in \Sigma$, for any finite $M > 0$ we have,

$$\int_E |f|\,d\mu = \int_{E\cap\{|f|\le M\}} |f|\,d\mu + \int_{E\cap\{|f|>M\}} |f|\,d\mu$$

$$\le M\mu(E) + \int_{E\cap\{|f|>M\}} |f|\,d\mu$$

$$\le M\mu(E) + \int_{\{|f|>M\}} |f|\,d\mu$$

$$\le M\mu(E) + \sup_{f\in\mathcal{F}} \int_{\{|f|>M\}} |f|\,d\mu.$$

Hence,

$$\sup_{f\in\mathcal{F}} \int_E |f|\,d\mu \le M\mu(E) + \sup_{f\in\mathcal{F}} \int_{\{|f|>M\}} |f|\,d\mu.$$

Given $\epsilon > 0$, let us pick $M = M(\epsilon)$ such that $\sup_{f \in \mathcal{F}} \int_{\{|f|>M\}} |f| \, d\mu < \frac{\epsilon}{2}$. Then if $\mu(E) < \frac{\epsilon}{2M}$ we obtain

$$\sup_{f \in \mathcal{F}} \int_E |f| \, d\mu \le M \frac{\epsilon}{2M} + \frac{\epsilon}{2} = \epsilon.$$

$\square$

Note that whenever $(f_n)_{n=1}^{\infty}$ is a sequence bounded above by an integrable function then, in particular, $(f_n)_{n=1}^{\infty}$ is equi-integrable. The next lemma establishes that, conversely, equi-integrability is a condition that can replace the existence of a dominating function in the Lebesgue Dominated Convergence theorem:

Lemma 5.2.7. *Suppose $(f_n)_{n=1}^{\infty}$ is an equi-integrable sequence in $L_1(\mu)$ that converges a.e. to some $g \in L_1(\mu)$. Then*

$$\lim_{n \to \infty} \int_{\Omega} f_n \, d\mu = \int_{\Omega} g \, d\mu.$$

Proof. For each $M > 0$ let us consider the truncations

$$f_n^{(M)} = \begin{cases} M & \text{if } f_n > M \\ f_n & \text{if } |f_n| \le M \ , \\ -M & \text{if } f_n < -M \end{cases} \qquad g^{(M)} = \begin{cases} M & \text{if } g > M \\ g & \text{if } |g| \le M \\ -M & \text{if } g < -M \end{cases}$$

and let us write

$$\left| \int_{\Omega} f_n \, d\mu - \int_{\Omega} g \, d\mu \right|$$
$$\le \left| \int_{\Omega} (f_n - f_n^{(M)}) \, d\mu \right| + \left| \int_{\Omega} f_n^{(M)} \, d\mu - \int_{\Omega} g^{(M)} \, d\mu \right| + \left| \int_{\Omega} (g - g^{(M)}) \, d\mu \right|.$$

Now,

$$\left| \int_{\Omega} (f_n - f_n^{(M)}) \, d\mu \right| \le \int_{\{|f_n|>M\}} (|f_n| - M) \, d\mu \le \int_{\{|f_n|>M\}} |f_n| \, d\mu \to 0$$

uniformly in n as $M \to \infty$ by Lemma 5.2.6. Analogously, since $g \in L_1(\mu)$

$$\left| \int_{\Omega} (g - g^{(M)}) \, d\mu \right| \le \int_{\{|g|>M\}} (|g| - M) \, d\mu \le \int_{\{|g|>M\}} |g| \, d\mu \xrightarrow{M \to \infty} 0.$$

And, finally, for each M we have

$$\lim_{n \to \infty} \int_{\Omega} f_n^{(M)} \, d\mu = \int_{\Omega} g^{(M)} \, d\mu$$

by the Bounded Convergence theorem. The combination of these three facts finishes the proof.

□

We now come to an important technical lemma which is often referred to as the *Subsequence Splitting Lemma*. This lemma enables us to take an arbitrary bounded sequence in $L_1(\mu)$ and extract a subsequence that can be split into two sequences, the first disjointly supported and the second equi-integrable. It is due to Kadets and Pełczyński and provides a very useful bridge between sequence space methods (gliding hump techniques) and function spaces.

Lemma 5.2.8 (Subsequence Splitting Lemma [98]). *Let $(f_n)_{n=1}^{\infty}$ be a bounded sequence in $L_1(\mu)$. Then there exists a subsequence $(g_n)_{n=1}^{\infty}$ of $(f_n)_{n=1}^{\infty}$ and a sequence of disjoint measurable sets $(A_n)_{n=1}^{\infty}$ such that if $B_n = \Omega \setminus A_n$ then $(g_n \chi_{B_n})_{n=1}^{\infty}$ is equi-integrable.*

Proof. Without loss of generality we can assume $\|f_n\|_1 \leq 1$ for all n.

We will first find a subsequence $(f_{n_s})_{s=1}^{\infty}$ and a sequence of measurable sets $(F_s)_{s=1}^{\infty}$ such that if $E_s = \Omega \setminus F_s$ then $(f_{n_s} \chi_{E_s})_{s=1}^{\infty}$ is equi-integrable and $\lim_{s \to \infty} f_{n_s} \chi_{F_s} = 0$ μ-a.e.

For every choice of $k \in \mathbb{N}$, Chebyshev's inequality gives

$$0 \leq \mu(|f_n| > k) \leq \frac{1}{k} \quad \text{for all} \quad n.$$

Then, since $\big(\mu(|f_n| > k)\big)_{n=1}^{\infty}$ is a bounded sequence, by passing to a subsequence we can assume that $\big(\mu(|f_n| > k)\big)_{n=1}^{\infty}$ converges for each k. Let us call α_k its limit. Our first goal is to see that the series $\sum_{k=1}^{\infty} \alpha_k$ is convergent with sum no bigger than 1.

For each n,

$$1 \geq \int_{\Omega} |f_n| \, d\mu = \int_0^{\infty} \mu(|f_n| > t) \, dt$$

$$= \sum_{k=1}^{\infty} \int_{k-1}^{k} \mu(|f_n| > t) \, dt$$

$$\geq \sum_{k=1}^{\infty} \mu(|f_n| > k).$$

Therefore the partial sums of $\sum_{k=1}^{\infty} \alpha_k$ are uniformly bounded:

$$\sum_{k=1}^{N} \alpha_k = \sum_{k=1}^{N} \lim_{n \to \infty} \mu(|f_n| > k) = \lim_{n \to \infty} \sum_{k=1}^{N} \mu(|f_n| > k) \leq 1.$$

Now, for each k we want to speed up the convergence of the sequence $\big(\mu(|f_n| > k)\big)_{n=1}^{\infty}$ to α_k. Let us extract a subsequence $(f_{n_s})_{s=1}^{\infty}$ of $(f_n)_{n=1}^{\infty}$ in such a way that for all $s \in \mathbb{N}$

$$\mu(|f_{n_s}| > k) < \alpha_k + 2^{-2s} \quad \text{if } 1 \le k \le 2^s. \tag{5.1}$$

For each s let us define

$$E_s = \{\omega \in \Omega : |f_{n_s}(\omega)| \le 2^s\}$$

and

$$F_s = \{\omega \in \Omega : |f_{n_s}(\omega)| > 2^s\}.$$

Notice that

$$\sum_{s=1}^{\infty} \mu(F_s) \le \sum_{s=1}^{\infty} \frac{\|f_{n_s}\|_1}{2^s} \le \sum_{s=1}^{\infty} \frac{1}{2^s} = 1.$$

This implies that for almost every $\omega \in \Omega$, there is just a finite number of sets such that $\omega \in F_s$. Thus $(f_{n_s} \chi_{F_s})_{s=1}^{\infty}$ converges to 0 μ-a.e.

Next we will prove that $(f_{n_s} \chi_{E_s})_{s=1}^{\infty}$ is equi-integrable. For the sake of simplicity in the notation we will denote $h_s = f_{n_s} \chi_{E_s}$, $s \in \mathbb{N}$. It suffices to show that

$$\sup_s \int_{\{|h_s|>2^r\}} |h_s|\, d\mu \overset{r \to \infty}{\longrightarrow} 0.$$

Clearly

$$\mu(|h_s| > k) = 0 \quad \text{if } k > 2^s,$$

which implies that for every fixed $r \in \mathbb{N}$, if $s < r$ then

$$\int_{\{|h_s|>2^r\}} |h_s|\, d\mu = 0.$$

For values of $s \ge r$,

$$\int_{\{|h_s|>2^r\}} |h_s|\, d\mu \le \int_{\{|h_s|>2^r\}} \left(|h_s| - 2^r\right) d\mu + 2^r \mu(|h_s| > 2^r).$$

By (5.1),

$$2^r \mu(|h_s| > 2^r) \le 2^r \alpha_{2^r} + 2^{r-2s}.$$

On the other hand,

$$\int_{\{|h_s|>2^r\}} \left(|h_s| - 2^r\right) d\mu = \int_0^{\infty} \mu\left(|h_s| - 2^r > t\right) dt$$

$$= \sum_{k=1}^{\infty} \int_{k-1}^{k} \mu\left(|h_s| - 2^r > t\right) dt$$

$$\le \sum_{k=1}^{\infty} \mu\left(|h_s| - 2^r > k - 1\right)$$

$$= \sum_{k=0}^{\infty} \mu\left(|h_s| > 2^r + k\right)$$

$$= \sum_{k=2^r}^{2^s} \mu(|h_s| > k)$$

$$\leq \sum_{k=2^r}^{2^s} (\alpha_k + 2^{-2s})$$

$$\leq 2^{-r} + \sum_{k=2^r}^{\infty} \alpha_k.$$

Summing up, if $s \geq r$ we get

$$\int_{\{|h_s| > 2^r\}} |h_s| \, d\mu \leq 2 \cdot 2^{-r} + 2^r \alpha_{2^r} + \sum_{k=2^r}^{\infty} \alpha_k \xrightarrow{r \to \infty} 0.$$

This establishes the equi-integrability of $(h_s)_{s \in \mathbb{N}}$.

Note that $\lim_{s \to \infty}(f_{n_s} - h_s) = 0$ μ-a.e. Thus we can apply Lemma 5.2.1 to the sequence $h'_s = f_{n_s} - h_s$ to deduce the existence of a further subsequence $(h'_{s_r})_{r=1}^{\infty}$ and a sequence of disjoint sets $(A_r)_{r=1}^{\infty}$ in Σ such that $\lim_{r \to \infty} \|h'_{s_r} \chi_{B_r}\| = 0$, where $B_r = \Omega \setminus A_r$. Clearly we may assume that $A_r \subset F_{s_r}$. Then the set $\{h'_{s_r} \chi_{B_r}\}_{r=1}^{\infty}$ is equi-integrable and so $\{h_{s_r} + h'_{s_r} \chi_{B_r}\}_{r=1}^{\infty}$ is also equi-integrable. If we write $g_r = f_{n_{s_r}}$ then the subsequence $(g_r)_{r=1}^{\infty}$ gives us the conclusion since $g_r \chi_{B_r} = h_{s_r} + h'_{s_r} \chi_{B_r}$.

$\square$

Now we come to our main result on weak compactness. The main equivalence, $(i) \Leftrightarrow (ii)$, is due to Dunford and Pettis [45].

Theorem 5.2.9. *Let $\mathcal{F}$ be a bounded set in $L_1(\mu)$. Then the following conditions on $\mathcal{F}$ are equivalent:*

(i) $\mathcal{F}$ is relatively weakly compact;

(ii) $\mathcal{F}$ is equi-integrable;

(iii) $\mathcal{F}$ does not contain a basic sequence equivalent to the canonical basis of ℓ_1;

(iv) $\mathcal{F}$ does not contain a complemented basic sequence equivalent to the canonical basis of ℓ_1;

(v) for every sequence $(A_n)_{n=1}^{\infty}$ of disjoint measurable sets,

$$\lim_{n \to \infty} \sup_{f \in \mathcal{F}} \int_{A_n} |f| \, d\mu = 0.$$

Proof. It is clear that $(i) \Rightarrow (iii)$ since the unit vector basis of ℓ_1 contains no weakly convergent subsequences. Trivially, $(iii) \Rightarrow (iv)$; $(ii) \Rightarrow (v)$ is also immediate since if (A_n) are disjoint measurable sets then $\mu(A_n) \to 0$ and so $\lim_{n \to 0} \sup_{f \in \mathcal{F}} \int_{A_n} |f| \, d\mu = 0$ by equi-integrability. We shall complete the circle by showing that $(iv) \Rightarrow (ii)$, $(v) \Rightarrow (ii)$, and $(ii) \Rightarrow (i)$.

If (ii) fails, by Lemma 5.2.6 there exists a sequence $(f_n)_{n=1}^\infty$ in $\mathcal{F}$ and some $\delta > 0$ such that for each $n \in \mathbb{N}$,

$$\int_{\{|f_n|>n\}} |f_n|\, d\mu \geq \delta. \tag{5.2}$$

We may suppose, using Lemma 5.2.8 and passing to a subsequence, that every f_n can be written as

$$f_n = f_n \chi_{A_n} + f_n \chi_{B_n},$$

where $(A_n)_{n=1}^\infty$ is a sequence of disjoint sets in Σ, $B_n = \Omega \backslash A_n$, and $(f_n \chi_{B_n})_{n=1}^\infty$ is equi-integrable. Then observe that, since $\mu(|f_n| > n) \to 0$, we must have

$$\lim_{n \to \infty} \int_{B_n \cap \{|f_n|>n\}} |f_n|\, d\mu = 0.$$

By deleting finitely many terms in the sequence (f_n), we can assume that

$$a_n = \int_{A_n} |f_n|\, d\mu \geq \frac{1}{2}\delta, \tag{5.3}$$

for all n.

By Lemma 5.1.1 the sequence $(a_n^{-1} f_n \chi_{A_n})_{n=1}^\infty$ is a norm-one complemented basic sequence in $L_1(\mu)$ isometrically equivalent to the canonical ℓ_1-basis. Let (h_n) in $L_\infty(\mu)$ be the norm-one biorthogonal functionals chosen in the proof of Lemma 5.1.1; each h_n is supported on A_n. Since $\mu A_n \to 0$ and the set $\{f\chi_{B_k}\}_{k=1}^\infty$ is equi-integrable we can pass to yet a further subsequence and assume that

$$\int_{A_n \cap B_m} |f_m|\, d\mu < \frac{1}{4} 2^{-n}\delta, \qquad m, n \in \mathbb{N}.$$

Define $T : L_1(\mu) \to \ell_1$ by

$$Tf = \left(\int_\Omega f h_n d\mu \right)_{n=1}^\infty$$

and $R : \ell_1 \to L_1(\mu)$ by

$$R(\xi) = \sum_{n=1}^\infty \xi_n a_n^{-1} f_n.$$

Then

$$TRe_k - e_k = \left(a_k^{-1} \int_{A_n \cap B_k} f_k h_n d\mu \right)_{n=1}^\infty$$

and we obtain the estimate

$$\left| a_k^{-1} \int_{A_n \cap B_k} f_k h_n d\mu \right| \leq 2^{-n-1}.$$

Hence

$$\|TRe_k - e_k\| \leq a_k^{-1} \sum_{n=1}^{\infty} \frac{1}{4} \delta 2^{-n} \leq \frac{1}{2},$$

which yields $\|TR - I\| \leq \frac{1}{2}$, where I is the identity operator on ℓ_1. This implies TR is invertible so there is $U : \ell_1 \to \ell_1$ such that $UTR = I$. RUT is a projection onto range of R, hence R maps ℓ_1 isomorphically onto a complemented subspace of $L_1(\mu)$; this shows that $(f_n)_{n=1}^{\infty}$ is a complemented basic sequence equivalent to the ℓ_1-basis. Thus (iv) is contradicted and so (iv) implies (ii).

Let us point out that equation (5.3), which we obtained with the only assumption that $\mathcal{F}$ failed to be equi-integrable, contradicts (v), hence in our way we also obtained the implication $(v) \Rightarrow (ii)$.

Finally, let us prove $(ii) \Rightarrow (i)$. We must show that $\overline{\mathcal{F}}^{w^*}$, the weak* closure of $\mathcal{F}$ in the bidual of $L_1(\mu)$, is contained in $L_1(\mu)$.

For each $M \in (0, \infty)$, let us consider the sets

$$\mathcal{F}_M = \{f \cdot \chi_{\{|f| \leq M\}} : f \in \mathcal{F}\}$$

and

$$\mathcal{F}^M = \{f \cdot \chi_{\{|f| > M\}} : f \in \mathcal{F}\}.$$

It is obvious that $\mathcal{F} \subset \mathcal{F}_M + \mathcal{F}^M$, therefore $\overline{\mathcal{F}}^{w^*} \subset \overline{\mathcal{F}_M}^{w^*} + \overline{\mathcal{F}^M}^{w^*}$.

Let us notice that if $f \in \mathcal{F}_M$, we have $\|f\|_2 \leq \|f\|_{\infty} \leq M$. Then,

$$\mathcal{F}_M \subset M B_{L_2(\mu)}.$$

Since $L_2(\mu)$ is reflexive, its closed unit ball is weakly compact. Therefore $M B_{L_2(\mu)}$ is weakly compact for each $M > 0$ and so $\mathcal{F}_M$ is a relatively weakly compact set in $L_2(\mu)$ for each $M > 0$. Being norm-to-norm continuous, the inclusion $\iota : L_2(\mu) \to L_1(\mu)$ is weak-to-weak continuous, so $\iota(\mathcal{F}_M) = \mathcal{F}_M$ is a relatively weakly compact set in $L_1(\mu)$ for each $M > 0$. This is equivalent to saying that for each positive M, the weak* closure of $\mathcal{F}_M$ in the bidual of $L_1(\mu)$ is a subset of $L_1(\mu)$, i.e.,

$$\overline{\mathcal{F}_M}^{w^*} \subset L_1(\mu) \text{ for all } M > 0.$$

On the other hand, if $f \in \mathcal{F}^M$, then $\|f\|_1 \leq \epsilon(M)$, where $\epsilon(M) = \sup_{f \in \mathcal{F}} \int_{\{|f| > M\}} |f| \, d\mu$. Hence, $\mathcal{F}^M \subset \epsilon(M) B_{L_1(\mu)}$. Using Goldstine's theorem we deduce that

$$\overline{\mathcal{F}^M}^{w^*} \subset \epsilon(M) B_{L_1(\mu)^{**}}.$$

Hence if $f \in \overline{\mathcal{F}}^{w^*}$ then we can write $f = \psi + \phi$, with $\psi \in L_1(\mu)$ and $\phi \in \epsilon(M) B_{L_1(\mu)^{**}}$. Therefore, for an arbitrary $M > 0$, $d(f, L_1(\mu)) \leq \epsilon(M)$. Since $\lim_{M \to \infty} \epsilon(M) = 0$ by Lemma 5.2.6, $d(f, L_1(\mu)) = 0$ and $f \in L_1(\mu)$.

$\square$

We conclude this section with a simple deduction from this theorem.

Theorem 5.2.10. $L_1(\mu)$ *is weakly sequentially complete.*

Proof. Let $(f_n)_{n=1}^{\infty} \subset L_1(\mu)$ be a weakly Cauchy sequence. Then, no subsequence of $(f_n)_{n=1}^{\infty}$ can be equivalent to the canonical ℓ_1-basis, which is not weakly Cauchy. Hence the set $\{f_n\}_{n=1}^{\infty}$ is relatively weakly compact by Theorem 5.2.9 and this implies the sequence must actually be weakly convergent. □

Corollary 5.2.11. *The space c_0 is not isomorphic to a subspace of $L_1(\mu)$.*

Proof. Since $L_1(\mu)$ is weakly sequentially complete, by Corollary 2.4.15 every WUC series in $L_1(\mu)$ is unconditionally convergent so, by Theorem 2.4.11, $L_1(\mu)$ does not contain a copy of c_0. □

5.3 Weak compactness in $\mathcal{M}(K)$

Suppose now that K is a compact Hausdorff space (not necessarily metrizable). The space $\mathcal{M}(K) = C(K)^*$ as a Banach space is a "very large" ℓ_1-sum of spaces $L_1(\mu)$ where μ is a probability measure on K. This fact has already been observed in the proof of Proposition 4.3.8 (*iii*). Using this it is possible to extend Theorem 5.2.9 to the spaces $\mathcal{M}(K)$; however, we need some additional characterizations of weak compactness in spaces of measures.

Definition 5.3.1. A subset $\mathcal{A}$ of $\mathcal{M}(K)$ is said to be *uniformly regular* if given any open set $U \subset K$ and $\epsilon > 0$, there is a compact set $H \subset U$ such that $|\mu|(U \setminus H) < \epsilon$ for all $\mu \in \mathcal{A}$.

The next equivalences are due to Grothendieck [75].

Theorem 5.3.2. *Let $\mathcal{A}$ be a bounded subset of $\mathcal{M}(K)$. The following are equivalent:*

(*i*) *$\mathcal{A}$ is relatively weakly compact;*
(*ii*) *$\mathcal{A}$ is uniformly regular;*
(*iii*) *for any sequence of disjoint Borel sets $(B_n)_{n=1}^{\infty}$ in K and any sequence of measures $(\mu_n)_{n=1}^{\infty}$ in $\mathcal{A}$, $\lim_{n\to\infty} |\mu_n|(B_n) = 0$;*
(*iv*) *for any sequence of disjoint open sets $(U_n)_{n=1}^{\infty}$ in K and any sequence of measures $(\mu_n)_{n=1}^{\infty}$ in $\mathcal{A}$, $\lim_{n\to\infty} \mu_n(U_n) = 0$;*
(*iv*)′ *for any sequence of disjoint open sets $(U_n)_{n=1}^{\infty}$ in K and any sequence of measures $(\mu_n)_{n=1}^{\infty}$ in $\mathcal{A}$, $\lim_{n\to\infty} |\mu_n|(U_n) = 0$.*

Remark 5.3.3. This theorem is true for either real or complex scalars. We give the proof in the real case. It is easy to extend this to the complex case by the simple procedure of splitting a complex measure into real and imaginary parts.

Proof. $(iii) \Rightarrow (iv)$ This is immediate because an open set is, in particular, a Borel set and

$$0 \le |\mu_n(U_n)| \le |\mu_n|(U_n) \overset{n\to\infty}{\longrightarrow} 0.$$

$(iv) \Rightarrow (iv)'$ Assume $(iv)'$ fails. Then there exist a sequence of open sets $(U_n)_{n=1}^{\infty}$ in K and a sequence of regular signed measures on K, $(\mu_n)_{n=1}^{\infty}$ such that $(|\mu_n|(U_n))_{n=1}^{\infty}$ does not converge to 0.

For each n we can write μ_n as the difference of its positive and negative parts: $\mu_n = \mu_n^+ - \mu_n^-$. Then the total variation of μ_n is the sum: $|\mu_n| = \mu_n^+ + \mu_n^-$. Therefore, without loss of generality we will suppose that the sequence $(\mu_n^+(U_n))_{n=1}^{\infty}$ does not converge to 0. By passing to a subsequence we can assume that there exists $\delta > 0$ such that $\mu_n^+(U_n) \ge \delta > 0$ for all n.

Let us fix $n \in \mathbb{N}$. Using the Hahn decomposition theorem, there is a Borel set $B_n \subset U_n$ such that $\mu_n(B_n) = \mu_n^+(U_n) \ge \delta$. Now, by the regularity of μ_n, there is an open set O_n such that $B_n \subset O_n \subset U_n$ and $\mu_n(O_n) \ge \frac{\delta}{2}$.

This way, we have a sequence of disjoint open sets $(O_n)_{n=1}^{\infty} \subset K$ such that $(\mu_n(O_n))_{n=1}^{\infty}$ does not converge to 0, contradicting (iv).

$(iv)' \Rightarrow (ii)$ Let us assume that $\mathcal{A}$ fails to be uniformly regular. Then there is an open set $U \subset K$ such that for some $\delta > 0$ we have

$$\sup_{\mu \in \mathcal{A}} |\mu|(U \setminus H) > \delta,$$

for all compact sets $H \subset K$.

Given $H_0 = \emptyset$, pick $\mu_1 \in \mathcal{A}$ so that $|\mu_1|(U \setminus H_0) > \delta$. By regularity of the measure μ_1 there exists a compact set $F_1 \subset U \setminus H_0$ such that $|\mu_1|(F_1) > \delta$. Using the T_4 separation property, we find an open set V_1 satisfying

$$F_1 \subset V_1 \subset \overline{V_1} \subset U \setminus H_0.$$

Now, given the compact set $H_1 = \overline{V_1}$ there is $\mu_2 \in \mathcal{A}$ such that $|\mu_2|(U \setminus H_1) > \delta$. By regularity of μ_2 there exists a compact set $F_2 \subset U \setminus H_1$ such that $|\mu_2|(F_2) > \delta$ and the T_4 separation property yields an open set V_2 such that

$$F_2 \subset V_2 \subset \overline{V_2} \subset U \setminus H_1.$$

For the next step in this recurrence argument we would pick $H_2 = \overline{V_1} \cup \overline{V_2}$ and repeat the previous procedure. This way, by induction we obtain a sequence of disjoint open sets $(V_n)_{n=1}^{\infty} \subset K$ and a sequence $(\mu_n)_{n=1}^{\infty} \subset \mathcal{A}$ such that $|\mu_n|(V_n) > \delta$ for all n, contradicting (ii).

$(ii) \Rightarrow (i)$ In order to prove that $\mathcal{A}$ is relatively weakly compact in $\mathcal{M}(K)$, by the Eberlein-Šmulian theorem it suffices to show that any sequence $(\mu_n) \subset \mathcal{A}$ is relatively weakly compact.

Let us consider the (positive) finite measure on the Borel sets of K given by

$$\mu = \sum_{n=1}^{\infty} \frac{1}{2^n} |\mu_n|.$$

Every μ_n is absolutely continuous with respect to μ. By the Radon-Nikodym theorem, for each n there exists a unique $f_n \in L_1(K, \mu)$ such that $d\mu_n = f_n \, d\mu$ and $\|\mu_n\| = \int_K |f_n| \, d\mu$. This provides an isometric isomorphism from $L_1(\mu)$ onto the closed subspace of $\mathcal{M}(K)$ consisting of the regular signed measures on K which are absolutely continuous with respect to μ. The isometry, in particular, takes each f_n in $L_1(K, \mu)$ to μ_n. Therefore we need only show that (f_n) is equi-integrable in $L_1(K, \mu)$.

If (f_n) is not equi-integrable, using (ii) we find a sequence (U_n) of open sets and some $\epsilon > 0$ such that $\mu(U_n) < 2^{-n}$ and $\sup_k \int_{U_n} |f_k| \, d\mu > \epsilon$. For each n put $V_n = \bigcup_{k>n} U_k$. (V_n) is a decreasing sequence of open sets such that $\mu(V_n) < 2^{-n}$ and

$$\sup_k \int_{V_n} |f_k| \, d\mu > \epsilon. \tag{5.4}$$

Now, for each n there exists E_n compact, $E_n \subset V_n$, for which

$$\sup_k \int_{V_n \setminus E_n} |f_k| \, d\mu < \frac{\epsilon}{2^{n+2}}.$$

Obviously, $\mu(\bigcap_{n=1}^{\infty} E_n) = 0$. The uniform regularity yields an open set W such that $\bigcap_{n=1}^{\infty} E_n \subset W$ and $\sup_k \int_W |f_k| \, d\mu < \frac{\epsilon}{2}$. By compactness there exists N such that $\bigcap_{n=1}^{N} E_n \subset W$ and so

$$\int_{\bigcap_{n=1}^{N} E_n} |f_k| \, d\mu < \frac{\epsilon}{2} \quad \text{for each } k.$$

Thus, for each k we have

$$\int_{V_{N+1}} |f_k| \, d\mu \leq \int_{\bigcap_{n=1}^{N} E_n} |f_k| \, d\mu + \sum_{n=1}^{N} \int_{V_n \setminus E_n} |f_k| \, d\mu < \frac{\epsilon}{2} + \sum_{k=1}^{N} \frac{\epsilon}{2^{k+2}} < \epsilon,$$

which contradicts (5.4).

$(i) \Rightarrow (iii)$ Let $(B_n)_{n=1}^{\infty}$ be an arbitrary sequence of disjoint Borel sets in K and $(\mu_n)_{n=1}^{\infty}$ be an arbitrary sequence of measures in $\mathcal{A}$. Put

$$\mu = \sum_{n=1}^{\infty} \frac{1}{2^n} |\mu_n|.$$

Reasoning as we did in the previous implication, for each n there exists a unique $g_n \in L_1(K, \mu)$ such that $d\mu_n = g_n \, d\mu$. If $\mathcal{A}$ is relatively weakly compact in $\mathcal{M}(K)$ the sequence $(g_n)_{n=1}^{\infty}$ is relatively weakly compact in $L_1(K, \mu)$, hence equi-integrable. Thus, since $\mu(B_n) \to 0$, we have

$$|\mu_n|(B_n) = \int_{B_n} |g_n| \, d\mu \to 0.$$

$\square$

5.4 The Dunford-Pettis property

Definition 5.4.1. Let X and Y be Banach spaces. A bounded linear operator $T : X \to Y$ is *completely continuous* or a *Dunford-Pettis* operator if whenever W is a weakly compact subset of X then $T(W)$ is a norm-compact subset of Y.

Clearly, if an operator is compact then it is Dunford-Pettis. If X is reflexive then an operator $T : X \to Y$ is compact if and only if T is Dunford-Pettis.

Proposition 5.4.2. *Suppose that X and Y are Banach spaces. A linear operator $T : X \to Y$ is Dunford-Pettis if and only if T is weak-to-norm sequentially continuous, i.e., whenever $(x_n)_{n=1}^{\infty} \subset X$ converges to x weakly then $(Tx_n)_{n=1}^{\infty}$ converges to Tx in norm.*

Proof. Let $T : X \to Y$ be Dunford-Pettis and suppose that there is a weakly null sequence $(x_n)_{n=1}^{\infty} \subset X$ such that $\|Tx_n\| \geq \delta > 0$ for some positive δ. Since the subset $W = \{x_n : n \in \mathbb{N}\} \cup \{0\}$ is weakly compact, its image under T is norm-compact, therefore it contains a subsequence $(T(x_{n_k}))_{k=1}^{\infty}$ that converges in norm to some $y \in Y$. From the fact that T, in particular, is weak-to-weak continuous, it follows that the sequence $(T(x_n))_{n=1}^{\infty}$ is weakly null, so y must be 0, which contradicts our assumption.

For the converse implication, suppose T is weak-to-norm sequentially continuous. Let W be a weakly compact subset of X and let $(y_n)_{n=1}^{\infty}$ be a sequence in $T(W)$. Pick (x_n) in X so that $y_n = Tx_n$ for all n. By the Eberlein-Šmulian theorem (x_n) contains a subsequence (x_{n_k}) that converges weakly to some x in W. Hence $(y_{n_k})_{k=1}^{\infty}$ converges in norm to Tx. We conclude that $T(W)$ is norm-compact.

$\square$

The following definition was introduced by Grothendieck [75] as an abstraction of ideas originally developed by Dunford and Pettis [45].

Definition 5.4.3. A Banach space X is said to have the *Dunford-Pettis property* (or, in short, X has (DPP)) if every weakly compact operator T from X into a Banach space Y is Dunford-Pettis.

For example c_0 has (DPP) because if Y is a Banach space and $T : c_0 \to Y$ is a weakly compact operator then T is compact, hence Dunford-Pettis. ℓ_1 has also (DPP) because ℓ_1 has the Schur property, which implies, as we saw, that weakly compact subsets in ℓ_1 are actually compact.

On the other hand, no infinite-dimensional reflexive Banach space X has (DPP) since the identity operator $I : X \to X$ is weakly compact but cannot be a Dunford-Pettis operator because the closed unit ball of X is not compact.

Theorem 5.4.4. *Suppose that X is a Banach space. Then X has (DPP) if and only if for every sequence $(x_n)_{n=1}^{\infty}$ in X converging weakly to 0 and every sequence $(x_n^*)_{n=1}^{\infty}$ in X^* converging weakly to 0, the sequence of scalars $(x_n^*(x_n))_{n=1}^{\infty}$ converges to 0.*

Proof. Let Y be a Banach space and $T : X \to Y$ a weakly compact operator. Let us suppose that T is not Dunford-Pettis. Then there is $(x_n)_{n=1}^{\infty}$ in X such that $x_n \xrightarrow{w} 0$ but $\|Tx_n\| \geq \delta > 0$ for all n.

Pick $(y_n^*)_{n=1}^{\infty} \subset Y^*$ such that $y_n^*(Tx_n) = \|Tx_n\|$ and $\|y_n^*\| = 1$ for all n. By Gantmacher's theorem T^* is weakly compact hence $T^*(B_{Y^*})$ is a relatively weakly compact subset of X^*. By the Eberlein-Šmulian theorem the sequence $(T^*y_n^*)_{n=1}^{\infty} \subset T^*(B_{Y^*})$ can be assumed weakly convergent to some x^* in X^*. Then $(T^*y_n^* - x^*)_{n=1}^{\infty}$ is weakly convergent to 0, which implies $(T^*y_n^* - x^*)(x_n) \to 0$. But, since $x^*(x_n) \to 0$, it would follow that $(T^*y_n^*(x_n))_{n=1}^{\infty} = (\|Tx_n\|)_{n=1}^{\infty}$ must converge to 0, which is absurd.

For the converse, let (x_n) in X be such that $x_n \xrightarrow{w} 0$ and (x_n^*) in X^* be such that $x_n^* \xrightarrow{w} 0$. Consider the operator

$$T : X \longrightarrow c_0, \quad Tx = (x_n^*(x)).$$

The adjoint operator T^* of T satisfies $T^*e_k = x_k^*$ for all $k \in \mathbb{N}$, where (e_k) denotes the canonical basis of ℓ_1. This implies that $T^*(B_{\ell_1})$ is contained in the convex hull of the weakly null sequence (x_n^*). Therefore T^* is weakly compact, hence by Gantmacher's theorem so is T.

As T is weakly compact, T is also Dunford-Pettis by the hypothesis. Then, by Proposition 5.4.2, $\|Tx_n\|_{\infty} \to 0$. Thus $(x_n^*(x_n))_{n=1}^{\infty}$ converges to 0 since, for all n,

$$|x_n^*(x_n)| \leq \max_k |x_k^*(x_n)| = \|Tx_n\|_{\infty}.$$

$\square$

We now reach the main result of the chapter. The fact that $L_1(\mu)$-spaces have (DPP) is due to Dunford and Pettis [45] (at least for the case when $L_1(\mu)$ is separable) and to Phillips [180]. The case of $\mathcal{C}(K)$-spaces was covered by Grothendieck in [75].

Theorem 5.4.5 (The Dunford-Pettis Theorem).

(i) If μ is a σ-finite measure then $L_1(\mu)$ has (DPP).
(ii) If K is a compact Hausdorff space then $\mathcal{C}(K)$ has (DPP).

Proof. Let us first prove part *(ii)*. Take any weakly null sequence $(f_n)_{n=1}^{\infty}$ in $\mathcal{C}(K)$ and any weakly null sequence $(\mu_n)_{n=1}^{\infty}$ in $\mathcal{M}(K)$. Without loss of generality both sequences can be assumed to lie inside the unit balls of the respective spaces. Define the (positive) measure

$$\nu = \sum_{n=1}^{\infty} \frac{1}{2^n} |\mu_n|.$$

Clearly each μ_n is absolutely continuous with respect to ν. By the Radon-Nikodym theorem, for each n there exists a nonnegative, Borel-measurable function g_n such that $d\mu_n = g_n \, d\nu$ and $\|\mu_n\| = \int_K g_n \, d\nu$. This provides an isometry from $L_1(\nu)$ onto the closed subspace of $\mathcal{M}(K)$ consisting of the regular signed measures on K which are absolutely continuous with respect to ν. This isometry in particular takes each g_n in $L_1(\nu)$ to μ_n. Therefore the sequence $(g_n)_{n=1}^{\infty}$ is weakly null. Thus the set $\{g_n \; ; \; n \in \mathbb{N}\}$ is relatively weakly compact in $L_1(\nu)$, hence equi-integrable.

Now for any $M > 0$, by the Bounded Convergence theorem, we have that

$$\lim_{n \to \infty} \int_{|g_n| \leq M} f_n g_n \, d\nu = 0.$$

Hence

$$\limsup_{n \to \infty} \int f_n g_n \, d\nu \leq \sup_n \int_{|g_n| > M} |g_n| \, d\nu.$$

Note that the right-hand side term tends to zero as $M \to \infty$ by Lemma 5.2.6. Then

$$\lim_{n \to \infty} \int f_n \, d\mu_n = 0$$

as required.

(i) follows from (ii) since the dual space of $L_1(\mu)$, $L_\infty(\mu)$, can be regarded as a $\mathcal{C}(K)$-space for a suitable compact Hausdorff space K. Hence if $(f_n)_{n=1}^{\infty}$ is weakly null in $L_1(\mu)$ and $(g_n)_{n=1}^{\infty}$ is weakly null in $L_\infty(\mu)$ then $\lim_{n \to \infty} \int f_n g_n \, d\mu = 0$ by the preceding argument. $\qquad \square$

Corollary 5.4.6. *If K is a compact Hausdorff space then $\mathcal{M}(K)$ has (DPP).*

The Dunford-Pettis theorem was a remarkable achievement in the early history of Banach spaces. The motivation of Dunford and Pettis came from the study of integral equations and their hope was to develop an understanding of linear operators $T : L_p(\mu) \to L_p(\mu)$ for $p \geq 1$. In fact the Dunford-Pettis theorem immediately gives the following application.

Theorem 5.4.7. *Let $T : L_1(\mu) \to L_1(\mu)$ or $T : \mathcal{C}(K) \to \mathcal{C}(K)$ be a weakly compact operator. Then T^2 is compact.*

Proof. This is immediate. For example, in the first case, $T(B_{L_1(\mu)})$ is relatively weakly compact hence $T^2(B_{L_1(\mu)})$ is relatively norm compact. $\qquad \square$

It is well known that compact operators have very nice spectral properties. For instance, any nonzero λ in the spectrum is an eigenvalue, and the only possible accumulation point of the spectrum is 0. These properties extend in a very simple way to an operator whose square is compact, so the previous result means that weakly compact operators on $L_1(\mu)$-spaces or $\mathcal{C}(K)$-spaces have similar properties. The Dunford-Pettis theorem was thus an important step in the development of the theory of linear operators in the first half of the twentieth century; this theory reached its apex in the publication of a three-volume treatise by Dunford and Schwartz between 1958 and 1971 ([46], [47], and [48]). The first of these volumes alone runs to more than 1000 pages!

The original proof of Dunford and Pettis relied heavily on the theory of representations for operators on L_1. In order to study an operator $T : L_1(\mu) \to X$ one can associate it to a *vector measure* $\nu : \Sigma \to X$ given by $\nu(E) = T\chi_E$. Thus $\|\nu(E)\| \leq \|T\|\mu(E)$. Dunford and Pettis [45] and Phillips [180] showed that if T is weakly compact one can prove a vector-valued Radon-Nikodym theorem and thus produce a Bochner integrable function $g : \Omega \to X$ so that

$$\mu(E) = \int_E g(\omega)\, d\mu(\omega).$$

This permits a representation for the operator T in the form

$$Tf = \int g(\omega) f(\omega) d\mu(\omega),$$

and they established the Dunford-Pettis theorem from this representation.

In particular if X is reflexive, every operator $T : L_1(\mu) \to X$ is weakly compact, and one has a Radon-Nikodym theorem for vector measures taking values in X. It was also shown by Dunford and Pettis [45] that this property is also enjoyed by any separable Banach space which is also a dual space (separable dual spaces). This was the springboard for the definition of the *Radon-Nikodym Property (RNP)* for Banach spaces, which led to a remarkable theory developed largely between 1965 and 1980. We will not follow up on this direction in this book. A very nice account of this theory is contained in the book of Diestel and Uhl from 1977 [42].

One of the surprising aspects of this theory is the connection between the Radon-Nikodym Property and the *Krein-Milman Property (KMP)*. A Banach space X has (KMP) if every closed bounded (not necessarily compact!) convex set is the closed convex hull of its extreme points. Obviously reflexive spaces have (KMP) but, remarkably, any space with (RNP) has (KMP) (Lindenstrauss [128]). The converse remains the major open problem in this area; the best results in this direction are due to Phelps [179] and Schachermayer [201]. It is probably fair to say that the subject has received relatively little attention since the 1980s and some really new ideas seem to be necessary to make further progress.

5.5 Weakly compact operators on $\mathcal{C}(K)$-spaces

Let us refer back again to Theorem 2.4.10. In that theorem it was shown that for operators $T : c_0 \to X$ the properties of being weakly compact, compact, or strictly singular are equivalent. For general $\mathcal{C}(K)$-spaces we have seen that weak compactness implies Dunford-Pettis. Next we turn to strict singularity.

Theorem 5.5.1. *Let K be a compact Hausdorff space. If $T : \mathcal{C}(K) \to X$ is weakly compact then T is strictly singular.*

Proof. Let Y be a subspace of $\mathcal{C}(K)$ such that $T|_Y$ is an isomorphism onto its image. Since T is weakly compact, $T(B_Y)$ is relatively weakly compact, which implies that B_Y is weakly compact. But $T(B_Y)$ is actually compact by the Dunford-Pettis theorem, Theorem 5.4.5. It follows that Y is finite-dimensional.

$\square$

Remark 5.5.2. Clearly, Theorem 5.5.1 also holds replacing $\mathcal{C}(K)$ by $L_1(\mu)$.

The following result by Pełczyński [171] is a much more precise statement than Theorem 5.5.1.

Theorem 5.5.3 (Pełczyński). *Suppose K is a compact Hausdorff space and X is a Banach space. Suppose that $T : \mathcal{C}(K) \to X$ is a bounded linear operator. If T fails to be weakly compact then there is a closed subspace E of $\mathcal{C}(K)$ isomorphic to c_0 such that $T|_E$ is an isomorphism.*

Proof. Suppose that $T : \mathcal{C}(K) \to X$ fails to be weakly compact. Then, by Gantmacher's theorem, its adjoint operator $T^* : X^* \to \mathcal{M}(K)$ also fails to be weakly compact, and so the subset $T^*(B_{X^*})$ of $\mathcal{M}(K)$ is not relatively weakly compact. By Theorem 5.3.2, there exists $\delta > 0$, a disjoint sequence of open sets $(U_n)_{n=1}^{\infty}$ in K, and a sequence $(x_n^*)_{n=1}^{\infty}$ in B_{X^*} such that if we call $\nu_n = T^* x_n^*$ then $\nu_n(U_n) > \delta$ for all n.

For each n there exists a compact subset F_n of U_n, such that $|\nu|(U_n \setminus F_n) < \frac{\delta}{2}$. By Urysohn's lemma there exists $f_n \in \mathcal{C}(K)$, $0 \le f_n \le 1$, such that $f_n = 0$ on $K \setminus U_n$ and $f_n = 1$ on F_n. Then $(f_n)_{n=1}^{\infty}$ is isometrically equivalent to the canonical basis of c_0, which implies that $[f_n]$, the closed linear span of the basic sequence (f_n), is isometrically isomorphic to c_0. Let $S : c_0 \to \mathcal{C}(K)$ be the isometric embedding defined by $Se_n = f_n$ where $(e_n)_{n=1}^{\infty}$ is the canonical basis of c_0.

Consider the operator $TS : c_0 \to X$. We claim that TS cannot be compact. Indeed since $(e_n)_{n=1}^{\infty}$ is weakly null, if TS were compact we would have $\lim_{n \to \infty} \|TSe_n\| = 0$. However,

$$
\begin{aligned}
x_n^*(TSe_n) &= x_n^*(Tf_n) \\
&= (T^* x_n^*)(f_n) \\
&= \int_K f_n \, d\nu_n \\
&= \int_{U_n} d\nu_n + \int_{U_n} (f_n - 1) \, d\nu_n \\
&\ge \delta - |\nu_n|(U_n \setminus E_n) \\
&\ge \frac{\delta}{2}.
\end{aligned}
$$

Thus TS is not compact and, by Theorem 2.4.10, it is also not strictly singular. In fact TS must be an isomorphism on a subspace isomorphic to c_0 (Proposition 2.2.1).

$\square$

Corollary 5.5.4. *Let X be a Banach space such that no closed subspace of X is isomorphic to c_0. Then any operator $T : \mathcal{C}(K) \to X$ is weakly compact.*

Using the above theorem we can now say a little bit more about injective Banach spaces.

Theorem 5.5.5. *Suppose X is an injective Banach space and $T : X \to Y$ is a bounded linear operator. If T fails to be weakly compact then there is a closed subspace F of ℓ_∞ such that F is isomorphic to ℓ_∞ and $T|_F$ is an isomorphism.*

Proof. We start by embedding X isometrically into an $\ell_\infty(\Gamma)$-space; this can be done by taking $\Gamma = B_{X^*}$ and using the embedding $x \mapsto \hat{x}$, where $\hat{x}(x^*) = x^*(x)$.

Since X is injective there is a projection $P : \ell_\infty(\Gamma) \to X$. Now the operator $TP : \ell_\infty(\Gamma) \to Y$ is not weakly compact; since $\ell_\infty(\Gamma)$ can be represented as a $\mathcal{C}(K)$-space we can find a subspace E of $\ell_\infty(\Gamma)$ which is isomorphic to c_0 and such that $TP|_E$ is an isomorphism. Let $J : c_0 \to E$ be any isomorphism. Since X is injective we can find a bounded linear extension $S : \ell_\infty \to X$ of the operator $PJ : c_0 \to X$. Note also that TPJ maps c_0 isomorphically onto a subspace G of Y and thus using the fact that ℓ_∞ is injective we can find a bounded linear operator $R : Y \to \ell_\infty$ which extends the operator $(TPJ)^{-1} : G \to c_0$. Thus we have the following commutative diagram:

$$
\begin{array}{ccccccc}
\ell_\infty & \xrightarrow{\;S\;} & X & \xrightarrow{\;T\;} & Y & \xrightarrow{\;R\;} & \ell_\infty \\
\uparrow & & \| & & \uparrow & & \uparrow \\
c_0 & \xrightarrow{\;PJ\;} & X & \xrightarrow{\;T\;} & G & \xrightarrow{\;R\;} & c_0
\end{array}
$$

The operator in the second row, namely, $RTPJ$, is the identity operator I on c_0 and $RTS : \ell_\infty \to \ell_\infty$ is an extension. Thus the operator $RTS - I$ on ℓ_∞ vanishes on c_0. We can now refer back to Theorem 2.5.4 to deduce the existence of a subset $\mathbb{A}$ of $\mathbb{N}$ so that $RTS - I$ vanishes on $\ell_\infty(\mathbb{A})$. In particular RTS is an isomorphism from $\ell_\infty(\mathbb{A})$ to its range. This requires that $F = S(\ell_\infty)$ is isomorphic to ℓ_∞, and $T|_F$ is an isomorphism.

$\square$

5.6 Subspaces of $L_1(\mu)$-spaces and $\mathcal{C}(K)$-spaces

Our first result in this section is a direct application of Theorem 5.4.7.

Proposition 5.6.1. *$L_1(\mu)$ and $\mathcal{C}(K)$ have no infinite-dimensional complemented reflexive subspaces.*

Proposition 5.6.2. *If X is a nonreflexive subspace of $L_1(\mu)$ then X contains a subspace isomorphic to ℓ_1 and complemented in $L_1(\mu)$.*

Proof. If X is nonreflexive, its closed unit ball B_X is not weakly compact, therefore B_X is not an equi-integrable set in $L_1(\mu)$. The proposition then follows from Theorem 5.2.9.

$\square$

Combining Proposition 5.6.1 and Proposition 5.6.2 gives us:

Proposition 5.6.3. *If X is an infinite-dimensional complemented subspace of $L_1(\mu)$ then X contains a complemented subspace isomorphic to ℓ_1.*

The analogous result for $\mathcal{C}(K)$-spaces is just as easy:

Proposition 5.6.4. *Let K be a compact metric space. If X is an infinite-dimensional complemented subspace of $\mathcal{C}(K)$ then X contains a complemented subspace isomorphic to c_0.*

Proof. Again by Proposition 5.6.1, X is nonreflexive and hence any projection P onto it fails to be weakly compact. By Theorem 5.5.3, X must contain a subspace isomorphic to c_0, and this subspace must be complemented because (since K is metrizable) X is separable (by Sobczyk's theorem, Theorem 2.5.8).

$\square$

Note here that if K is not metrizable we can obtain a subspace isomorphic to c_0, but it need not be complemented. In the case of ℓ_∞ we can use these techniques to add this space to our list of prime spaces. This result is due to Lindenstrauss [129] and it completes our list of classical prime spaces. We remind the reader of Pełczyński's result that the sequence spaces ℓ_p for $1 \le p < \infty$ and c_0 are prime (Theorem 2.2.4).

Theorem 5.6.5. *ℓ_∞ is prime.*

Proof. Let X be an infinite-dimensional complemented subspace of ℓ_∞. We have already seen that X cannot be reflexive (Proposition 5.6.1) and hence a projection P onto X cannot be weakly compact. In this case we can use Theorem 5.5.5 to deduce that X contains a copy of ℓ_∞. Since ℓ_∞ is injective, X actually contains a complemented copy of ℓ_∞ (Proposition 2.5.2). We are now ready to use Proposition 2.2.3 (b) in the case $p = \infty$ and we deduce that $X \approx \ell_\infty$.

$\square$

Corollary 5.6.6. *There are no infinite-dimensional separable injective Banach spaces.*

PROOF. Suppose that X is a separable injective space. X embeds isometrically into ℓ_∞ by Theorem 2.5.7. Since X is injective, it embeds complementably into ℓ_∞, which is a prime space. That forces X to be isomorphic to ℓ_∞, a contradiction because ℓ_∞ is nonseparable.

$\square$

It is quite clear that the spaces L_1 and $\mathcal{C}[0,1]$ cannot be prime; the former contains a complemented subspace isomorphic to ℓ_1 and the latter contains a complemented subspace isomorphic to c_0. However, the classification of the complemented subspaces of these classical function spaces remains a very intriguing and important open question.

In the case of L_1 the following conjecture remains open:

Conjecture 5.6.7. *Every infinite-dimensional complemented subspace of L_1 is isomorphic to L_1 or ℓ_1.*

The best result known in this direction is the Lewis-Stegall theorem from 1973 that any complemented subspace of L_1 which is a dual space is isomorphic to ℓ_1 [125]. (More generally, we can replace the dual space assumption by the Radon-Nikodym property.) Later we will develop techniques which show that any complemented subspace with an unconditional basis is isomorphic to ℓ_1 (an earlier result which is due to Lindenstrauss and Pełczyński [131]).

The corresponding conjecture for $\mathcal{C}[0,1]$ is:

Conjecture 5.6.8. *Every infinite-dimensional complemented subspace of $\mathcal{C}[0,1]$ is isomorphic to a $\mathcal{C}(K)$-space for some compact metric space K.*

Here the best positive result known is due to Rosenthal [195] who proved that if X is a complemented subspace of $\mathcal{C}[0,1]$ with nonseparable dual then $X \approx \mathcal{C}[0,1]$. We refer to the survey article of Rosenthal [199] for a fuller discussion of this problem,

Since both these spaces fail to be prime, it is natural to weaken the notion:

Definition 5.6.9. A Banach space X is *primary* if whenever $X \approx Y \oplus Z$ then either $X \approx Y$ or $X \approx Z$.

The spaces L_1 and $\mathcal{C}[0,1]$ are both primary. In the case of L_1 this result is due to Enflo and Starbird [55] (for an alternative approach see [103]). In the case of $\mathcal{C}[0,1]$ this was proved by Lindenstrauss and Pełczyński in 1971 [132], but of course it follows from Rosenthal's result cited above [195], which was proved slightly later, since one factor must have nonseparable dual.

Problems

5.1. Show that there is a sequence $(a_n)_{n\in\mathbb{Z}} \in c_0(\mathbb{Z})$ which is not the Fourier transform of any $f \in L_1(\mathbb{T})$.

5.2. Let X be a Banach space that does not contain a copy of ℓ_1. Show that every Dunford-Pettis operator $T : X \to Y$, with Y any Banach space, is compact.

5.3. Show that the identity operator $I_{\ell_1} : \ell_1 \to \ell_1$ is Dunford-Pettis.

5.4. Let X be a Banach space that does not contain a copy of ℓ_1; show that every operator $T : X \to L_1$ is weakly compact.

5.5. Let μ be a probability measure. Show that an operator $T : L_1(\mu) \to X$ is Dunford-Pettis if and only if T restricted to $L_2(\mu)$ is compact.

5.6. In this exercise we work in the *complex* space $L_p(\mathbb{T})$ $(1 \le p < \infty)$, where $\mathbb{T}$ is the unit circle with the normalized Haar measure $d\theta/2\pi$. We identify functions f on $\mathbb{T}$ with 2π-periodic functions on $\mathbb{R}$. The Fourier coefficients of f in $L_1(\mathbb{T})$ are given by

$$\hat{f}(n) = \int_{-\pi}^{\pi} f(\theta)e^{-in\theta} \frac{d\theta}{2\pi}, \quad n \in \mathbb{Z}.$$

For measures $\mu \in \mathcal{M}(\mathbb{T})$ we write

$$\hat{\mu}(n) = \int_{-\pi}^{\pi} e^{-in\theta} d\mu(\theta).$$

(a) Let μ be a Borel measure on the unit circle $\mathbb{T}$ so that $\mu \in \mathcal{M}(\mathbb{T})$. Show that for $1 \le p < \infty$ the map $T_\mu : L_p(\mathbb{T}, d\theta/2\pi) \to L_p(\mathbb{T}, d\theta/2\pi)$ defined by

$$T_\mu f(s) = \mu * f(s) = \int f(s-t)d\mu(t) \quad \text{a.e.}$$

is a well-defined bounded operator with $\|T_\mu\| \le \|\mu\|$. [Note that T_μ maps continuous functions and can be extended to $L_p(\mu)$ by continuity.]

(b) Show that $T_\mu e_n = \hat{\mu}(n)e_n$, where $e_n(t) = e^{int}$. Deduce that T_μ is Dunford-Pettis if and only if $\lim_{n\to\infty} \hat{\mu}(n) = 0$.

(c) Show that $T_\mu : L_1(\mathbb{T}) \to L_1(\mathbb{T})$ is weakly compact if and only if μ is absolutely continuous with respect to Lebesgue measure. [*Hint:* To show that μ is absolutely continuous, consider $T_\mu f_n$ where f_n is a sequence of nonnegative continuous functions with $\int f_n(t)dt/2\pi = 1$ and whose supports shrink to 0.]

5.7. Let $T : \ell_\infty \to X$ be a weakly compact operator which vanishes on c_0. Show that there exists an infinite subset $\mathbb{A}$ of $\mathbb{N}$ so that $T|_{\ell_\infty(\mathbb{A})} = 0$. [*Hint:* Mimic the argument in Theorem 2.5.4.]

5.8. If $T : \ell_\infty \to X$ is a weakly compact operator show that, for any $\epsilon > 0$, there exists an infinite subset $\mathbb{A}$ of $\mathbb{N}$ so that $T : \ell_\infty(\mathbb{A}) \to X$ is compact and $\|T|_{\ell_\infty(\mathbb{A})}\| < \epsilon$.

5.9. Show that if X is a Banach space containing ℓ_∞ and E is a closed subspace of X then either E contains ℓ_∞ or X/E contains ℓ_∞.

5.10. Show that every injective Banach space X contains a copy of ℓ_∞.

5.11. Suppose X is a Banach space with a closed subspace E so that X/E is isomorphic to L_1. Show that $E^{\perp\perp}$ is complemented in X^{**}. [*Hint:* Use the injectivity of L_∞.]

5.12 (Lindenstrauss [127]). Show that ℓ_1 has a subspace E which is not complemented in its bidual. [*Hint:* Use the kernel of a quotient map onto L_1.] Show that this subspace also has no unconditional basis.

The L_p-Spaces for $1 \leq p < \infty$

In this chapter we will initiate the study of the Banach space structure of the spaces $L_p(\mu)$ where $1 \leq p < \infty$. We will be interested in some natural questions which ask which Banach spaces can be isomorphic to a subspace of a space $L_p(\mu)$. Questions of this type were called problems of *linear dimension* by Banach in his book [8].

If $1 < p < \infty$ the Banach space $L_p(\mu)$ is reflexive while $L_1(\mu)$ is nonreflexive; we will see that this is just an example of a discontinuity in behavior when $p = 1$. We will also show certain critical differences between the cases $1 < p < 2$ and $2 < p < \infty$.

Before proceeding we note that, just as with $L_1(\mu)$-spaces, any space $L_p(\nu)$ with ν a σ-finite measure is isometric to a space $L_p(\mu)$ where μ is a probability measure. We also note that if K is a Polish space and μ is nonatomic probability measure defined on the Borel sets of K then $L_p(K, \mu)$ is isometric to $L_p[0, 1]$ and the isometry is implemented by a map of the form $f \mapsto f \circ \sigma$, where $\sigma : K \to [0, 1]$ is a Borel isomorphism that preserves measure. We refer the reader to the discussion in Section 5.1. For this reason it is natural to restrict our study to the spaces $L_p[0, 1]$ in many cases. From now on we will use the abbreviation L_p for the space $L_p[0, 1]$.

6.1 Conditional expectations and the Haar basis

Let (Ω, Σ, μ) be a probability measure space, and Σ' a sub-σ-algebra of Σ. Given $f \in L_1(\Omega, \Sigma, \mu)$ we define a (signed) measure, ν, on Σ':

$$\nu(E) = \int_E f \, d\mu, \quad E \in \Sigma'.$$

ν is absolutely continuous with respect to $\mu|_{\Sigma'}$, hence by the Radon-Nikodym theorem, there is a (unique, up to sets of measure zero) function $\psi \in L_1(\Omega, \Sigma', \mu)$ such that

$$\nu(E) = \int_E \psi \, d\mu, \quad E \in \Sigma'.$$

Then ψ is the (unique) function that satisfies

$$\int_E f \, d\mu = \int_E \psi \, d\mu, \quad E \in \Sigma'.$$

ψ is called the *conditional expectation of f on the σ-algebra* Σ' and will be denoted by $\mathcal{E}(f \,|\, \Sigma')$.

Let us notice that if Σ' consists of countably many disjoint atoms $(A_n)_{n=1}^\infty$, the definition of $\mathcal{E}(f \,|\, \Sigma')$ is specially simple:

$$\mathcal{E}(f \,|\, \Sigma')(t) = \sum_{j=1}^\infty \frac{1}{\mu(A_j)} \left(\int_{A_j} f \, d\mu \right) \chi_{A_j}(t).$$

We also observe that if $f \in L_p(\mu)$ where $1 \leq p < \infty$ and $g \in L_q(\Omega, \Sigma', \mu)$ where $\frac{1}{p} + \frac{1}{q}$ then

$$\int_E g \, d\nu = \int fg \, d\mu, \quad E \in \Sigma',$$

and

$$\mathcal{E}(fg \,|\, \Sigma') = g\mathcal{E}(f \,|\, \Sigma').$$

Lemma 6.1.1. *Let (Ω, Σ, μ) be a probability measure space and suppose Σ' is a sub-σ-algebra of Σ. Then for every $1 \leq p \leq \infty$, $\mathcal{E}(\cdot \,|\, \Sigma')$ is a norm-one linear projection from $L_p(\Omega, \Sigma, \mu)$ onto $L_p(\Omega, \Sigma', \mu)$.*

Proof. We denote $\mathcal{E} = \mathcal{E}(\cdot \,|\, \Sigma')$. It is immediate to check that $\mathcal{E}^2 = \mathcal{E}$ for all $1 \leq p \leq \infty$.

Fix $1 \leq p < \infty$ (we leave the case $p = \infty$ to the reader). If $f \in L_p(\mu)$,

$$\|\mathcal{E}(f)\|_p = \sup \left\{ \int_\Omega \mathcal{E}(f) g \, d\mu : g \in L_q(\Omega, \Sigma', \mu), \|g\|_q \leq 1 \right\}$$

$$= \sup \left\{ \int_\Omega \mathcal{E}(fg) \, d\mu : g \in L_q(\Omega, \Sigma', \mu), \|g\|_q \leq 1 \right\}$$

$$= \sup \left\{ \int_\Omega fg \, d\mu : g \in L_q(\Omega, \Sigma', \mu), \|g\|_q \leq 1 \right\}$$

$$\leq \|f\|_p.$$

$\square$

Definition 6.1.2. The sequence of functions on $[0,1]$, $(h_n)_{n=1}^\infty$, defined by $h_1 = 1$ and for $n = 2^k + s$ (where $k = 0, 1, 2, \ldots$, and $s = 1, 2, \ldots, 2^k$),

$$h_n(t) = \begin{cases} 1 & \text{if } t \in [\frac{2s-2}{2^{k+1}}, \frac{2s-1}{2^{k+1}}) \\ -1 & \text{if } t \in [\frac{2s-1}{2^{k+1}}, \frac{2s}{2^{k+1}}) \\ 0 & \text{otherwise} \end{cases}$$

$$= \chi_{[\frac{2s-2}{2^{k+1}}, \frac{2s-1}{2^{k+1}})}(t) - \chi_{[\frac{2s-1}{2^{k+1}}, \frac{2s}{2^{k+1}})}(t)$$

is called the *Haar system*.

Given $k = 0, 1, 2, \ldots$ and $1 \le s \le 2^k$, each interval of the form $[\frac{s-1}{2^k}, \frac{s}{2^k})$ is called *dyadic*. It is often useful to label the elements of the Haar system by their supports; thus we write h_I to denote h_n when I is the dyadic interval support of h_n.

Proposition 6.1.3. *The Haar system is a monotone basis in L_p for $1 \le p < \infty$.*

Proof. Let us consider an increasing sequence of σ-algebras, $(\mathcal{B}_n)_{n=1}^\infty$, contained in the Borel σ-algebra of $[0, 1]$ defined as follows: we let $\mathcal{B}_1$ be the trivial σ-algebra, $\{\emptyset, [0, 1]\}$, and for $n = 2^k + s$ $(k = 0, 1, 2, \ldots, 1 \le s \le 2^k)$ we let $\mathcal{B}_n$ be the finite subalgebra of the Borel sets of $[0, 1]$ whose atoms are the dyadic intervals of the family

$$\mathcal{F}_n = \begin{cases} [\frac{j-1}{2^{k+1}}, \frac{j}{2^{k+1}}) & \text{for } j = 1, \ldots, 2s \\ [\frac{j-1}{2^k}, \frac{j}{2^k}) & \text{for } j = s+1, \ldots, 2^k. \end{cases}$$

Fix $1 \le p < \infty$. For each n, $\mathcal{E}_n$ will denote the conditional expectation operator on the σ-algebra $\mathcal{B}_n$. By Lemma 6.1.1, $\mathcal{E}_n$ is a norm-one projection from L_p onto $L_p([0, 1], \mathcal{B}_n, \lambda)$, the space of functions which are constant on intervals of the family $\mathcal{F}_n$. We will denote this space by $L_p(\mathcal{B}_n)$. Clearly, rank $\mathcal{E}_n = n$. Furthermore, $\mathcal{E}_n \mathcal{E}_m = \mathcal{E}_m \mathcal{E}_n = \mathcal{E}_{\min\{m,n\}}$ for any two positive integers m, n.

On the other hand, the set

$$\{f \in L_p : \|\mathcal{E}_n(f) - f\|_p \to 0\}$$

is closed (using the partial converse of the Banach-Steinhaus theorem, see the Appendix) and contains the set $\cup_{k=1}^\infty L_p(\mathcal{B}_k)$, which is dense in L_p. Therefore $\|\mathcal{E}_n(f) - f\|_p \to 0$ for all $f \in L_p$. By Proposition 1.1.7, L_p has a basis whose natural projections are $(\mathcal{E}_n)_{n=1}^\infty$. This basis is actually the Haar system because for each $n \in \mathbb{N}$, $\mathcal{E}_m(h_n) = h_n$ for $m \ge n$ and $\mathcal{E}_m h_n = 0$ for $m < n$. The basis constant is $\sup_n \|\mathcal{E}_n\| = 1$.

$\square$

The Haar system as we have defined it is not normalized in L_p for $1 \le p < \infty$ (it is normalized in L_∞). To normalize in L_p one should take $h_n/\|h_n\|_p = |I_n|^{-1/p} h_n$, where I_n denotes the support of the Haar function h_n.

Let us observe that if $f \in L_p$ $(1 \le p < \infty)$, then

$$\mathcal{E}_n f - \mathcal{E}_{n-1} f = \left(\frac{1}{|I_n|} \int f(t) h_n(t) \, dt \right) h_n.$$

We deduce that the dual functionals associated to the Haar system are given by

$$h_n^* = \frac{1}{|I_n|} h_n, \quad n \in \mathbb{N},$$

and the series expansion of $f \in L_p$ in terms of the Haar basis is

$$f = \sum_{n=1}^{\infty} \left(\frac{1}{|I_n|} \int f(t) h_n(t)\, dt \right) h_n.$$

Notice that if $p = 2$ then $(h_n / \|h_n\|_2)_{n=1}^{\infty}$ is an orthonormal basis for the Hilbert space L_2 and is thus unconditional.

It is an important fact that, actually, the Haar basis is an unconditional basis in L_p for $1 < p < \infty$. This was first proved by Paley [165] in 1932. Much more recently, Burkholder [20] established the best constant.

We are going to present another proof of Burkholder from 1988 [21]. We will only treat the real case here, although, remarkably, the same proof works for complex scalars with the same constant; however, the calculations needed for the complex case are a little harder to follow. For our purposes the constant is not so important, and we simply note that if the Haar basis is unconditional for real scalars, one readily checks it is also unconditional for complex scalars. There is one drawback to Burkholder's argument: it is simply too clever in the sense that the proof looks very like magic.

We start with some elementary calculus.

Lemma 6.1.4. *Suppose $p > 2$ and $\frac{1}{p} + \frac{1}{q} = 1$. Then for $0 \leq t \leq 1$ we have*

$$t^p - p^p q^{-p}(1-t)^p \leq p^2 q^{1-p}\left(t - \frac{1}{q}\right). \tag{6.1}$$

Proof. For $0 \leq t \leq 1$ put

$$f(t) = t^p - p^p q^{-p}(1-t)^p - p^2 q^{1-p}\left(t - \frac{1}{q}\right).$$

Then

$$f'(t) = pt^{p-1} + p^{p+1} q^{-p}(1-t)^{p-1} - p^2 q^{1-p}$$

and

$$f''(t) = p(p-1)t^{p-2} - p^{p+1}(p-1)q^{-p}(1-t)^{p-2}.$$

Observe that $f(0) = -p^p q^{-p} + p^2 q^{-p} < 0$ and $f(1) = 1 - pq^{1-p}$. Since $p > 2$ we have $(1 - \frac{1}{p})^{p-1} > \frac{1}{p}$, i.e., $pq^{1-p} > 1$; thus $f(1) < 0$.

Next note that $f(\frac{1}{q}) = 0$ and

$$f'\left(\frac{1}{q}\right) = pq^{1-p} + p^2 q^{-p} - p^2 q^{1-p} = 0.$$

We also have

$$f''\left(\frac{1}{q}\right) = (p-1)(pq^{2-p} - p^3q^{-p}) = (p-1)pq^{-p}(q^2 - p^2) < 0.$$

Assume that there exists some $0 < s < 1$ with $f(s) > 0$. Then there must exist at least *three* solutions of $f'(t) = 0$ in the open interval $(0,1)$, including $1/q$. By Rolle's theorem this means there are at least two solutions of $f''(t) = 0$, which is clearly false.

□

In the next lemma we introduce a mysterious function which will enable us to prove Burkholder's theorem. This function appears to be plucked out of the air although there are sound reasons behind its selection. The use of such functions to prove sharp inequalities has been developed extensively by Nazarov, Treil, and Volberg who term them *Bellman functions*. We refer to [156] for a discussion of this technique.

Lemma 6.1.5. *Suppose $p > 2$ and define $\varphi : \mathbb{R}^2 \to \mathbb{R}$ by*

$$\varphi(x,y) = (|x| + |y|)^{p-1}\big((p-1)|x| - |y|\big).$$

(i) If $1/p + 1/q = 1$, the following inequality holds for all $(x,y) \in \mathbb{R}^2$

$$(p-1)^p|x|^p - |y|^p \geq pq^{1-p}\varphi(x,y). \tag{6.2}$$

(ii) φ is twice continuous differentiable and satisfies the condition

$$\frac{\partial^2\varphi}{\partial y^2} + \frac{\partial^2\varphi}{\partial x^2} = 2\left|\frac{\partial^2\varphi}{\partial x \partial y}\right| \geq 0. \tag{6.3}$$

Proof. (i) If we substitute $t = |y|(|x| + |y|)^{-1}$ (for $(x,y) \neq (0,0)$) in equation (6.1) we have

$$|y|^p - p^pq^{-p}|x|^p \leq pq^{1-p}(|y| - (p-1)|x|)(|x| + |y|)^{p-1}.$$

Thus

$$p^pq^{-p}|x|^p - |y|^p \geq pq^{1-p}\varphi(x,y).$$

Note that $p^pq^{-p} = (p-1)^p$.

(ii) The fact that φ is twice continuously differentiable is immediate since $p > 2$.

Clearly, it suffices to prove (6.3) in the first quadrant, where $x > 0$, $y > 0$. Let $u = x + y$ and $v = (p-1)x - y$. Then $\varphi(x,y) = u^{p-1}v$. Hence

$$\frac{\partial^2\varphi}{\partial y^2} = (p-1)(p-2)u^{p-3}v - 2(p-1)u^{p-2}$$

while

$$\frac{\partial^2\varphi}{\partial x^2} = (p-1)(p-2)u^{p-3}v + 2(p-1)^2u^{p-2}.$$

Hence

$$\frac{\partial^2 \varphi}{\partial x^2} + \frac{\partial^2 \varphi}{\partial y^2} = 2(p-1)(p-2)u^{p-3}(u+v) \ge 0.$$

On the other hand, since φ is linear on any line of slope one (or by routine calculation) we must also have

$$\frac{\partial^2 \varphi}{\partial x \partial y} = (p-1)(p-2)u^{p-3}(u+v).$$

$\square$

Theorem 6.1.6. *Suppose $1 < p < \infty$ and $\frac{1}{p} + \frac{1}{q} = 1$. Let $p^* = \max(p,q)$. The Haar basis $(h_k)_{k=1}^{\infty}$ in L_p is unconditional with unconditional constant at most $p^* - 1$. That is,*

$$\left\| \sum_{j=1}^{n} \epsilon_j a_j h_j \right\|_p \le (p^* - 1) \left\| \sum_{j=1}^{n} a_j h_j \right\|_p,$$

whenever $n \in \mathbb{N}$, for any real scalars $a_1, \ldots, a_n$ and any signs $\epsilon_1, \ldots, \epsilon_n$.

Proof. Suppose $p > 2$, in which case $p^* = p$. For each fixed $n \in \mathbb{N}$, let $f_0 = g_0 = 0$ and for $1 \le k \le n$ put

$$f_k = \sum_{j=1}^{k} a_j h_j \quad \text{and} \quad g_k = \sum_{j=1}^{k} \epsilon_j a_j h_j.$$

We will prove by induction on k that

$$\int_0^1 \varphi(f_k(s), g_k(s)) \, ds \ge 0, \quad 1 \le k \le n, \tag{6.4}$$

where φ is the function defined in Lemma 6.1.5. This is trivial when $k = 0$. In order to establish the inductive step, for a given k let us consider the function $F : [0,1] \to \mathbb{R}$ defined by

$$F(t) = \int_0^1 \varphi\big((1-t)f_{k-1}(s) + tf_k(s), (1-t)g_{k-1}(s) + tg_k(s)\big) \, ds,$$

and show that $F(1) \ge 0$ assuming that $F(0) \ge 0$.

Let $u_t = (1-t)f_{k-1} + tf_k$ and $v_t = (1-t)g_{k-1} + tg_k$. Then

$$F'(t) = a_k \int_0^1 \frac{\partial \varphi}{\partial x}(u_t, v_t) h_k \, ds + \epsilon_k a_k \int_0^1 \frac{\partial \varphi}{\partial y}(u_t, v_t) h_k \, ds.$$

Observe that $F'(0) = 0$ since $\frac{\partial \varphi}{\partial x}(u_0, v_0)$ and $\frac{\partial \varphi}{\partial y}(u_0, v_0)$ are constant on the support interval of h_k.

Differentiating again gives

$$F''(t) = a_k^2 \int_{I_k} \left(\frac{\partial^2 \varphi}{\partial^2 x}(u_t, v_t) + \frac{\partial^2 \varphi}{\partial^2 y}(u_t, v_t) + 2\epsilon_k \frac{\partial^2 \varphi}{\partial x \partial y}(u_t, v_t) \right) ds.$$

By Lemma 6.1.5 (ii), $F''(t) \geq 0$. Hence $F(1) \geq F(0) \geq 0$ and thus(6.4) holds.
To complete the proof when $p > 2$ we plug $x = f_n$ and $y = g_n$ in (6.2).
Integrating both sides of this inequality and using (6.4) we obtain

$$\int_0^1 (p-1)^p |f_n(s)|^p - |g_n(s)|^p ds \geq 0.$$

The case when $1 < p < 2$ now follows by duality: with f_n, g_n as before
choose $g_n' \in L_q(\mathcal{B}_n)$ so that $\|g_n'\|_q = 1$ and

$$\int_0^1 g_n'(s)g_n(s)\, ds = \|g_n\|_p.$$

Then $g_n' = \sum_{j=1}^n b_j h_j$ for some $(b_j)_{j=1}^n$ and

$$\|g_n\|_p = \sum_{j=1}^n |I_j| \epsilon_j a_j b_j \leq \|f_n\|_p \left\| \sum_{j=1}^n \epsilon_j b_j h_j \right\|_q \leq (q-1)\|f_n\|_p.$$

□

The constant $p^* - 1$ in Burkholder's theorem is sharp, although we will
not prove this here.

6.2 Averaging in Banach spaces

In discussing unconditional bases and unconditional convergence of series in a
Banach space X we have frequently met the problem of estimating expressions
of the type

$$\max \left\{ \left\| \sum_{i=1}^n \epsilon_i x_i \right\| \; : \; (\epsilon_i) \in \{-1, 1\}^n \right\},$$

where $\{x_i\}_{i=1}^n$ are vectors in X. In many situations it is much easier to replace
the maximum by the average over all choices of signs $\epsilon_i = \pm 1$.

It turns out to be helpful to consider such averages using the Rademacher
functions $(r_i)_{i=1}^\infty$ since the sequence $(r_i(t))_{i=1}^n$ gives us all possible choices of
signs $(\epsilon_i)_{i=1}^n$ when t ranges over $[0, 1]$. Thus,

$$\operatorname*{Average}_{\epsilon_i = \pm 1} \left\| \sum_{i=1}^n \epsilon_i x_i \right\| = 2^{-n} \sum_{\epsilon_i = \pm 1} \left\| \sum_{i=1}^n \epsilon_i x_i \right\| = \int_0^1 \left\| \sum_{i=1}^n r_i(t) x_i \right\| dt.$$

For reference let us recall the definition of the Rademacher functions and their
basic properties.

Definition 6.2.1. The *Rademacher functions* $(r_k)_{k=1}^\infty$ are defined on $[0,1]$ by

$$r_k(t) = \text{sgn } (\sin 2^k \pi t).$$

Alternatively, the sequence $(r_k)_{k=1}^\infty$ can be described as

$$r_1(t) = \begin{cases} 1 & \text{if } t \in [0, \frac{1}{2}) \\ -1 & \text{if } t \in [\frac{1}{2}, 1) \end{cases}$$

$$r_2(t) = \begin{cases} 1 & \text{if } t \in [0, \frac{1}{4}) \cup [\frac{1}{2}, \frac{3}{4}) \\ -1 & \text{if } t \in [\frac{1}{4}, \frac{1}{2}) \cup [\frac{3}{4}, 1) \end{cases}$$

$$\vdots$$

$$r_{k+1}(t) = \begin{cases} 1 & \text{if } t \in \bigcup_{s=1}^{2^k} [\frac{2s-2}{2^{k+1}}, \frac{2s-1}{2^{k+1}}) \\ -1 & \text{if } t \in \bigcup_{s=1}^{2^k} [\frac{2s-1}{2^{k+1}}, \frac{2s}{2^{k+1}}). \end{cases}$$

That is,

$$r_{k+1} = \sum_{s=1}^{2^k} h_{2^k+s}, \qquad k = 0, 1, 2, \dots$$

Thus $(r_k)_{k=1}^\infty$ is a block-basic sequence with respect to the Haar basis in every L_p for $1 \leq p < \infty$. The key properties we need are the following:

- $r_k(t) = \pm 1$ a.e. for all k,
- $\int r_{k_1} r_{k_2}(t) \dots r_{k_m}(t) dt = 0$, whenever $k_1 < k_2 < \cdots < k_m$.

The Rademacher functions were first introduced by Rademacher in 1922 [191] with the idea of studying the problem of finding conditions under which a series of real numbers $\sum \pm a_n$, where the signs were assigned randomly, would converge almost surely. Rademacher showed that if $\sum |a_n|^2 < \infty$ then indeed $\sum \pm a_n$ converges almost surely. The converse was proved in 1925 by Khintchine and Kolmogoroff [111].

For our purposes it will be convenient to replace the concrete Rademacher functions by an abstract model. To that end we will use the language and methods of probability theory.

Let us recall that a *random variable* is a real-valued measurable function on some probability space $(\Omega, \Sigma, \mathbb{P})$. The *expectation* (or *mean*) of a random variable f is defined by

$$\mathbb{E}f = \int_\Omega f(\omega)\, d\mathbb{P}(\omega).$$

A finite set of random variables $\{f_j\}_{j=1}^n$ on the same probability space is *independent* if

$$\mathbb{P} \bigcap_{j=1}^n (f_j \in B_j) = \prod_{j=1}^n \mathbb{P}(f_j \in B_j)$$

for all Borel sets B_j. Therefore if $(f_j)_{j=1}^n$ are independent,

$$\mathbb{E}(f_1 f_2 \cdots f_n) = \mathbb{E}(f_1)\mathbb{E}(f_2)\cdots\mathbb{E}(f_n).$$

An arbitrary set of random variables is said to be independent if any finite subcollection of the set is independent.

Definition 6.2.2. A *Rademacher sequence* is a sequence of mutually independent random variables $(\varepsilon_n)_{n=1}^\infty$ defined on some probability space $(\Omega, \mathbb{P})$ such that $\mathbb{P}(\varepsilon_n = 1) = \mathbb{P}(\varepsilon_n = -1) = \frac{1}{2}$ for every n.

The terminology is justified by the fact that the Rademacher functions $(r_n)_{n=1}^\infty$ are a Rademacher sequence on $[0, 1]$. Thus,

$$\int_0^1 \left\| \sum_{i=1}^n r_i(t) x_i \right\| dt = \mathbb{E}\left\| \sum_{i=1}^n \varepsilon_i x_i \right\| = \int_\Omega \left\| \sum_{i=1}^n \varepsilon_i(\omega) x_i \right\| d\mathbb{P}.$$

Historically, the subject of finding estimates for averages over all choices of signs was initiated in 1923 by the classical Khintchine inequality [110], but the usefulness of a probabilistic viewpoint in studying the L_p-spaces seems to have been fully appreciated quite late (around 1970).

Theorem 6.2.3 (Khintchine's Inequality). *There exist constants* A_p, B_p *$(1 \le p < \infty)$ such that for any finite sequence of scalars $(a_i)_{i=1}^n$ and any $n \in \mathbb{N}$ we have*

$$A_p \Big(\sum_{i=1}^n |a_i|^2 \Big)^{1/2} \le \left\| \sum_{i=1}^n a_i r_i \right\|_p \le \Big(\sum_{i=1}^n |a_i|^2 \Big)^{1/2} \quad \text{if } 1 \le p < 2,$$

and

$$\Big(\sum_{i=1}^n |a_i|^2 \Big)^{1/2} \le \left\| \sum_{i=1}^n a_i r_i \right\|_p \le B_p \Big(\sum_{i=1}^n |a_i|^2 \Big)^{1/2} \quad \text{if } p > 2.$$

We will not prove this here but it will be derived as a consequence of a more general result below. Theorem 6.2.3 was first given in the stated form by Littlewood in 1930 [141] but Khintchine's earlier work (of which Littlewood was unaware) implied these inequalities as a consequence.

Remark 6.2.4. (a) Khintchine's inequality says that $(r_i)_{i=1}^\infty$ is a basic sequence equivalent to the ℓ_2-basis in every L_p for $1 \le p < \infty$. In L_∞, though, one readily checks that $(r_i)_{i=1}^\infty$ is isometrically equivalent to the canonical ℓ_1-basis.
(b) $(r_i)_{i=1}^\infty$ is an orthonormal sequence in L_2, which yields the identity

$$\left\| \sum_{i=1}^n a_i r_i \right\|_2 = \Big(\sum_{i=1}^n |a_i|^2 \Big)^{1/2},$$

for any choice of scalars (a_i). But $(r_i)_{i=1}^{\infty}$ is not a complete system in L_2, that is, $[r_i] \neq L_2$ (for instance, notice that the function $r_1 r_2$ is orthogonal to the subspace $[r_i]$). However, one can obtain a complete orthonormal system for L_2 using the Rademacher functions by adding to (r_n) the constant function $r_0 = 1$ and the functions of the form $r_{k_1} r_{k_2} \dots r_{k_n}$ for any $k_1 < k_2 < \dots < k_n$. This collection of functions are the *Walsh functions*.

Thus we can also interpret Khintchine's inequality as stating that all the norms $\{\|\cdot\|_p : 1 \leq p < \infty\}$ are equivalent on the linear span of the Rademacher functions in L_p. It turns out that in this form the statement can be generalized to an arbitrary Banach space. This generalization was first obtained by Kahane in 1964 [101].

Theorem 6.2.5 (Kahane-Khintchine Inequality). *For each $1 \leq p < \infty$ there exists a constant C_p such that, for every Banach space X and for any finite sequence $(x_i)_{i=1}^{n}$ in X, the following inequality holds:*

$$\mathbb{E}\left\| \sum_{i=1}^{n} \varepsilon_i x_i \right\| \leq \left(\mathbb{E}\left\| \sum_{i=1}^{n} \varepsilon_i x_i \right\|^p \right)^{1/p} \leq C_p \, \mathbb{E}\left\| \sum_{i=1}^{n} \varepsilon_i x_i \right\|.$$

We will prove the Kahane-Khintchine inequality (and this will imply the Khintchine inequality by taking $X = \mathbb{R}$ or $X = \mathbb{C}$) but first we shall establish three lemmas on our way to the proof. To avoid repetitions, in all three lemmas $(\Omega, \Sigma, \mathbb{P})$ will be a probability space and X will be a Banach space. Let us recall that an *X-valued random variable* on Ω is a function $f : \Omega \to X$ such that $f^{-1}(B) \in \Sigma$ for every Borel set $B \subset X$. f is *symmetric* if $\mathbb{P}(f \in B) = \mathbb{P}(-f \in B)$ for all Borel subsets B of X.

Lemma 6.2.6. *Let $f : \Omega \to X$ be a symmetric random variable. Then for all $x \in X$ we have*
$$\mathbb{P}\big(\|f + x\| \geq \|x\| \big) \geq 1/2.$$

Proof. Let us take any $x \in X$. For every $\omega \in \Omega$, using the convexity of the norm of X, clearly $\|f(\omega) + x\| + \|x - f(\omega)\| \geq 2\|x\|$. Then, either $\|f(\omega) + x\| \geq \|x\|$ or $\|x - f(\omega)\| \geq \|x\|$. Hence
$$1 \leq \mathbb{P}\big(\|f + x\| \geq \|x\| \big) + \mathbb{P}\big(\|x - f\| \geq \|x\| \big).$$

Since f is symmetric, $x + f$ and $x - f$ have the same distribution and so the lemma follows.

$\square$

Let $(\varepsilon_i)_{i=1}^{\infty}$ be a Rademacher sequence on Ω. Given $n \in \mathbb{N}$ and vectors $x_1, \dots, x_n$ in X, we shall consider $\Lambda_m : \Omega \longrightarrow X$ ($1 \leq m \leq n$) defined by

$$\Lambda_m(\omega) = \sum_{i=1}^{m} \varepsilon_i(\omega) x_i.$$

Lemma 6.2.7. *For all* $\lambda > 0$,

$$\mathbb{P}\big(\max_{m \le n} \|\Lambda_m\| > \lambda \big) \le 2\mathbb{P}\big(\|\Lambda_n\| > \lambda \big).$$

Proof. Given $\lambda > 0$, for $m = 1, \ldots, n$ put

$$\Omega_m^{(\lambda)} = \{\omega \in \Omega : \|\Lambda_m(\omega)\| > \lambda \text{ and } \|\Lambda_j(\omega)\| \le \lambda \text{ for all } j = 1, \ldots, m-1\}.$$

Since $\{\omega \in \Omega : \max_{m \le n} \|\Lambda_m(\omega)\| > \lambda\} = \cup_{m=1}^n \Omega_m^{(\lambda)}$, by the disjointedness of the sets $\Omega_m^{(\lambda)}$ it follows that

$$\mathbb{P}\big(\max_{m \le n} \|\Lambda_m\| > \lambda \big) = \sum_{m=1}^n \mathbb{P}(\Omega_m^{(\lambda)}). \tag{6.5}$$

Therefore,

$$\mathbb{P}\big(\|\Lambda_n\| > \lambda \big) = \sum_{m=1}^n \mathbb{P}\big(\Omega_m^{(\lambda)} \cap (\|\Lambda_n\| > \lambda)\big). \tag{6.6}$$

Notice that every $\Omega_m^{(\lambda)}$ can be written as the union of sets of the type

$$\{\omega \in \Omega : \varepsilon_j(\omega) = \delta_j \text{ for } 1 \le j \le m\}$$

for some choices of signs $\delta_j = \pm 1$. For each of these choices of signs $\delta_1, \ldots, \delta_m$ we observe that by Lemma 6.2.6,

$$\mathbb{P}\Big(\Big\| \sum_{j=1}^m \delta_j x_j + \sum_{j=m+1}^n \varepsilon_j x_j \Big\| > \Big\| \sum_{j=1}^m \delta_j x_j \Big\| \Big) \ge \frac{1}{2}.$$

Summing over the appropriate signs $(\delta_1, \ldots, \delta_m)$ it follows that

$$\mathbb{P}\big(\Omega_m^{(\lambda)} \cap (\|\Lambda_n\| \ge \|\Lambda_m\|)\big) \ge \frac{1}{2}\mathbb{P}(\Omega_m^{(\lambda)}).$$

Thus,

$$\mathbb{P}\big(\Omega_m^{(\lambda)} \cap (\|\Lambda_n\| > \lambda)\big) \ge \frac{1}{2}\mathbb{P}(\Omega_m^{(\lambda)}).$$

Summing in m and combining (6.5) and (6.6) we finish the proof.

$\square$

Lemma 6.2.8. *For all* $\lambda > 0$,

$$\mathbb{P}\big(\|\Lambda_n\| > 2\lambda \big) \le 4\big(\mathbb{P}\big(\|\Lambda_n\| > \lambda \big)\big)^2.$$

Proof. We will keep the notation that we introduced in the previous lemma. Notice that for each $1 \le m \le n$, the random variable $\|\sum_{i=m}^n \varepsilon_i x_i\|$ is independent of each of $\varepsilon_1, \ldots, \varepsilon_m$ and hence for all $\lambda > 0$ the events $\{\omega : \|\sum_{i=m}^n \varepsilon_i(\omega)x_i\| > \lambda\}$ and $\Omega_m^{(\lambda)}$ are independent. Observe as well that if some

$\omega \in \Omega_m^{(\lambda)}$ further satisfies $\|\Lambda_n(w)\| > 2\lambda$, then $\|\Lambda_n(\omega) - \Lambda_{m-1}(\omega)\| > \lambda$ (for $m = 1$, take $\Lambda_0 = 0$). Therefore, since $\mathbb{P}(\| \sum_{i=m}^n \varepsilon_i x_i\| > \lambda) \le 2\mathbb{P}(\|\Lambda_n\| > \lambda)$ for each $m = 1, \dots, n$ by Lemma 6.2.7, we have

$$\mathbb{P}\big(\Omega_m^{(\lambda)} \cap (\|\Lambda_n\| > 2\lambda)\big) \le \mathbb{P}(\Omega_m^{(\lambda)})\mathbb{P}\big(\| \sum_{i=m}^n \varepsilon_i x_i\| > \lambda\big)$$

$$\le 2\mathbb{P}(\Omega_m^{(\lambda)})\mathbb{P}\big(\|\Lambda_n\| > \lambda\big).$$

Summing in m and using again Lemma 6.2.7 we obtain

$$\mathbb{P}\big(\|\Lambda_n\| > 2\lambda\big) \le \mathbb{P}\big(\max_{m \le n}\|\Lambda_m\| > \lambda\big)\mathbb{P}\big(\|\Lambda_n\| > \lambda\big) \le 4\big(\mathbb{P}\big(\|\Lambda_n\| > \lambda\big)\big)^2.$$

$\square$

Proof of Theorem 6.2.5. Fix $1 \le p < \infty$ and let $\{x_i\}_{i=1}^n$ be any finite set of vectors in X. Without loss of generality we will suppose that $\mathbb{E}\| \sum_{i=1}^n \varepsilon_i x_i\| = 1$. Then, by Chebyshev's inequality,

$$\mathbb{P}\big(\|\Lambda_n\| > 8\big) \le \frac{1}{8}. \tag{6.7}$$

Using Lemma 6.2.8 repeatedly we obtain

$$\mathbb{P}\big(\|\Lambda_n\| > 2 \cdot 8\big) \le 4(1/8)^2,$$
$$\mathbb{P}\big(\|\Lambda_n\| > 2^2 \cdot 8\big) \le 4^3(1/8)^4,$$
$$\mathbb{P}\big(\|\Lambda_n\| > 2^3 \cdot 8\big) \le 4^7(1/8)^8,$$

and so on. Hence, by induction, we deduce that

$$\mathbb{P}\big(\|\Lambda_n\| > 2^n \cdot 8\big) \le 4^{2^n-1}(1/8)^{2^n} \le 4^{2^n}(1/8)^{2^n} = (1/2)^{2^n}.$$

Therefore,

$$\mathbb{E}\Big\| \sum_{i=1}^n \varepsilon_i x_i\Big\|^p = \int_0^\infty \mathbb{P}\big(\|\Lambda_n\|^p > t\big)\, dt$$

$$= \int_0^\infty p\, t^{p-1}\mathbb{P}\big(\|\Lambda_n\| > t\big)\, dt$$

$$= \int_0^8 p\, t^{p-1}\mathbb{P}\big(\|\Lambda_n\| > t\big)\, dt + \sum_{n=1}^\infty \int_{2^{n-1}\cdot 8}^{2^n\cdot 8} p\, t^{p-1}\mathbb{P}\big(\|\Lambda_n\| > t\big)\, dt$$

$$\le \int_0^8 p\, t^{p-1}\, dt + \sum_{n=1}^\infty (1/2)^{2^n-1} \int_{2^{n-1}\cdot 8}^{2^n\cdot 8} p\, t^{p-1}\, dt$$

$$\le 8^p \Big(1 + \sum_{n=1}^\infty (1/2)^{2^n-1}\, 2^{np}\Big)$$

$$= C_p^p.$$

□

Suppose that H is a Hilbert space. The well-known Parallelogram Law states that for any two vectors x, y in H we have

$$\frac{\|x+y\|^2 + \|x-y\|^2}{2} = \|x\|^2 + \|y\|^2 .$$

This identity is a simple example of the power of averaging over signs and has an elementary generalization:

Proposition 6.2.9 (Generalized Parallelogram Law). *Suppose that H is a Hilbert space. Then for every finite sequence $(x_i)_i^n$ in H,*

$$\mathbb{E}\Big\|\sum_{i=1}^n \varepsilon_i x_i\Big\|^2 = \sum_{i=1}^n \|x_i\|^2 .$$

Proof. For any vectors $(x_i)_{i=1}^n$ in H we have

$$\mathbb{E}\Big\|\sum_{i=1}^n \varepsilon_i x_i\Big\|^2 = \mathbb{E}\Big\langle \sum_{i=1}^n \varepsilon_i x_i, \sum_{i=1}^n \varepsilon_i x_i\Big\rangle$$
$$= \sum_{i,j=1}^n \langle x_i, x_j\rangle \mathbb{E}(\varepsilon_i\varepsilon_j)$$
$$= \sum_{i=1}^n \|x_i\|^2 .$$

□

Next we are going to study how the averages $(\mathbb{E}\|\sum_{i=1}^n \varepsilon_i x_i\|^p)^{1/p}$ are situated with respect to the sums $(\sum_{i=1}^n \|x_i\|^p)^{1/p}$ using the concepts of *type* and *cotype* of a Banach space. These were introduced into Banach space theory by Hoffmann-Jørgensen [79] and their basic theory was developed in the early 1970s by Maurey and Pisier [147]; see [146] for historical comments. However, it should be said that the origin of these ideas was in two very early papers of Orlicz in 1933, [163] and [164]. Orlicz essentially introduced the notion of cotype for the spaces L_p although he did not use the more modern terminology.

Definition 6.2.10. A Banach space X is said to have *Rademacher type p* (in short, *type p*) for some $1 \le p \le 2$ if there is a constant C such that for every finite set of vectors $\{x_i\}_{i=1}^n$ in X,

$$\Big(\mathbb{E}\Big\|\sum_{i=1}^n \varepsilon_i x_i\Big\|^p\Big)^{1/p} \le C\Big(\sum_{i=1}^n \|x_i\|^p\Big)^{1/p}. \tag{6.8}$$

The smallest constant for which (6.8) holds is called the *type-p constant of X* and is denoted $T_p(X)$.

Similarly, a Banach space X is said to have *Rademacher cotype q* (in short, *cotype q*) for some $2 \leq q \leq \infty$ if there is a constant C such that for every finite sequence $x_1, x_2, \ldots, x_n$ in X,

$$\Big(\sum_{i=1}^{n} \|x_i\|^q \Big)^{1/q} \leq C \Big(\mathbb{E} \Big\| \sum_{i=1}^{n} \varepsilon_i x_i \Big\|^q \Big)^{1/q}, \tag{6.9}$$

with the usual modification of $\max_{1 \leq i \leq n} \|x_i\|$ replacing $\big(\sum_{i=1}^{n} \|x_i\|^q \big)^{1/q}$ when $q = \infty$. The smallest constant for which (6.9) holds is called the *cotype-q constant of X* and is denoted $C_q(X)$.

Remark 6.2.11. (*a*) The restrictions on p and q in the definitions of type and cotype respectively are natural since it is impossible to have type $p > 2$ or cotype $q < 2$ even in a one-dimensional space. To see this, for each n take vectors $\{x_i\}_{i=1}^{n}$ all equal to some $x \in X$ with $\|x\| = 1$. The combination of Khintchine's inequality with (6.8) and (6.9) gives us the range of eligible values for p and q.

(*b*) Every Banach space X has type 1 with $T_1(X) = 1$ and cotype ∞ with $C_\infty(X) = 1$ by the triangle law. Thus X is said to have *nontrivial type* if it has type p for some $1 < p \leq 2$; similarly X is said to have *nontrivial cotype* if it has cotype q for some $2 \leq q < \infty$.

(*c*) The generalized Parallelogram Law (Proposition 6.2.9) says that a Hilbert space H has type 2 and cotype 2 with $T_2(H) = C_2(H) = 1$. In particular a one-dimensional space has type 2 and cotype 2. But the Parallelogram Law is also a characterization of Banach spaces which are linearly isometric to Hilbert spaces, hence we deduce that a Banach space X is isometric to a Hilbert space if and only if $T_2(X) = C_2(X) = 1$ (see Problem 7.6).

(*d*) By Theorem 6.2.5, the L_p-average $(\mathbb{E} \| \sum_{i=1}^{n} \varepsilon_i x_i \|^p)^{1/p}$ in the definition of type can be replaced by any other L_r-average $(\mathbb{E} \| \sum_{i=1}^{n} \varepsilon_i x_i \|^r)^{1/r}$ $(1 \leq r < \infty)$ and this has the effect only of changing the constant. The same comment applies to the L_q-average in the definition of cotype.

(*e*) If X has type p then X has type r for $r < p$ and if X has cotype q then X has cotype s for $s > q$.

(*f*) The type and cotype of a Banach space are isomorphic invariants and are inherited by subspaces.

(*g*) Consider the unit vector basis $(e_n)_{n=1}^{\infty}$ in ℓ_p $(1 \leq p < \infty)$ or c_0. Then for any signs (ϵ_k) we have

$$\|\epsilon_1 e_1 + \cdots + \epsilon_n e_n\|_p = n^{\frac{1}{p}}$$

and

$$\|\epsilon_1 e_1 + \cdots + \epsilon_n e_n\|_\infty = 1.$$

Thus ℓ_p cannot have type greater than p if $1 \leq p \leq 2$ or cotype less than p if $2 \leq p \leq \infty$.

Proposition 6.2.12. *If a Banach space X has type p then X^* has cotype q, where $\frac{1}{p} + \frac{1}{q} = 1$ and $C_q(X^*) \leq T_p(X)$.*

Proof. Let us pick an arbitrary finite set $\{x_i^*\}_{i=1}^n$ in X^*. Given $\epsilon > 0$ we can find $x_1, \ldots, x_n$ in X such that $\|x_i\| = 1$ and $|x_i^*(x_i)| \geq (1 - \epsilon)\|x_i^*\|$ for all $i = 1, \ldots, n$. Thus

$$\Big(\sum_{i=1}^n |x_i^*(x_i)|^q\Big)^{1/q} \geq (1 - \epsilon)\Big(\sum_{i=1}^n \|x_i^*\|^q\Big)^{1/q}.$$

On the other hand,

$$\Big(\sum_{i=1}^n |x_i^*(x_i)|^q\Big)^{\frac{1}{q}} = \sup\Big\{\Big|\sum_{i=1}^n a_i x_i^*(x_i)\Big| : \sum_{i=1}^n |a_i|^p \leq 1\Big\}.$$

For any scalars $(a_i)_{i=1}^n$ with $\sum_{i=1}^n |a_i|^p \leq 1$ we have

$$\sum_{i=1}^n a_i x_i^*(x_i) = \int_\Omega \Big(\sum_{i=1}^n \varepsilon_i x_i^*\Big)\Big(\sum_{i=1}^n \varepsilon_i a_i x_i\Big) \, d\mathbb{P}$$

$$\leq \int_\Omega \Big\|\sum_{i=1}^n \varepsilon_i x_i^*\Big\|\Big\|\sum_{i=1}^n \varepsilon_i a_i x_i\Big\| \, d\mathbb{P}$$

$$\leq \Big(\int_\Omega \Big\|\sum_{i=1}^n \varepsilon_i x_i^*\Big\|^q \, d\mathbb{P}\Big)^{\frac{1}{q}}\Big(\int_\Omega \Big\|\sum_{i=1}^n \varepsilon_i a_i x_i\Big\|^p \, d\mathbb{P}\Big)^{\frac{1}{p}}$$

$$\leq \Big(\int_\Omega \Big\|\sum_{i=1}^n \varepsilon_i x_i^*\Big\|^q \, d\mathbb{P}\Big)^{\frac{1}{q}} T_p(X)\Big(\sum_{i=1}^n |a_i|^p\Big)^{\frac{1}{p}}.$$

Therefore,

$$\Big(\sum_{i=1}^n \|x_i^*\|^q\Big)^{\frac{1}{q}} \leq (1 - \epsilon)^{-1} T_p(X)\Big(\mathbb{E}\Big\|\sum_{i=1}^n \varepsilon_i x_i^*\Big\|^q\Big)^{\frac{1}{q}}.$$

Since ϵ was arbitrary, this shows $C_q(X^*) \leq T_p(X)$. $\square$

Curiously, Proposition 6.2.12 does not have a converse statement. At the end of the section we shall give an example showing that if X has cotype q for $q < \infty$ then X^* may not have type p where $\frac{1}{p} + \frac{1}{q}$.

Next we want to investigate the type and cotype of L_p for $1 \leq p < \infty$. To do so we will estimate $\|(\sum_{i=1}^n |f_i|^2)^{1/2}\|_p$ in relation with the Rademacher averages $(\mathbb{E}\|\sum_{j=1}^n \varepsilon_j f_j\|_p^p)^{1/p}$ on a generic $L_p(\mu)$-space.

Theorem 6.2.13. *For every finite set of functions $\{f_i\}_{i=1}^n$ in $L_p(\mu)$ $(1 \leq p < \infty)$,*

$$A_p \left\| \left(\sum_{i=1}^{n} |f_i|^2 \right)^{\frac{1}{2}} \right\|_p \leq \left(\mathbb{E} \left\| \sum_{i=1}^{n} \varepsilon_i f_i \right\|_p^p \right)^{1/p} \leq B_p \left\| \left(\sum_{i=1}^{n} |f_i|^2 \right)^{\frac{1}{2}} \right\|_p,$$

where A_p, B_p are the constants in Khintchine's inequality (in particular $A_p = 1$ for $2 \leq p < \infty$ and $B_p = 1$ for $1 \leq p \leq 2$).

Proof. For each $\omega \in \Omega$, from Khintchine's inequality

$$A_p \left(\sum_{i=1}^{n} |f_i(\omega)|^2 \right)^{1/2} \leq \left(\mathbb{E} \left| \sum_{i=1}^{n} \varepsilon_i f_i(\omega) \right|^p \right)^{1/p},$$

where $A_p = 1$ for $2 \leq p < \infty$. Now, using Fubini's theorem

$$A_p^p \left\| \left(\sum_{i=1}^{n} |f_i|^2 \right)^{1/2} \right\|_p^p \leq \int_\Omega \mathbb{E} \left| \sum_{i=1}^{n} \varepsilon_i f_i(\omega) \right|^p d\mu$$

$$= \mathbb{E} \left(\int_\Omega \left| \sum_{i=1}^{n} \varepsilon_i f_i(\omega) \right|^p d\mu \right)$$

$$= \mathbb{E} \left\| \sum_{i=1}^{n} \varepsilon_i f_i \right\|_p^p.$$

The converse estimate is obtained similarly.

$\square$

The next theorem is due to Orlicz for cotype [163, 164] and Nordlander for type [159]. Obviously, the language of type and cotype did not exist before the 1970s and their results were stated differently. Note the difference in behavior of the L_p-spaces when $p > 2$ or $p < 2$. This is the first example where we meet some fundamental change around the index $p = 2$ and, as the reader will see, it is really because when $p/2 < 1$ the triangle law for positive functions in $L_{p/2}$ reverses.

Theorem 6.2.14.

(a) If $1 \leq p \leq 2$, $L_p(\mu)$ has type p and cotype 2.
(b) If $2 < p < \infty$, $L_p(\mu)$ has type 2 and cotype p.

Moreover, (a) and (b) are optimal.

Proof. (a) Let us prove first that if $1 \leq p \leq 2$, then $L_p(\mu)$ has type p. We recall this elementary inequality:

Lemma 6.2.15. *Let $0 < r \leq 1$. Then for any nonnegative scalars $(\alpha_i)_{i=1}^{n}$ we have*

$$(\alpha_1 + \cdots + \alpha_n)^r \leq \alpha_1^r + \cdots + \alpha_n^r. \tag{6.10}$$

This way, combining Theorem 6.2.13 with (6.10) we obtain

$$
\begin{aligned}
\left(\mathbb{E}\Big\|\sum_{i=1}^{n}\varepsilon_i f_i\Big\|_p^p\right)^{\frac{1}{p}} &\le \Big\|\big(\sum_{i=1}^{n}|f_i|^2\big)^{1/2}\Big\|_p \\
&= \Big\|\sum_{i=1}^{n}|f_i|^2\Big\|_{p/2}^{1/2} \\
&\le \Big\|\big(\sum_{i=1}^{n}|f_i|^p\big)^{2/p}\Big\|_{p/2}^{1/2} \\
&= \Big(\int_\Omega \sum_{i=1}^{n}|f_i|^p\,d\mu\Big)^{1/p} \\
&= \Big(\sum_{i=1}^{n}\|f\|_p^p\Big)^{1/p}.
\end{aligned}
$$

To show that $L_p(\mu)$ has cotype 2 when $1 \le p \le 2$ we need the reverse of Minkowski's inequality:

Lemma 6.2.16. *Let* $0 < r < 1$. *Then*

$$\|f+g\|_r \ge \|f\|_r + \|g\|_r,$$

whenever f and g are nonnegative functions in $L_r(\mu)$.

Proof. Without loss of generality we can assume that $\|f+g\|_r = 1$ and so $d\nu = (f+g)^r\,d\mu$ is a probability measure. This implies

$$
\|f\|_r = \Big(\int_\Omega f^r\,d\mu\Big)^{1/r} = \Big(\int_{\{f+g>0\}}\frac{f^r}{(f+g)^r}(f+g)^r d\mu\Big)^{1/r} \\
\le \int_{\{f+g>0\}}\frac{f}{f+g}(f+g)^r d\mu.
$$

Analogously,

$$\|g\|_r \le \int_{\{f+g>0\}}\frac{g}{f+g}(f+g)^r d\mu.$$

Therefore $\|f\|_r + \|g\|_r \le 1 = \|f+g\|_r$. $\square$

Now, combining Theorem 6.2.13 with Lemma 6.2.16,

$$
\begin{aligned}
A_p^{-1}\Big(\mathbb{E}\Big\|\sum_{i=1}^{n}\varepsilon_i f_i\Big\|_p^p\Big)^{1/p} &\ge \Big\|\big(\sum_{i=1}^{n}|f_i|^2\big)^{\frac{1}{2}}\Big\|_p \\
&= \Big\|\sum_{i=1}^{n}|f_i|^2\Big\|_{p/2}^{\frac{1}{2}}
\end{aligned}
$$

$$\ge \Big(\sum_{i=1}^{n} \|f_i^2\|_{p/2}\Big)^{1/2}$$

$$= \Big(\sum_{i=1}^{n} \|f_i\|_p^2\Big)^{1/2}.$$

To obtain the cotype-2 estimate we just have to replace the L_p-average $(\mathbb{E}\|\sum_{j=1}^{n} \varepsilon_j f_j\|_p^p)^{1/p}$ by $(\mathbb{E}\|\sum_{j=1}^{n} \varepsilon_j f_j\|_p^2)^{1/2}$ using Kahane's inequality (at the small cost of a constant) .

(b) For each $2 < p < \infty$, from Theorem 6.2.13 in combination with Kahane's inequality there exists a constant $C = C(p)$ so that

$$\Big(\mathbb{E}\Big\|\sum_{i=1}^{n} \varepsilon_i f_i\Big\|_p^2\Big)^{1/2} \le C\Big\|\Big(\sum_{i=1}^{n} |f_i|^2\Big)^{\frac{1}{2}}\Big\|_p.$$

Since $p/2 > 1$, the triangle law now holds in $L_{p/2}(\mu)$ and hence

$$\Big\|\Big(\sum_{i=1}^{n} |f_i|^2\Big)^{\frac{1}{2}}\Big\|_p = \Big\|\sum_{i=1}^{n} |f_i|^2\Big\|_{p/2}^{1/2} \le \Big(\sum_{i=1}^{n} \|f_i^2\|_{p/2}\Big)^{1/2} = \Big(\sum_{i=1}^{n} \|f_i\|_p^2\Big)^{1/2}.$$

This shows that $L_p(\mu)$ has type 2. Therefore, from part (a) and Proposition 6.2.12 it follows that $L_p(\mu)$ has cotype p.

The last statement of the theorem follows from Remark 6.2.11 and the fact that $L_p(\mu)$ contains ℓ_p as a subspace.

$\square$

Example 6.2.17. To finish the section let us give an example showing that the concepts of type and cotype are not in duality, in the sense that the converse of Proposition 6.2.12 need not hold. The space $\mathcal{C}[0,1]$ fails to have nontrivial type because it contains a copy of L_1, whereas its dual, $\mathcal{M}(K)$, has cotype 2 (we leave the verification of this fact to the reader).

6.3 Properties of L_1

In Section 6.1 we saw that the Haar basis is unconditional in L_p when $1 < p < \infty$. It is, however, not unconditional in L_1 and this highlights an important difference between the cases $p = 1$ and $p > 1$.

Proposition 6.3.1. *The Haar basis is not unconditional in L_1.*

Proof. Let us use the device of labeling the elements of the Haar system by their supports and let f_N denote the characteristic function of the interval $[0, 2^{1-2N})$. Then expanding with respect to the Haar basis gives

$$f_N = 2^{1-2N}\chi_{[0,1]} + \sum_{j=0}^{2N} 2^{j+1-2N} h_{[0,2^{-j})}.$$

Put

$$g_N = \sum_{j=0}^{N} 2^{2j+1-2N} h_{[0,2^{-2j})}.$$

It is clear that

$$g_N(t) = -2^{2j+1-2N} \quad \text{for } 2^{-2j-1} \le t < 2^{-2j} \text{ and } 0 \le j \le N.$$

Thus

$$\|g_N\|_1 \ge \sum_{j=0}^{N} 2^{2j+1-2N} 2^{-2j-1} = (N+1)2^{-2N} = (N+1)\|f_N\|_1.$$

This shows immediately that the Haar system cannot be unconditional.

□

In fact we will show that L_1 cannot be embedded in a space with an unconditional basis; this result is due to Pełczyński (1961) [170]. In Theorem 4.5.2 we showed, by the technique of testing property (u), that $\mathcal{C}(K)$ embeds in a space with unconditional basis if and only if $\mathcal{C}(K) \approx c_0$. For L_1 this approach does not work because L_1 is weakly sequentially complete and therefore has property (u). A more sophisticated argument is therefore required. The argument we use was originally discovered by Milman [151]; first we need a lemma:

Lemma 6.3.2. *For every $f \in L_1$ we have*

$$\lim_{n \to \infty} \int f(t)r_n(t)dt = 0.$$

Thus $(fr_n)_{n=1}^{\infty}$ is weakly null for every $f \in L_1$.

Proof. $(r_n)_{n=1}^{\infty}$ is an orthonormal sequence in L_2, which implies (by Bessel's inequality) that

$$\lim_{n \to \infty} \int f(t)r_n(t)dt = 0 \quad \text{for all } f \in L_2.$$

Since $(r_n)_{n=1}^{\infty}$ is uniformly bounded in L_∞, and L_2 is dense in L_1 we deduce

$$\lim_{n \to \infty} \int f(t)r_n(t)dt = 0 \quad \text{for all } f \in L_1.$$

Thus if $f \in L_1$ and $g \in L_\infty$, since $fg \in L_1$ we obtain

$$\lim_{n \to \infty} \int g(t)f(t)r_n(t)dt = 0,$$

which gives the latter statement in the lemma.

$\square$

Theorem 6.3.3. L_1 *cannot be embedded in a Banach space with unconditional basis.*

Proof. Let X be a Banach space with K-unconditional basis $(e_n)_{n=1}^{\infty}$ and suppose that $T : L_1 \to X$ is an embedding. We can assume that for some constant $M \geq 1$,

$$\|f\|_1 \leq \|Tf\| \leq M\|f\|_1, \qquad f \in L_1.$$

By exploiting the unconditionality of $(e_n)_{n=1}^{\infty}$ we are going to build an unconditional basic sequence in L_1 using a gliding-hump type argument.

Take $(\delta_k)_{k=1}^{\infty}$ a sequence of positive real numbers with $\sum_{k=1}^{\infty} \delta_k < 1$. Let $f_0 = 1 = \chi_{[0,1]}$, $n_1 = 1$, $s_0 = 0$ and pick $s_1 \in \mathbb{N}$ such that

$$\left\| \sum_{j=s_1+1}^{\infty} e_j^*(T(f_0 r_{n_1}))e_j \right\| < \frac{1}{2}\delta_1.$$

Put

$$x_1 = \sum_{j=s_0+1}^{s_1} e_j^*(T(f_0 r_{n_1}))e_j.$$

Next take $f_1 = (1 + r_{n_1})f_0$. Since the sequence $(f_1 r_k)_{k=1}^{\infty}$ is weakly null by Lemma 6.3.2, $(T(f_1 r_k))_{k=1}^{\infty}$ is also weakly null, hence we can find $n_2 \in \mathbb{N}$, $n_2 > n_1$, so that

$$\left\| \sum_{j=1}^{s_1} e_j^*(T(f_1 r_{n_2}))e_j \right\| < \frac{1}{2}\delta_2.$$

Now, pick $s_2 \in \mathbb{N}$, $s_2 > s_1$, for which

$$\left\| \sum_{j=s_2+1}^{\infty} e_j^*(T(f_1 r_{n_2}))e_j \right\| < \frac{1}{2}\delta_2.$$

Continuing in this way we will inductively select two strictly increasing sequences of natural numbers $(n_k)_{k=1}^{\infty}$ and $(s_k)_{k=0}^{\infty}$, a sequence of functions $(f_k)_{k=0}^{\infty}$ in L_1 given by

$$f_k = (1 + r_{n_k})f_{k-1} \quad \text{for } k \geq 1,$$

and a block basic sequence $(x_k)_{k=1}^{\infty}$ of $(e_n)_{n=1}^{\infty}$ defined by

$$x_k = \sum_{j=s_{k-1}+1}^{s_k} e_j^*(T(f_{k-1} r_{n_k}))e_j, \qquad k = 1, 2, \ldots.$$

This is how the inductive step goes: suppose $n_1, n_2, \ldots, n_{l-1}, s_0, s_1, \ldots, s_{l-1}$, and therefore $f_1, \ldots, f_{l-1}$ have been determined. Since $(T(f_{l-1} r_k))_{k=1}^{\infty}$ is weakly null we can find $n_l > n_{l-1}$ so that

$$\left\| \sum_{j=1}^{s_{l-1}} e_j^*(T(f_{l-1}r_{n_l}))e_j \right\| < \frac{1}{2}\delta_l,$$

and then we choose $s_l > s_{l-1}$ so that

$$\left\| \sum_{j=s_l+1}^{\infty} e_j^*(T(f_{l-1}r_{n_l}))e_j \right\| < \frac{1}{2}\delta_l.$$

Note that for $k \geq 1$ we have

$$f_k = \prod_{j=1}^{k}(1 + r_{n_j}), \tag{6.11}$$

which yields $f_k \geq 0$ for all k. Expanding out (6.11), it is also clear that for each k,

$$\|f_k\|_1 = \int_0^1 f_k(t)\, dt = 1.$$

On the other hand, for $k \geq 1$ we have

$$\left\| Tf_k - Tf_{k-1} - x_k \right\| < \delta_k,$$

and hence the estimate

$$\left\| \sum_{j=1}^{n} x_j \right\| < M + \sum_{j=1}^{n} \delta_j < M + 1$$

holds for all n.

Since it is a block basic sequence with respect to $(e_n)_{n=1}^{\infty}$, $(x_k)_{n=1}^{\infty}$ is an unconditional basic sequence in X with unconditional constant $\leq K$ (see Problem 3.1). Therefore for all choices of signs $\epsilon_j = \pm 1$ and all $n = 1, 2, \ldots$ we have a bound:

$$\left\| \sum_{j=1}^{n} \epsilon_j x_j \right\| \leq K(M + 1),$$

which implies

$$\left\| \sum_{j=1}^{n} \epsilon_j(Tf_j - Tf_{j-1}) \right\| \leq K(M + 1) + 1,$$

and thus

$$\left\| \sum_{j=1}^{n} \epsilon_j(f_j - f_{j-1}) \right\|_1 \leq K(M + 1) + 1.$$

This shows that $\sum_{j=1}^{\infty}(f_j - f_{j-1})$ in L_1 is a WUC series in L_1 (see Lemma 2.4.6). Since L_1 is weakly sequentially complete (Theorem 5.2.10), by Corollary 2.4.15 the series $\sum_{j=1}^{\infty}(f_j - f_{j-1})$ must converge (unconditionally) in norm in L_1 and, in particular, $\lim_{j \to \infty} \|f_j - f_{j-1}\|_1 = 0$. But for $j \geq 1$ we have $\|f_j - f_{j-1}\|_1 = \|r_{n_j} f_{j-1}\|_1 = 1$, a contradiction.

□

In Corollary 2.5.6 we saw that c_0 is not a dual space. We will show that L_1 is also not a dual space and, even more generally, that it cannot be embedded in a separable dual space. We know that c_0 is not isomorphic to a dual space because c_0 is uncomplemented in its bidual. This is not the case for L_1 as we shall see below. Thus to show L_1 is not a dual space requires another type of argument and we will use some rather more delicate geometrical properties of separable dual spaces.

Lemma 6.3.4. *Let X be a Banach space such that X^* is separable. Assume that K is a weak* compact set in X^*. Then K has a point of weak*-to-norm continuity. That is, there is $x^* \in K$ such that whenever a sequence $(x_n^*)_{n=1}^\infty \subset K$ converges to x^* with respect to the weak* topology of X^*, then $(x_n^*)_{n=1}^\infty$ converges to x^* in the norm topology of X^*.*

Proof. Let $(\epsilon_n)_{n=1}^\infty$ be a sequence of scalars converging to zero. Using that X^* is separable for the norm topology, for each ϵ_n there is a sequence of points $(x_k^{(n)})_{k=1}^\infty \subset X^*$ such that

$$K \subset \bigcup_{k=1}^\infty \left(B(x_k^{(n)}, \epsilon_n) \cap K \right).$$

Observe that for all integers n and k, $B(x_k^{(n)}, \epsilon_n)$ (the closed ball centered in $x_k^{(n)}$ of radius ϵ_n) is weak* compact by Banach-Alaoglu's theorem, so the sets $B(x_k^{(n)}, \epsilon_n) \cap K$ are weak* closed. Let us call $B_k^{(n)}$ the weak* interior of $B(x_k^{(n)}, \epsilon_n) \cap K$. Hence

$$V_n = \bigcup_{k=1}^\infty B_k^{(n)}$$

is dense and open.

Since X^* is separable, the weak* topology of X^* relative to K is metrizable. Then K is compact metric, therefore complete. By the Baire Category theorem, the set $V = \bigcap_{n=1}^\infty V_n$ is a dense G_δ-set. We are going to see that all of the elements in V are points of weak*-to-norm continuity. Indeed, take $v^* \in V$. Then for each ϵ_n there exists a weak* neighborhood of v^* relative to K of diameter at most $2\epsilon_n$. Since $(\epsilon_n)_{n=1}^\infty$ converges to zero, the identity operator

$$I : (K, w^*) \longrightarrow (K, \|\cdot\|)$$

is continuous at v^*.

□

Lemma 6.3.5. *Suppose X is a Banach space which embeds in a separable dual space. Then every closed bounded subset F of X has a point of weak-to-norm continuity.*

Proof. Let F be a closed bounded subset of X. Suppose $T : X \to Y^*$ is an embedding in Y^*, where Y is a Banach space with separable dual. We can assume that $\|x\| \leq \|Tx\| \leq M\|x\|$ for $x \in X$ where M is a constant independent of x. Let W be the weak* closure of $T(F)$. Then by Lemma 6.3.4 there is $y^* \in W$ which is a point of weak*-to-norm continuity. In particular there is a sequence (y_n^*) in $T(F)$ with $\|y_n^* - y^*\| \to 0$. If we let $y_n^* = Tx_n$ with $x_n \in F$ for each n, then $(x_n)_{n=1}^\infty$ is Cauchy in X and so converges to some $x \in F$, hence $Tx = y^*$. Now for any $\epsilon > 0$ we can find a weak* neighborhood U_ϵ of y^* so that $w^* \in U_\epsilon \cap W$ implies $\|w^* - y^*\| < \epsilon$. In particular if $v \in T^{-1}(U_\epsilon) \cap C$ then $\|v - x\| < \epsilon$. Clearly $T^{-1}(U_\epsilon)$ is weakly open since the map $T : X \to Y^*$ is weak-to-weak* continuous. This shows x is a point of weak-to-norm continuity.

□

Lemma 6.3.6. *Suppose X is a Banach space which embeds in a separable dual space and let $x \in B_X$ be a point of weak-to-norm continuity. If (x_n) is a weakly null sequence in X such that $\limsup \|x + x_n\| \leq 1$ then $\lim_{n\to\infty} \|x_n\| = 0$.*

Proof. Put
$$u_n = \begin{cases} x + x_n & \text{if } \|x + x_n\| \leq 1 \\ \frac{x+x_n}{\|x+x_n\|} & \text{if } \|x + x_n\| > 1 \end{cases}$$
and observe that
$$u_n - x = x_n + (1 - \alpha_n)(x + x_n),$$
where $\alpha_n = (\|x + x_n\| - 1)_+ \to 0$. Thus $\lim_{n\to\infty} u_n = x$ weakly and so $\lim_{n\to\infty} \|u_n - x\| = 0$. This implies that $\lim_{n\to\infty} \|x_n\| = 0$.

□

Theorem 6.3.7. *Neither of the Banach spaces L_1 and c_0 can be embedded in a separable dual space.*

Proof. If L_1 embeds in a separable dual space, Lemma 6.3.5 yields a function $f \in B_{L_1}$ that is a point of weak-to-norm continuity. By Lemma 6.3.2 the sequence $(r_n f)_{n=1}^\infty$ is weakly null in L_1 and satisfies
$$\|f + fr_n\|_1 = \int (1 + r_n(t))|f(t)|\, dt \longrightarrow 1.$$

Therefore by Lemma 6.3.6 it must be $\lim_{n\to\infty} \|r_n f\|_1 = 0$, which implies $f = 0$. This is absurd since $(r_n)_{n=1}^\infty$ is a weakly null sequence and $\|r_n\|_1 = 1$.

For c_0 the argument is similar. Let ξ be a point of weak-to-norm continuity in B_{c_0}. Then if $(e_n)_{n=1}^\infty$ is the canonical basis we have $\lim_{n\to\infty} \|\xi + e_n\| = 1$ and so $\lim_{n\to\infty} \|e_n\| = 0$, which is again absurd.

□

Remark 6.3.8. The fact that c_0 cannot be embedded in a separable dual space can be proved in many ways, and we have already seen this in Problems 2.6 and 2.9.

Corollary 6.3.9. L_1 *does not have a boundedly-complete basis.*

Proof. We need only recall that, by Theorem 3.2.10, a space with a boundedly-complete basis is (isomorphic to) a separable dual space.

$\square$

Theorem 6.3.7 is rather classical: it is due to Gelfand [66]. In fact the argument we have given is somewhat *ad hoc*; to be more precise, one should use the concept of the Radon-Nikodym Property which we discussed earlier in Section 5.4. The main point here is that neither L_1 nor c_0 have the Radon-Nikodym Property while separable dual spaces do. Gelfand approaches this through differentiability of Lipschitz maps: a Banach space X has (RNP) if and only if every Lipschitz map $F : [0,1] \to X$ is differentiable a.e. In L_1 the Lipschitz map

$$F(t) = \chi_{(0,t)}, \qquad 0 \le t \le 1$$

is nowhere differentiable. In c_0 we can consider the map

$$F(t) = \left(\frac{1}{n} \sin nt \right)_{n=1}^{\infty}, \qquad 0 \le t \le 1$$

which is again nowhere differentiable (note that formally differentiating takes us into the bidual!). These examples are due to Clarkson [30].

Let us conclude this section with the promised result that L_1 is complemented in its bidual.

Proposition 6.3.10. *There is a norm-one linear projection* $P : L_1^{**} \to L_1$.

Proof. Let us first define $R : L_1^{**} \to \mathcal{M}[0,1]$ to be the restriction map $\varphi \mapsto \varphi|_{\mathcal{M}[0,1]}$. Clearly $\|R\varphi\| \le \|\varphi\|$. Next we define a map $S : \mathcal{M}[0,1] \to L_1$ by $S\mu = f$ where

$$d\mu = d\nu + f \, dt$$

is the Lebesgue decomposition of μ (i.e., ν is singular with respect to the Lebesgue measure). Then $\|S\| = 1$. We conclude that $P = SR$ is a norm-one projection of L_1^{**} onto L_1.

$\square$

6.4 Subspaces of L_p

In Chapter 2 we studied the subspace structure and the complemented subspace structure of the spaces ℓ_p for $1 \le p < \infty$ (see particularly Corollary 2.1.6 and Theorem 2.2.4). Now we would like to analyze the function space analogues, the L_p-spaces for $1 \le p < \infty$, in the same way. This is a more delicate problem and the subspace structure is much richer, with the exception of the case $p = 2$ which is trivial since L_2 is isometric to ℓ_2. We will also see some fundamental differences between the cases $1 < p < 2$ and $2 < p < \infty$.

Proposition 6.4.1. *Let $(f_n)_{n=1}^{\infty}$ be a sequence of norm-one, disjointly supported functions in L_p. Then $(f_n)_{n=1}^{\infty}$ is a complemented basic sequence isometrically equivalent to the canonical basis of ℓ_p.*

Proof. The case $p = 1$ was seen in Lemma 5.1.1. Let us fix $1 < p < \infty$. For any sequence of scalars $(a_i)_{i=1}^{\infty} \in c_{00}$, by the disjointness of the f_i's we have

$$\left\| \sum_{i=1}^{\infty} a_i f_i \right\|_p^p = \int \left| \sum_{i=1}^{\infty} a_i f_i(t) \right|^p dt$$

$$= \int \sum_{i=1}^{\infty} |a_i f_i(t)|^p dt$$

$$= \sum_{i=1}^{\infty} |a_i|^p \int |f_i(t)|^p dt$$

$$= \sum_{i=1}^{\infty} |a_i|^p.$$

By the Hahn-Banach theorem, for each $i \in \mathbb{N}$ there exists $g_i \in L_q$ (q the conjugate exponent of p) with $\|g_i\|_q = 1$ so that $1 = \|f_i\|_p = \int f_i(t) g_i(t) \, dt$. Furthermore, without loss of generality, we can assume g_i to have the same support as f_i for all i. Let us define the linear operator from L_p onto $[f_i]$ given by

$$P(f) = \sum_{i=1}^{\infty} \left(\int f(t) g_i(t) \, dt \right) f_i, \quad f \in L_p.$$

Then,

$$\|P(f)\|_p = \left(\sum_{i=1}^{\infty} \left| \int f(t) g_i(t) \, dt \right|^p \right)^{1/p}$$

$$= \left(\sum_{i=1}^{\infty} \left| \int_{\{|f_i|>0\}} f(t) g_i(t) \, dt \right|^p \right)^{1/p}$$

$$\leq \left(\sum_{i=1}^{\infty} \int_{\{|f_i|>0\}} |f(t)|^p \, dt \right)^{1/p}$$

$$\leq \left(\int |f(t)|^p \, dt \right)^{1/p}.$$

$\square$

The following proposition allows us to deduce that L_p is not isomorphic to ℓ_p for $p \neq 2$, and already hints at the fact that the the L_p-spaces have a more complicated structure than the spaces ℓ_p.

Proposition 6.4.2. ℓ_2 *embeds in L_p for all $1 \leq p < \infty$. Furthermore, ℓ_2 embeds complementably in L_p if and only if $1 < p < \infty$.*

Proof. For each $1 \leq p < \infty$ let R_p be the closed subspace spanned in L_p by the Rademacher functions $(r_n)_{n=1}^{\infty}$. By Khintchine's inequality, $(r_n)_{n=1}^{\infty}$ is equivalent to the canonical basis of ℓ_2, then R_p is isomorphic to ℓ_2.

By Proposition 5.6.1, L_1 has no infinite-dimensional complemented reflexive subspaces, so R_1 is not complemented in L_1. Let us prove that if $1 < p < \infty$, R_p is complemented in L_p.

Assume first that $2 \leq p < \infty$. Consider the map from L_p onto R_p given by

$$P(f) = \sum_{n=1}^{\infty} \left(\int f(t) r_n(t) \, dt \right) r_n, \quad f \in L_p.$$

P is linear and well defined. Indeed, the series is convergent in L_p because $f \in L_p \subset L_2$ implies $\sum_{n=1}^{\infty} (\int f(t) r_n(t) \, dt)^2 < \infty$. Now, Khintchine's inequality and Bessel's inequality yield

$$\|P(f)\|_p^2 = \left\| \sum_{n=1}^{\infty} \left(\int f(t) r_n(t) \, dt \right) r_n \right\|_p^2$$

$$\leq B_p^2 \sum_{n=1}^{\infty} \left| \int f(t) r_n(t) \, dt \right|^2$$

$$\leq B_p^2 \|f\|_2^2$$

$$\leq B_p^2 \|f\|_p^2.$$

If $1 \leq p < 2$ we define P as before for each $f \in L_p \cap L_2$ (which is a dense subspace in L_p). Then, using Khintchine's inequality, we obtain

$$\|P(f)\|_p \leq \left(\sum_{n=1}^{\infty} \left| \int f(t) r_n(t) \, dt \right|^2 \right)^{1/2}$$

$$= \sup \left\{ \sum_{n=1}^{\infty} \left(\alpha_n \int f(t) r_n(t) \, dt \right) : \sum_{n=1}^{\infty} \alpha_n^2 = 1 \right\}$$

$$= \sup \left\{ \int f(t) \left(\sum_{n=1}^{\infty} \alpha_n r_n(t) \right) dt : \sum_{n=1}^{\infty} \alpha_n^2 = 1 \right\}$$

$$\leq \sup \left\{ \|f\|_p \left\| \sum_{n=1}^{\infty} \alpha_n r_n(t) \right\|_q : \sum_{n=1}^{\infty} \alpha_n^2 = 1 \right\}$$

$$\leq \sup \left\{ \|f\|_p B_q \left\| \sum_{n=1}^{\infty} \alpha_n r_n(t) \right\|_2 : \sum_{n=1}^{\infty} \alpha_n^2 = 1 \right\}$$

$$= B_q \|f\|_p.$$

By density, P extends continuously to L_p with preservation of norm.

□

Proposition 6.4.3. *If ℓ_q embeds in L_p then either $p \leq q \leq 2$ or $2 \leq q \leq p$.*

Proof. Let us start by noticing that if $(e_i)_{i=1}^{\infty}$ is the canonical basis of ℓ_q, for each n we have

$$\mathbb{E}\left\|\sum_{i=1}^{n} \varepsilon_i e_i\right\|_q = n^{1/q}.$$

If ℓ_q embeds in L_p for some $p < 2$, by Theorem 6.2.14 there exist constants c_1 and c_2 (given by the embedding and the type and cotype constants) such that

$$c_1 \, n^{\frac{1}{2}} \leq n^{\frac{1}{q}} \leq c_2 \, n^{\frac{1}{p}}.$$

For these inequalities to hold for all $n \in \mathbb{N}$ it is necessary that $q \in [p, 2]$. If ℓ_q embeds in L_p for some $2 < p < \infty$, with the same kind of argument we deduce that q must belong to the interval $[2, p]$.

□

Definition 6.4.4. Suppose (Ω, Σ, μ) is a probability measure space and let X be a closed subspace of $L_p(\mu)$ for some $1 \leq p < \infty$. X is said to be *strongly embedded* in $L_p(\mu)$ if, in X, convergence in measure is equivalent to convergence in the $L_p(\mu)$-norm; that is, a sequence of functions $(f_n)_{n=1}^{\infty}$ in X converges to 0 in measure if and only if $\|f_n\|_p \to 0$.

Proposition 6.4.5. *Suppose (Ω, Σ, μ) is a probability measure space and let $1 \leq p < \infty$. Suppose X is an infinite-dimensional closed subspace of $L_p(\mu)$. Then the following are equivalent:*

(i) X is strongly embedded in $L_p(\mu)$;
(ii) For each $0 < q < p$ there exists a constant C_q such that

$$\|f\|_q \leq \|f\|_p \leq C_q \, \|f\|_q \quad \text{for all } f \in X;$$

(iii) For some $0 < q < p$ there exists a constant C_q such that

$$\|f\|_q \leq \|f\|_p \leq C_q \, \|f\|_q \quad \text{for all } f \in X.$$

Proof. Let us suppose that X is strongly embedded in $L_p(\mu)$ but (ii) fails. Then there would exist a sequence $(f_n)_{n=1}^{\infty}$ in X such that $\|f_n\|_p = 1$ and $\|f_n\|_q \to 0$ for some $0 < q < p$. Obviously, this implies that $(f_n)_{n=1}^{\infty}$ converges to 0 in measure, which would force $(\|f_n\|_p)_{n=1}^{\infty}$ to converge to 0. This contradiction shows that $(i) \Rightarrow (ii)$.

Suppose now that (iii) holds and there is a sequence of functions $(f_n)_{n=1}^{\infty}$ in X such that $(f_n)_{n=1}^{\infty}$ converges to 0 in measure but $(\|f_n\|_p)_{n=1}^{\infty}$ does not tend to 0. By passing to a subsequence we can assume that $(f_n)_{n=1}^{\infty}$ converges to 0 almost everywhere and $\|f_n\|_p = 1$ for all n.

For each $M > 0$, since $q < p$ we have

$$\int_\Omega |f_n|^q \, d\mu = \int_{\{|f_n| \ge M\}} |f_n|^q \, d\mu + \int_{\{|f_n| < M\}} |f_n|^q \, d\mu$$

$$\le \int_{\{|f_n| \ge M\}} M^{q-p} |f_n|^p \, d\mu + \int_{\{|f_n| < M\}} |f_n|^q \, d\mu$$

$$\le \frac{1}{M^{p-q}} + \int_{\{|f_n| < M\}} |f_n|^q \, d\mu.$$

Let $\epsilon > 0$. By the Lebesgue Bounded Convergence theorem, there is $N_0 \in \mathbb{N}$ such that $\int_{\{|f_n| < M\}} |f_n|^q \, d\mu < \epsilon/2$ for all $n > N_0$. So, if we pick $M > (2\epsilon^{-1})^{\frac{1}{p-q}}$ we get

$$\int_\Omega |f_n|^q \, d\mu < \epsilon,$$

contradicting (iii). Hence $(iii) \Rightarrow (i)$, and so the proof is over because, trivially, $(ii) \Rightarrow (iii)$. $\qquad\square$

Example 6.4.6. For each $1 \le p < \infty$ the closed subspace spanned in L_p by the Rademacher functions, R_p, is strongly embedded in L_p since, using Khintchine's inequality, the L_q-norm and the L_p-norm are equivalent in R_p for all $1 \le q < \infty$.

Theorem 6.4.7. *Suppose that X is an infinite-dimensional closed subspace of L_p for some $1 \le p < \infty$. If X is not strongly embedded in L_p then X contains a subspace isomorphic to ℓ_p and complemented in L_p.*

Proof. If X is not strongly embedded in L_p, by Proposition 6.4.5 there is a sequence $(f_n)_{n=1}^\infty$ in X, $\|f_n\|_p = 1$ for all n, such that $f_n \to 0$ a.e. By Lemma 5.2.1 there is a subsequence $(f_{n_k})_{k=1}^\infty$ of $(f_n)_{n=1}^\infty$ and a sequence of disjoint subsets $(A_k)_{k=1}^\infty$ of $[0,1]$ such that if $B_k = [0,1] \setminus A_k$, then $(|f_{n_k}|^p \chi_{B_k})_{k=1}^\infty$ is equi-integrable. Lemma 5.2.7 implies $\int |f_{n_k}|^p \chi_{B_k} \, d\mu \to 0$. That is, $\|f_{n_k} - f_{n_k} \chi_{A_k}\|_p \to 0$. Now, by standard perturbation arguments we obtain a subsequence $(f_{n_{k_j}})_{j=1}^\infty$ of $(f_{n_k})_{k=1}^\infty$ such that $(f_{n_{k_j}})_{j=1}^\infty$ is equivalent to the canonical basis of ℓ_p and $[f_{n_{k_j}}]$ is complemented in L_p. $\qquad\square$

The following theorem was proved in 1962 by Kadets and Pełczyński [98] in a paper which really initiated the study of L_p-spaces by basic sequence techniques. We will see that the case $p > 2$ is quite different from the case $p < 2$ and this theorem emphasizes this point.

Theorem 6.4.8 (Kadets-Pełczyński). *Suppose that X is an infinite-dimensional closed subspace of L_p for some $2 < p < \infty$. Then the following are equivalent:*

(i) ℓ_p does not embed in X;
(ii) ℓ_p does not embed complementably in X;

(iii) X *is strongly embedded in* L_p*;*
(iv) X *is isomorphic to a Hilbert space and is complemented in* L_p*;*
(v) X *is isomorphic to a Hilbert space.*

Proof. $(i) \Rightarrow (ii)$ and $(iv) \Rightarrow (v)$ are obvious, and $(ii) \Rightarrow (iii)$ was proved in Theorem 6.4.7. Let us complete the circle by showing that $(iii) \Rightarrow (iv)$ and that $(v) \Rightarrow (i)$.

$(iii) \Rightarrow (iv)$ If X is strongly embedded in L_p, Proposition 6.4.5 yields a constant C_2 such that $\|f\|_2 \leq \|f\|_p \leq C_2 \|f\|_2$ for all $f \in X$. This shows that X embeds in L_2 and hence it is isomorphic to a Hilbert space. Let us see that X is complemented in L_2.

Since $p > 2$, L_p is contained in L_2 and the inclusion $\iota : L_p \to L_2$ is norm decreasing. The restriction of ι to X is an isomorphism onto the subspace $\iota(X)$ of L_2, and $\iota(X)$ is complemented in L_2 by an orthogonal projection P:

Then $\iota^{-1} P \iota$ is a projection of L_p onto X (this projection is simply the restriction of P to L_p).

$(v) \Rightarrow (i)$ If $X \approx \ell_2$ then X cannot contain an isomorphic copy of ℓ_p for any $p \neq 2$ because the classical sequence spaces are totally incomparable (Corollary 2.1.6).

$\square$

The Kadets-Pełczyński theorem establishes a dichotomy for subspaces of L_p when $2 < p < \infty$:

Corollary 6.4.9. *Suppose X is a closed subspace of L_p for some $2 < p < \infty$. Then either*

(i) X is isomorphic to ℓ_2, in which case X is complemented in L_p, or
(ii) X contains a subspace that is isomorphic to ℓ_p and complemented in L_p.

Notice that, in particular, this settles the question of which L_q-spaces for $1 \leq q < \infty$ embed in L_p for $p > 2$:

Corollary 6.4.10. *For $2 < p < \infty$ and $1 \leq q < \infty$ with $q \neq p, 2$, L_p does not have any subspace isomorphic to L_q or ℓ_q.*

We are now ready to find a more efficient embedding of ℓ_2 into the L_p-spaces, replacing the Rademacher sequences by sequences of independent Gaussians. We consider only the real case, although modifications can be made to handle complex functions. In order to introduce these ideas, we will require some more basic notions from probability theory.

If f is a real random variable, its *distribution* is the probability measure μ_f on $\mathbb{R}$ given by

$$\mu_f(B) = \mathbb{P}(f^{-1}B)$$

for any Borel set B. f is called *symmetric* if f and $-f$ have the same distribution.

Conversely, for each probability measure μ on $\mathbb{R}$ there exist real random variables f with $\mu_f = \mu$, and the formula

$$\int_\Omega F(f(\omega))\, d\mathbb{P}(\omega) = \int_{-\infty}^{\infty} F(x)\, d\mu_f(x) \tag{6.12}$$

holds for any positive Borel function $F : \mathbb{R} \to \mathbb{R}$.

The *characteristic function* ϕ_f of a random variable f is the function $\phi_f : \mathbb{R} \to \mathbb{C}$ defined by

$$\phi_f(t) = \mathbb{E}(e^{itf}).$$

This is related to μ_f via the Fourier transform:

$$\hat{\mu}_f(-t) = \int_{\mathbb{R}} e^{itx} d\mu_f(x) = \phi_f(t).$$

In particular ϕ_f *determines* μ_f, i.e., if f and g are two random variables (possibly on different probability spaces) with $\phi_f = \phi_g$ then $\mu_f = \mu_g$. Other basic useful properties of characteristic functions are:

- $\phi_f(-t) = \overline{\phi_f(t)}$;
- $\phi_{cf+d}(-t) = e^{idt}\phi_f(ct)$, for c, d constants;
- $\phi_{f+g} = \phi_f \phi_g$ if f and g are independent.

Remark 6.4.11. If $f_1, \ldots, f_n$ are independent random variables (not necessarily equally distributed) on some probability space, then we can exploit independence to compute the characteristic function of any linear combination $\sum_{j=1}^{n} a_j f_j$:

$$\mathbb{E}\left(e^{it \sum_{j=1}^{n} a_j f_j}\right) = \prod_{j=1}^{n} \mathbb{E}\left(e^{ita_j f_j}\right) = \prod_{j=1}^{n} \phi_{f_j}(a_j t). \tag{6.13}$$

Suppose we are given a probability measure μ on $\mathbb{R}$. The random variable $f(x) = x$ has distribution μ with respect to the probability space $(\mathbb{R}, \mu)$. Next consider the countable product space $\mathbb{R}^{\mathbb{N}}$ with the product measure $\mathbb{P} = \mu \times \mu \times \cdots$. $(\mathbb{R}^{\mathbb{N}}, \mathbb{P})$ is also a probability space and the coordinate maps $f_j : \mathbb{R}^{\mathbb{N}} \to \mathbb{R}$,

$$f_j(x_1, \ldots, x_n, \ldots) = x_j,$$

are identically distributed random variables on $\mathbb{R}^{\mathbb{N}}$ with distribution μ. Moreover, the random variables $(f_j)_{j=1}^{\infty}$ are independent.

Although we created the sequence of functions $(f_j)_{j=1}^\infty$ on $(\mathbb{R}^\mathbb{N}, \mathbb{P})$ we might just as well have worked on $([0,1], \mathcal{B}, \lambda)$. As we discussed in Section 5.1 there is a Borel isomorphism $\sigma : \mathbb{R}^\mathbb{N} \to [0,1]$ which preserves measure, that is,

$$\lambda(B) = \mathbb{P}(\sigma^{-1}B), \qquad B \in \mathcal{B}$$

and the functions $(f_j \circ \sigma^{-1})_{j=1}^\infty$ have exactly the same properties on $[0,1]$.

This remark, in particular, allows us to pick an infinite sequence of independent identically distributed random variables on $[0,1]$ with a given distribution.

The *standard normal distribution* is given by the measure on $\mathbb{R}$

$$d\mu_g = \frac{1}{\sqrt{2\pi}} e^{-x^2/2}\, dx.$$

We will call any random variable with this distribution a *(normalized) Gaussian*. In this case we have

$$\hat{\mu}_g(-t) = \frac{1}{\sqrt{2\pi}} \int_{-\infty}^\infty e^{itx - x^2/2}\, dx = e^{-t^2/2},$$

so the characteristic function of a Gaussian is $e^{-t^2/2}$.

Proposition 6.4.12. *If g is a Gaussian on some probability measure space (Ω, Σ, μ) then $g \in L_p(\mu)$ for every $1 \le p < \infty$.*

Proof. This is because

$$\int_\Omega |g(\omega)|^p\, d\omega = \frac{1}{\sqrt{2\pi}} \int_{-\infty}^\infty |x|^p e^{-\frac{1}{2}x^2}\, dx,$$

and the last integral is finite and indeed computable in terms of the Γ function as

$$\frac{2^{p/2}}{\sqrt{\pi}} \Gamma\left(\frac{p+1}{2}\right).$$

$\square$

Proposition 6.4.13. *ℓ_2 embeds isometrically in L_p for all $1 \le p < \infty$.*

Proof. Take $(g_j)_{j=1}^\infty$, a sequence of independent Gaussians on $[0,1]$. By Proposition 6.4.12, $(g_j)_{j=1}^\infty \subset L_p$. We will show that $[g_j]$ is isometrically isomorphic to ℓ_2.

For every $n \in \mathbb{N}$ and scalars $(a_j)_{j=1}^n$ such that $\sum_{j=1}^n a_j^2 = 1$, put

$$h_n = \sum_{j=1}^n a_j g_j.$$

By (6.13) we have

$$\phi_{h_n}(t) = e^{-(a_1^2 + \cdots + a_n^2)t^2/2} = e^{-t^2/2}.$$

This means that $\mu_{h_n} = \mu_{g_1}$ and so by (6.12)

$$\|h_n\|_p = \|g_1\|_p.$$

It follows that for any $a_1, \ldots, a_n$ in $\mathbb{R}$,

$$\left\| \sum_{j=1}^{n} a_j g_j \right\|_p = \|g_1\|_p \left(\sum_{j=1}^{n} |a_j|^2 \right)^{1/2}.$$

Thus the mapping $e_n \mapsto \|g_1\|_p^{-1} g_n$ linearly extends to an isometry from ℓ_2 onto the subspace $[g_n]$ of L_p.

□

The connection between the Gaussians and ℓ_2 is encoded in the characteristic function. We are now going to dig a little deeper to try to make copies of ℓ_q for other values of q in the L_p-spaces. A moment's thought shows that we need a random variable f with characteristic function

$$\phi_f(t) = e^{-c|t|^q}$$

for some constant $c = c(q)$. It turns out that if (and only if) $0 < q < 2$ we can construct such a random variable. This has long been known to Probabilists; here we give a treatment based on some unpublished notes of Ben Garling.

We will need the following classical lemma due to Paul Lévy (see, for instance, [57]).

Lemma 6.4.14. *Suppose $(\mu_n)_{n=1}^{\infty}$ is a sequence of probability measures on $\mathbb{R}$ such that*

$$\lim_{n \to \infty} \hat{\mu}_n(-t) = F(t)$$

exists for all $t \in \mathbb{R}$. If F is continuous then there is a probability measure μ on $\mathbb{R}$ such that $\hat{\mu}(-t) = F(t)$.

Proof. It is convenient to compactify the real line by adding one point at ∞ to make the one-point compactification $K = \mathbb{R} \cup \infty$. We can then regard each μ_n as a Borel measure on K which assigns zero mass to $\{\infty\}$. Let μ be any weak* cluster point of this sequence (viewed as elements of $\mathcal{C}(K)^*$; such a measure then exists by Banach-Alaoglu's theorem). The functions $x \mapsto e^{itx}$ cannot be extended continuously to K. However, for $t \neq 0$ the functions

$$h_t(x) = \begin{cases} t & \text{if } x = 0 \\ \frac{e^{itx}-1}{ix} & \text{if } x \in \mathbb{R} \setminus \{0\} \\ 0 & \text{if } x = \infty \end{cases}$$

are continuous on K.

If $t > 0$,

$$\int_K h_t(x)d\mu_n(x) = \int_{\mathbb{R}} \left(\int_0^t e^{isx}ds \right) d\mu_n(x)$$
$$= \int_0^t \left(\int_{\mathbb{R}} e^{isx} d\mu_n(x) \right) ds$$
$$= \int_0^t \hat{\mu}_n(-s) \, ds.$$

Thus

$$\int_K h_t(x)d\mu(x) = \int_0^t F(s) \, ds.$$

If $t < 0$ the same calculation works to give

$$\int_K h_t(x)d\mu(x) = -\int_t^0 F(s) \, ds.$$

Note that for $t > 0$, $|h_t(x)| \leq t$ for all x and vanishes at ∞. Thus

$$\left| \int_K h_t(x) \, d\mu(x) \right| \leq t\mu(\mathbb{R}).$$

Hence, for $t > 0$

$$\frac{1}{t} \int_0^t F(s) \, ds \leq \mu(\mathbb{R}).$$

F is *continuous* and, obviously, $F(1) = 1$. Thus the left-hand side converges to 1. We conclude that $\mu(\mathbb{R}) = 1$, i.e., μ is actually a Borel measure on $\mathbb{R}$. Now $\hat{\mu}(-t)$ is a continuous function of t and if $t > 0$,

$$\int_0^t \hat{\mu}(-s)ds = \int_{\mathbb{R}} h_t(x)d\mu(x) = \int_0^t F(s) \, ds.$$

By the Fundamental Theorem of Calculus, since both $\hat{\mu}(-t)$ and $F(t)$ are continuous, $\hat{\mu}(-t) = F(t)$ for $t > 0$. A similar calculation works if $t < 0$. $\quad\square$

Theorem 6.4.15. *For every $0 < p \leq 2$ there is a probability measure μ_p on $(\mathbb{R}, dx)$ such that*

$$\int_{-\infty}^{\infty} e^{itx} \, d\mu_p(x) = e^{-|t|^p}, \quad t \in \mathbb{R}.$$

Proof. It obviously suffices to show the existence of μ_p with

$$\int_{-\infty}^{\infty} e^{itx} \, d\mu_p(x) = e^{-c_p|t|^p}, \quad t \in \mathbb{R},$$

where c_p is some positive constant. For the case $p = 2$ this is achieved by using a Gaussian.

Now suppose $0 < p < 2$. Let f be a random variable on some probability space with probability distribution

$$d\mu_f = \frac{p}{2|x|^{p+1}} \left[\chi_{(-\infty,-1)}(x) + \chi_{(1,+\infty)}(x) \right] dx.$$

The characteristic function of f is the following:

$$
\begin{aligned}
\mathbb{E}\left(e^{itf}\right) &= \int_{-\infty}^{\infty} e^{itx} \, d\mu_f(x) \\
&= \frac{p}{2} \int_{-\infty}^{-1} \frac{e^{itx}}{(-x)^{p+1}} \, dx + \frac{p}{2} \int_1^{\infty} \frac{e^{itx}}{x^{p+1}} \, dx \\
&= p \int_1^{\infty} \frac{e^{itx} + e^{-itx}}{2} \frac{dx}{x^{p+1}} \\
&= p \int_1^{\infty} \frac{\cos(tx)}{x^{p+1}} \, dx.
\end{aligned}
$$

Then, if $t > 0$ the substitution $u = tx$ in the last integral yields

$$
\begin{aligned}
1 - \mathbb{E}\left(e^{itf}\right) &= p \int_1^{\infty} \frac{dx}{x^{p+1}} - p \int_1^{\infty} \frac{\cos(tx)}{x^{p+1}} \, dx \\
&= p \int_1^{\infty} \frac{1 - \cos(tx)}{x^{p+1}} \, dx \\
&= pt^p \int_t^{\infty} \frac{1 - \cos u}{u^{p+1}} \, du.
\end{aligned}
$$

Let

$$\omega_p(t) = p \int_t^{\infty} \frac{1 - \cos u}{u^{p+1}} \, du$$

and

$$c_p = \lim_{t \to 0^+} \omega_p(t) = p \int_0^{\infty} \frac{1 - \cos u}{u^{p+1}} \, du.$$

Note that $\int_0^{\infty} \frac{1-\cos u}{u^{p+1}} \, du$ is finite and positive for every $0 < p < 2$.

Since f is symmetric, its characteristic function is even and therefore the equality

$$\mathbb{E}\left(e^{itf}\right) = 1 - |t|^p \omega_p(t)$$

holds for all $t \in \mathbb{R}$.

Let $(f_j)_{j=1}^{\infty}$ be a sequence of independent random variables with the same distribution as f. Then, for every n the characteristic function of the random variable $\frac{f_1 + \cdots + f_n}{n^{1/p}}$ is

$$\mathbb{E}\left(e^{it\frac{f_1+\cdots+f_n}{n^{1/p}}}\right) = \prod_{i=1}^{n} \mathbb{E}\left(e^{it\frac{f_i}{n^{1/p}}}\right) = \left(\mathbb{E}\left(e^{it\frac{f}{n^{1/p}}}\right)\right)^n = \left(1 - \frac{|t|^p}{n}\, \omega_p\left(\frac{|t|}{n^{1/p}}\right)\right)^n.$$

Since

$$\lim_{n\to\infty} \left(1 - \frac{|t|^p}{n}\, \omega_p\left(\frac{|t|}{n^{1/p}}\right)\right)^n = e^{-c_p|t|^p},$$

we can apply the preceding lemma to obtain the required measure μ_p.

$\square$

Definition 6.4.16. A random variable f on a probability space is called *p-stable* $(0 < p < 2)$ if

$$\hat{\mu}_f(-t) = e^{-c|t|^p}, \quad t \in \mathbb{R}$$

for some positive constant $c = c(p)$. f is called *normalized p-stable* if $c = 1$.

Note that the normalization for Gaussians is somewhat different, i.e., the characteristic function of a normalized Gaussian would correspond to the case $c = 1/2$ in the previous definition.

Theorem 6.4.17. *Let f be a p-stable random variable on a probability measure space (Ω, Σ, μ) for some $0 < p < 2$. Then*

(i) $f \in L_q(\mu)$ for all $0 < q < p$;
(ii) $f \notin L_p(\mu)$.

Proof. Suppose that f is normalized p-stable for some $0 < p < 2$ with distribution of probability μ_p. Then

$$\int_{\Omega} |f(\omega)|^q \, d\omega = \int_{-\infty}^{\infty} |x|^q \, d\mu_p(x).$$

For every $x \in \mathbb{R}$ the substitution $u = |x|t$ in the integral $\int_0^\infty \frac{1-\cos tx}{t^{1+q}}\, dt$ yields

$$\int_0^\infty \frac{1 - \cos tx}{t^{1+q}}\, dt = |x|^q \alpha_q,$$

where $\alpha_q = \int_0^\infty \frac{1-\cos u}{u^{1+q}}\, du$ is a positive constant for $0 < q < 2$. Hence,

$$\int_{-\infty}^{\infty} |x|^q \, d\mu_p(x) = \alpha_q^{-1} \int_{-\infty}^{\infty} \left(\int_0^\infty \frac{1 - \cos tx}{t^{1+q}}\, dt\right) d\mu_p(x)$$

$$= \alpha_q^{-1} \int_0^\infty \frac{1}{t^{q+1}} \left(\int_{-\infty}^{\infty} (1 - \cos tx)\, d\mu_p(x)\right) dt$$

$$= \alpha_q^{-1} \int_0^\infty \frac{1}{t^{q+1}} \left(\int_{-\infty}^{\infty} (1 - \Re e^{ixt})\, d\mu_p(x)\right) dt$$

$$= \alpha_q^{-1} \int_0^\infty \frac{1}{t^{q+1}} (1 - e^{-t^p})\, dt.$$

The last integral is finite for $0 < q < p$ and fails to converge when $q = p$.

$\square$

Theorem 6.4.18. *If $1 \leq p < 2$ and $p \leq q \leq 2$, then ℓ_q embeds isometrically in L_p.*

Proof. We have already seen the cases when $q = p$ and $q = 2$. For $1 \leq p < q < 2$, let $(f_j)_{j=1}^\infty$ be a sequence of independent normalized q-stable random variables on $[0,1]$. Then we can repeat the argument we used in Proposition 6.4.13 to prove that $[f_j]$ is isometric to ℓ_q in L_p. The only constraint is that the sequence (f_j) must belong to L_p, which requires that $p < q$.

$\square$

We can summarize our discussion by stating:

Theorem 6.4.19 (ℓ_q-subspaces of L_p).
(i) For $1 \leq p \leq 2$, ℓ_q embeds in L_p if and only if $p \leq q \leq 2$;
(ii) For $2 < p < \infty$, ℓ_q embeds in L_p if and only if $q = 2$ or $q = p$.
Moreover, if ℓ_q embeds in L_p then it embeds isometrically.

Remark 6.4.20. The alert reader will wonder for which values of q, the function space L_q can be embedded in L_p. In fact, the answer is exactly the same as for the sequence space ℓ_q, but we will postpone the proof of this until Chapter 11. A direct proof of this facts can be based on a discussion of stochastic integrals (see [106]).

Theorem 6.4.21. *Let $1 < p, q < \infty$. Then ℓ_q embeds complementably in L_p if and only $q = p$ or $q = 2$.*

Proof. We know (Proposition 6.4.2 and Proposition 6.4.1) that both $q = p$ and $q = 2$ allow complemented embeddings. Suppose ℓ_q embeds in L_p complementably and $q \notin \{2, p\}$. By Theorem 6.4.19 we must have $p < q < 2$. Taking duals it follows that $\ell_{q'}$ embeds complementably in $L_{p'}$, where q', p' are the conjugate indices of q and p. This is impossible.

$\square$

The L_p-spaces ($1 \leq p < \infty$) are primary. Alspach, Enflo, and Odell [3] proved the result for $1 < p < \infty$ in 1977. The case $p = 1$ was established in 1979 by Enflo and Starbird [55] as we already mentioned in Chapter 5.

The problem of classifying the complemented subspaces of L_p when $1 < p < \infty$ received a great deal of attention during the 1970s. At this stage we know of three isomorphism classes that we can find as complemented subspaces inside any L_p: ℓ_2, ℓ_p, and L_p, and it is easily seen that we can add $\ell_p \oplus \ell_2$ and $\ell_p(\ell_2)$ to that list. In fact, it turns out that L_p has a very rich class of complemented subspaces and the classification of them seems beyond reach. In 1981, Bourgain, Rosenthal, and Schechtman [17] showed the existence of uncountably many mutually nonisomorphic complemented subspaces of L_p; curiously it seems unknown (unless we assume the Continuum Hypothesis) whether there is a continuum of such spaces!

Problems

6.1. This exercise can be considered as a continuation of Problem 5.6.

(a) A closed subspace X of $L_p(\mathbb{T})$ is called *translation-invariant* if $f \in X$ implies $\tau_\phi(f) \in X$, where $\tau_\phi(f) = f(\theta - \phi)$. Show that if X is translation-invariant and $E = \{n \in \mathbb{Z} : e^{in\theta} \in X\}$, then X is the closed linear span of $\{e^{in\theta} : n \in E\}$. In this case we put $X = L_{p,E}(\mathbb{T})$.

(b) E is called a $\Lambda(p)$-set if $L_{p,E}(\mathbb{T})$ is strongly embedded in $L_p(\mathbb{T})$. Show that if E is a $\Lambda(p)$-set then it is a $\Lambda(q)$-set for $q < p$.

(c) Show that if $p > 2$, E is a $\Lambda(p)$-set if and only if $\{e^{in\theta} : n \in E\}$ is an unconditional basis of $L_{p,E}(\mathbb{T})$.

(d) Prove that $E = \{4^n : n \in \mathbb{N}\}$ is a $\Lambda(4)$-set. [*Hint*: Expand $|\sum_{n \in E} a_n e^{in\theta}|^4$.]

(e) E is called a *Sidon set* if for any $(a_n)_{n \in E} \in \ell_\infty(E)$ there exists $\mu \in M(\mathbb{T})$ with $\hat{\mu}(n) = a_n$. Show that the following are equivalent:

 (i) E is a Sidon set;
 (ii) $(e^{in\theta})_{n \in E}$ is an unconditional basic sequence in $C(\mathbb{T})$;
 (iii) $(e^{in\theta})_{n \in E}$ is a basic sequence equivalent to the canonical ℓ_1-basis in $C(\mathbb{T})$.

(f) Show that a Sidon set is a $\Lambda(p)$-set for every $1 \le p < \infty$.

(g) Show that $E = \{4^n : n \in \mathbb{N}\}$ is a Sidon set. [*Hint*: For $-1 \le a_n \le 1$, consider the functions $f_n(\theta) = \prod_{k=1}^n (1 + a_k \cos 4^k \theta)$, and let μ be a weak* cluster point of the measures $f_n \frac{d\theta}{2\pi}$.]

6.2. In this problem we aim to obtain **Khintchine's inequality** directly, not as a consequence of Kahane's inequality.

(a) Prove that $\cosh t \le e^{t^2/2}$ for all $t \in \mathbb{R}$.

(b) Show that if $p \ge 1$ then $t^p \le p^p e^{-p} e^t$.

(c) Let $(\epsilon_n)_{n=1}^\infty$ be a sequence of Rademachers and suppose $f = \sum_{k=1}^n a_k \epsilon_k$ where $\sum_{k=1}^n a_k^2 = 1$. Show that

$$\mathbb{E}(e^f) \le e$$

and deduce that

$$\mathbb{E}(e^{|f|}) \le 2e.$$

Hence show that

$$(\mathbb{E}(|f|^p))^{1/p} \le 2^{1/p} e^{1/p} \frac{p}{e}.$$

Finally obtain Khintchine's inequality for $p > 2$.

(d) Show by using Hölder's inequality that (c) implies Khintchine's inequality for $p < 2$.

6.3. The classical proof of Khintchine's inequality.

(a) Let $(\Omega, \mathbb{P})$ be a probability space and (ε_k) be a Rademacher sequence on it. Suppose $\sum_{k=1}^{n} a_k^2 = 1$. If $p = 2m$ is an even integer expand $\mathbb{E}(\sum_{k=1}^{n} a_k \varepsilon_k)^{2m}$ using the multinomial theorem and compare with $(\sum_{k=1}^{n} a_k^2)^m$.

(b) Deduce that

$$\mathbb{E}\left(\sum_{k=1}^{n} a_k \varepsilon_k\right)^{2m} \le \frac{(2m)!}{2^m m!}.$$

(c) Obtain Khintchine's inequality.

6.4. Let $(\Omega, \mathbb{P})$ be a probability space and (ε_k) be a Rademacher sequence on it. Consider a finite series $f = \sum_{k=1}^{N} a_k \varepsilon_k$ and let

$$M(t) = \max_{1 \le n \le N} \left| \sum_{k=1}^{n} a_k \varepsilon_k(t) \right|.$$

(a) Show that $\mathbb{P}(M > \lambda) \le 2\mathbb{P}(|f| > \lambda)$.

(b) Deduce that $\mathbb{E}(M^2) \le 2 \sum_{k=1}^{N} a_k^2$.

6.5. Suppose $\sum_{k=1}^{\infty} a_k^2 < \infty$. Let

$$M_m(t) = \sup_{n > m} \left| \sum_{j=m+1}^{n} a_k \varepsilon_k(t) \right|.$$

Show that $M_m(t) < \infty$ almost everywhere and $\lim_{m \to \infty} \mathbb{E}(M_m^2) = 0$. Deduce that $\sum_{k=1}^{\infty} a_k \varepsilon_k$ converges a.e.

6.6. Suppose the series $\sum_{k=1}^{\infty} a_k \varepsilon_k$ converges on a set of positive measure.

(a) Argue that there is a measurable set E with $\mathbb{P}(E) > 0$ and a constant C so that

$$\left| \sum_{j=m+1}^{n} a_j \varepsilon_k(\omega) \right| \le C, \qquad \omega \in E, \ 1 \le m < n < \infty.$$

(b) Let $b_{jk} = \mathbb{E}(\chi_E \varepsilon_j \varepsilon_k)$ for $j < k$. Show that

$$\sum_{j<k} b_{jk}^2 \le \mathbb{P}(E).$$

(c) Deduce the existence of m so that

$$\sum_{m \le j < k} b_{jk}^2 \le \frac{1}{100} (\mathbb{P}(E))^2.$$

(d) Deduce that $\sum_{k=1}^{\infty} a_k^2 < \infty$. [Hint: Estimate $\mathbb{E}|\chi_E \sum_{j=m+1}^{n} a_j \varepsilon_j|^2$.]

6.7. A Banach space X has the *Orlicz property* if whenever a series $\sum_{n=1}^{\infty} x_n$ is unconditionally convergent in X implies $\sum_{n=1}^{\infty} \|x_n\|^2 < \infty$.

(a) **Orlicz's Theorem.** Prove that the spaces L_p for $1 \le p \le 2$ have the Orlicz property.

(b) Show that for $2 \le p < \infty$, if $\sum_{n=1}^{\infty} f_n$ is unconditionally convergent in L_p then $\sum_{n=1}^{\infty} \|f_n\|^p < \infty$.

(c) Prove that if a series $\sum_{n=1}^{\infty} f_n$ is unconditionally convergent in L_p for $1 \le p < \infty$, then $f_n \to 0$ almost everywhere.

6.8. Prove that for $1 < p < \infty$, ℓ_2 embeds isometrically and complementably in L_p.

6.9. Show that a quotient of a space with type p also has type p. Is the same statement valid for cotype?

6.10. Show that every operator from $\mathcal{C}(K)$ into ℓ_p, $1 \le p < 2$, is compact.

7

Factorization Theory

This chapter is devoted to some important results on factorization of operators. Suppose X, Y are Banach spaces and that $T : X \to Y$ is a continuous operator. T *factorizes* through a Banach space E if there are continuous operators $R : X \to E$ and $S : E \to Y$ so that $T = SR$. Pictorially we have:

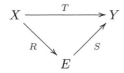

To illustrate the importance of such theorems, consider the case when we can factor the identity operator $I_X : X \to X$ through E. Then X is isomorphic to a complemented subspace of E. Another classical example: if an operator $T : X \to Y$ between Banach spaces factors through a reflexive Banach space then T is weakly compact (actually, this property characterizes weakly compact operators as Davis, Figiel, Johnson, and Pełczyński proved in [35]).

Most of the results of this chapter were obtained during the period 1970-74 by Maurey, Rosenthal, and Nikishin. Factorization builds on the theory of type and cotype as we will see. In fact, some of the work which preceded the results of this chapter and provided much of the impetus for factorization theory will only be developed in the following chapter. This particularly includes the fundamental work of Grothendieck [75] and Lindenstrauss and Pełczyński [131].

7.1 Maurey-Nikishin factorization theorems

In this section we shall discuss factorization theory of operators with values in the L_p-spaces. Here factorization is related to the notion of change of density.

The first factorization result of this type, essentially discovered by Nikishin [157], establishes a criterion for an operator with values in an $L_p(\mu)$-space to factor through $L_q(\nu)$ ($q > p$), where ν is obtained from μ after a suitable change of density. Nikishin's motivation came from harmonic analysis rather than Banach space theory, where versions of this result for translation-invariant operators had been known for some time (e.g., in the work of Stein [210]). However, it was the work of Maurey [144] that combined the ideas of Nikishin with the newly evolving theory of Rademacher type to create a very powerful tool.

The proof given below is based on one presented in [221] but is similar to the proof given by Maurey.

Definition 7.1.1. If (Ω, Σ, μ) is a σ-finite measure space then a *density function* h on Ω is a measurable function such that $h \geq 0$ a.e. and $\int h \, d\mu = 1$.

Theorem 7.1.2. *Let μ be a σ-finite measure on some measurable space (Ω, Σ). Suppose that T is an operator from a Banach space X into $L_p(\mu)$ and that $1 \leq p < q < \infty$. Suppose $0 < C < \infty$. Then the following conditions are equivalent:*

(a) There exists a density function h on Ω such that

$$\left(\int_{\{h>0\}} |Tx|^q h^{1-q/p} \, d\mu \right)^{1/q} \leq C \, \|x\|, \qquad x \in X, \qquad (7.1)$$

and

$$\mu\{\omega : |Tx(\omega)| > 0, \ h(\omega) = 0\} = 0, \qquad x \in X. \qquad (7.2)$$

(b) For every finite sequence $(x_k)_{k=1}^n$ in X,

$$\left\| \left(\sum_{k=1}^n |Tx_k|^q \right)^{1/q} \right\|_p \leq C \left(\sum_{k=1}^n \|x_k\|^q \right)^{1/q}. \qquad (7.3)$$

Interpretation. Condition (a) is to be interpreted in the sense that each function Tx is essentially supported on $A = \{\omega \in \Omega : h(\omega) > 0\}$. Thus the operator $Sx := h^{-1/p}Tx$ maps into $L_p(\Omega, h \, d\mu)$. However, (a) asserts that S actually maps boundedly into the *smaller* space $L_q(\Omega, h \, d\mu)$. This is the diagram depicting the situation:

$$
\begin{array}{ccc}
X & \xrightarrow{\ \ T\ \ } & L_p(\mu) \\
\big\downarrow & & \big\uparrow{\scriptstyle j} \\
L_q(hd\mu) & \lhook\joinrel\longrightarrow & L_p(hd\mu)
\end{array}
$$

Here j is an isometric embedding of $L_p(h \, d\mu)$ onto the subspace $L_p(A, \mu)$ of $L_p(\mu)$, defined by $j(f) = fh^{1/p}$.

Of course, at a very small cost we could insist that h is a strictly positive density (i.e., $h > 0$ a.e.) and drop equation (7.2): simply replace h by $(1 + \epsilon v)^{-1}(h + \epsilon v)$ where $\epsilon > 0$ and v is any strictly positive density. Then j becomes a genuine isometric isomorphism. In this case, however, the norm of $S = h^{-1/p}T$ is a little greater than C. Since the precise value of $\|S\|$ is rarely of interest we will often use the theorem is this form. In fact, in a formal sense we could replace (7.1) and (7.2) by

$$\left(\int_\Omega |Tx|^q h^{1-q/p} \, d\mu \right)^{1/q} \leq C \|x\|, \qquad x \in X,$$

with the implicit understanding that $Tx = 0$ a.e. on the set $\{\omega \in \Omega : h(\omega) = 0\}$ (i.e., where $h^{-q/p} = 0$). We will use this convention later.

Before continuing let us notice that, although we have stated this for general σ-finite measures, it is enough to prove the theorem under our usual convention that μ is a probability measure. If μ is not a probability measure we choose some strictly positive density v and set $d\mu' = v \, d\mu$; then we define $T' : X \to L_p(\mu')$ by $T'x = v^{-1}Tx$. A quick inspection will show the reader that the statement of the theorem for T' implies exactly the same statements for T. Thus we can and do resume our convention that μ is a probability measure.

Proof. $(a) \Rightarrow (b)$ Since $(\Omega, h \, d\mu)$ is a probability measure space and $p < q$, the $L_p(hd\mu)$-norm is smaller than the $L_q(hd\mu)$-norm and thus we have

$$\left(\int_\Omega \left(\sum_{k=1}^n |Tx_k|^q \right)^{p/q} d\mu \right)^{1/p} = \left(\int_{\{h>0\}} \left(\sum_{k=1}^n |Tx_k|^q \, h^{-q/p} \right)^{p/q} h \, d\mu \right)^{1/p}$$

$$\leq \left(\int_{\{h>0\}} \sum_{k=1}^n |Tx_k|^q h^{-\frac{q}{p}} h \, d\mu \right)^{1/q}$$

$$= \left(\sum_{k=1}^n \int_{\{h>0\}} |Tx_k|^q h^{-\frac{q}{p}} h d\mu \right)^{1/q}$$

$$\leq C \left(\sum_{k=1}^n \|x_k\|^q \right)^{1/q}.$$

$(b) \Rightarrow (a)$ Let us assume that C is the best constant so that (7.3) holds. Then

$$\sup \left\{ \left\| \left(\sum_{k=1}^n |Tx_k|^q \right)^{1/q} \right\|_p : (x_k)_{k=1}^n \subset X, \ \sum_{k=1}^n \|x_k\|^q \leq C^{-q}, \ n \in \mathbb{N} \right\} = 1.$$

Let W_0 be the set of all nonnegative functions in L_1 that are bounded above by functions of the form $(\sum_{k=1}^n |Tx_k|^q)^{p/q}$, where $n \in \mathbb{N}$ and $(x_k)_{k=1}^n \subset X$ with $\sum_{k=1}^n \|x_k\|^q \leq C^{-q}$, i.e.,

$$0 \le f \le \Big(\sum_{k=1}^{n} |Tx_k|^q \Big)^{p/q};$$

let W be the norm closure of W_0.

W_0 and W have the following property:

(*) Let $r = q/p > 1$. Given $f_1, \ldots, f_n \in W_0$ [respectively, W] and $c_1, \ldots, c_n \ge 0$ with $c_1 + \cdots + c_n \le 1$ then $(c_1 f_1^r + \cdots + c_n f_n^r)^{1/r} \in W_0$ [respectively, W].

To prove (*) it suffices to consider the case of W_0. Suppose

$$0 \le f_k \le \Big(\sum_{j=1}^{m_k} |Tx_{jk}|^q \Big)^{p/q} , \quad 1 \le k \le n,$$

where $\sum_{j=1}^{m_k} \|x_{jk}\|^q \le C^{-q}$ for $1 \le k \le n$. Then we also have

$$0 \le \Big(\sum_{k=1}^{n} c_k f_k^r \Big)^{1/r} \le \Big(\sum_{k=1}^{n} \sum_{j=1}^{m_k} |T(c_k^{\frac{1}{q}} x_{jk})|^q \Big)^{p/q} ,$$

with

$$\sum_{k=1}^{n} c_k \sum_{j=1}^{m_k} \|x_{jk}\|^q \le C^{-q},$$

and this establishes (*).

Property (*) immediately yields that W_0 (and hence its norm-closure W) is convex. Indeed, if $f_1, \ldots, f_n \in W_0$ and $c_1, \ldots, c_n \ge 0$ with $c_1 + \cdots + c_n = 1$ then by (*) we obtain

$$\sum_{j=1}^{n} c_j f_j \le \Big(\sum_{j=1}^{n} c_j f_j^r \Big)^{1/r} \in W_0.$$

Using Mazur's theorem, W is therefore weakly closed. Note that from the choice of C, we have

$$\sup_{f \in W_0} \int f \, d\mu = \sup_{f \in W} \int f \, d\mu = 1,$$

so in particular W is bounded. We next show that W is weakly compact. This requires to show that it is equi-integrable.

Suppose W is not equi-integrable. Then there is some $\delta > 0$, a sequence $(f_n)_{n=1}^{\infty}$ in W, and a sequence of disjoint measurable sets $(E_n)_{n=1}^{\infty}$ such that

$$\int_{E_n} f_n \, d\mu > \delta > 0, \quad n \in \mathbb{N}.$$

Thus for any N we have

$$\delta N^{1-\frac{1}{r}} \leq N^{-\frac{1}{r}} \int \max(f_1, f_2, \ldots, f_N) d\mu$$

$$\leq \int \left(\frac{1}{N} \sum_{j=1}^{N} f_j^r \right)^{1/r} d\mu$$

$$\leq 1,$$

by using (*). This is a contradiction for large enough N.

Hence W is weakly compact and, since integration is a weakly continuous functional on $L_1(\mu)$, it follows that there exists $h \in W$ with

$$\int h \, d\mu = 1. \tag{7.4}$$

Now suppose $f \in W$. On the one hand, for any $\tau > 0$ we have

$$(1+\tau)^{-\frac{1}{r}} (h^r + \tau f^r)^{\frac{1}{r}} \in W,$$

therefore, by property (*),

$$\int (h^r + \tau f^r)^{\frac{1}{r}} d\mu \leq (1+\tau)^{\frac{1}{r}}. \tag{7.5}$$

On the other hand,

$$\int (h^r + \tau f^r)^{\frac{1}{r}} d\mu \geq 1 + \tau^{\frac{1}{r}} \int_{h=0} f \, d\mu. \tag{7.6}$$

Since $1/r < 1$, combining (7.5) and (7.6) yields

$$\int_{\{h=0\}} f \, d\mu = 0. \tag{7.7}$$

From (7.6) and (7.4) we have

$$\int_{\{h>0\}} h \frac{(1+\tau f^r h^{-r})^{\frac{1}{r}} - 1}{\tau} d\mu \leq \frac{(1+\tau)^{\frac{1}{r}} - 1}{\tau}, \qquad \tau > 0.$$

Letting $\tau \to 0$ and using Fatou's lemma we obtain

$$\int_{\{h>0\}} f^r h^{1-r} d\mu \leq 1, \qquad f \in W. \tag{7.8}$$

In particular (7.7) and (7.8) hold for $f = C^{-p} \|x\|^{-p} |Tx|^p$ when $0 \neq x \in X$. This immediately gives (7.2) and (7.1).

$\square$

Theorem 7.1.3. *Let* $1 \le p < \infty$. *Suppose that* T *is an operator from a Banach space* X *into* $L_p(\mu)$. *If* X *has type 2 then there exists a constant* $C = C(p)$ *such that for every finite sequence* $(x_k)_{k=1}^n$ *in* X *we have*

$$\left\|\left(\sum_{k=1}^n |Tx_k|^2\right)^{1/2}\right\|_p \le C\left(\sum_{k=1}^n \|x_k\|^2\right)^{1/2}.$$

Proof. By Theorem 6.2.13, for every $1 \le p < \infty$ there is a constant $c = c(p)$ such that for any finite set of vectors $(x_k)_{k=1}^n$ in X,

$$\left\|\left(\sum_{k=1}^n |Tx_k|^2\right)^{\frac{1}{2}}\right\|_p \le c\,\mathbb{E}\left\|\sum_{k=1}^n \varepsilon_k Tx_k\right\|_p \le c\,\|T\|\,\mathbb{E}\left\|\sum_{k=1}^n \varepsilon_k x_k\right\|.$$

Using Kahane's inequality and the type 2 of X,

$$\mathbb{E}\left\|\sum_{k=1}^n \varepsilon_k x_k\right\| \le \left(\mathbb{E}\left\|\sum_{k=1}^n \varepsilon_k x_k\right\|^2\right)^{1/2} \le T_2(X)\left(\sum_{k=1}^n \|x_k\|^2\right)^{\frac{1}{2}}.$$

$\square$

Since $L_r(\mu)$ for $r \ge 2$ are type-2 spaces, we immediately obtain:

Corollary 7.1.4.

(a) *Every operator from a subspace of* $L_r(\mu)$ $(2 \le r < \infty)$ *into* $L_p(\mu)$ $(1 \le p < 2)$ *factors through a Hilbert space.*

(b) *If a Banach space* X *is isomorphic to a closed subspace of both* $L_p(\mu)$ *for some* $1 \le p < 2$ *and* $L_r(\mu)$ *for some* $2 < r < \infty$, *then* X *is isomorphic to a Hilbert space.*

Corollary 7.1.4 follows immediately from Theorems 7.1.2 and 7.1.3. Curiously, the isometric version of (b) does not hold. That is, if X is isometric to a subspace of L_p $(1 \le p < 2)$ and isometric to a subspace of L_r $(2 < r < \infty)$, it is not true that X must be isometric to a Hilbert space. Finite-dimensional counterexamples were given by Koldobsky [112]; however, the following problem is still open (see [114]):

Problem 7.1.5. *If an infinite-dimensional Banach space* X *is isometric to a closed subspace of both* L_p *for some* $1 \le p < 2$ *and* L_r *for some* $2 < r < \infty$, *must* X *be isometric to a Hilbert space?*

To push our results further we need a replacement for Theorem 6.2.13 for exponents other than 2. If $1 \le q < 2$ then it turns out that the q-stable random variables constructed in the previous chapter do very nicely. Indeed, we could have used Gaussians in place of Rademachers in the preceding argument.

Lemma 7.1.6. *Let* $1 \le p < q < 2$. *Suppose that* $\gamma = (\gamma_j)_{j=1}^\infty$ *is a sequence of independent normalized* q-stable random variables. Then for any finite sequence of functions* $(f_j)_{j=1}^n$ *in* $L_p(\mu)$,

$$\left\| \Big(\sum_{j=1}^{n} |f_j|^q \Big)^{1/q} \right\|_p = c \Big(\mathbb{E} \Big\| \sum_{j=1}^{n} \gamma_j f_j \Big\|_p^p \Big)^{1/p},$$

where $c = c(p,q) > 0$.

Proof. We recall from Theorem 6.4.18 that there is a constant $c = c(p,q)$ so that

$$\Big(\mathbb{E} \Big| \sum_{j=1}^{n} a_j \gamma_j \Big|^p \Big)^{1/p} = c^{-1} \Big(\sum_{j=1}^{n} |a_j|^q \Big)^{1/q}, \qquad (a_j)_{j=1}^{n} \subset \mathbb{R}.$$

Using Fubini's theorem,

$$\int \Big(\sum_{j=1}^{n} |f_j|^q \Big)^{\frac{p}{q}} d\mu = c^p \, \mathbb{E} \int \Big| \sum_{j=1}^{n} \gamma_j f_j \Big|^p d\mu,$$

and the lemma follows.

$\square$

Theorem 7.1.7. *Let $1 \le p < 2$. Suppose that T is an operator from a Banach space X into $L_p(\mu)$. If X has type r for some $p < r < 2$, then for each $q \in (p, r)$ there exists a constant C such that*

$$\left\| \Big(\sum_{j=1}^{n} |Tx_j|^q \Big)^{1/q} \right\|_p \le C \Big(\sum_{j=1}^{n} \|x_j\|^q \Big)^{1/q},$$

for every finite sequence $(x_j)_{j=1}^{n}$ in X.

Proof. In this proof we will require three mutually independent sequences of independent identically distributed random variables: a sequence $(\varepsilon_j)_{j=1}^{\infty}$ of Rademachers, a sequence $(\gamma_j)_{j=1}^{\infty}$ of normalized q-stable random variables and a sequence $(\eta_j)_{j=1}^{\infty}$ of normalized r-stable random variables.

Let $(x_j)_{j=1}^{n}$ be a finite sequence in X. By the previous lemma, for a certain constant $c = c(p,q)$ we have

$$\left\| \Big(\sum_{j=1}^{n} |Tx_j|^q \Big)^{1/q} \right\|_p = c \Big(\mathbb{E}_\gamma \Big\| \sum_{j=1}^{n} \gamma_j Tx_j \Big\|_p^p \Big)^{1/p} \le c \, \|T\| \, \Big(\mathbb{E}_\gamma \Big\| \sum_{j=1}^{n} \gamma_j x_j \Big\|^p \Big)^{1/p}.$$

Since the normalized q-stables are symmetric and X has type r,

$$\Big(\mathbb{E}_\gamma \Big\| \sum_{j=1}^{n} \gamma_j x_j \Big\|^p \Big)^{1/p} = \Big(\mathbb{E}_\gamma \mathbb{E}_\varepsilon \Big\| \sum_{j=1}^{n} \varepsilon_j \gamma_j x_j \Big\|^p \Big)^{1/p}$$

$$\le \Big(\mathbb{E}_\gamma \Big(\mathbb{E}_\varepsilon \Big\| \sum_{j=1}^{n} \varepsilon_j \gamma_j x_j \Big\|^r \Big)^{p/r} \Big)^{1/p}$$

$$\leq T_r(X) \left(\mathbb{E}_\gamma \big(\sum_{j=1}^n |\gamma_j|^r \, \|x_j\|^r \big)^{p/r} \right)^{1/p}.$$

Now notice that

$$\mathbb{E} \left| \sum_{j=1}^n a_j \eta_j \right|^p = c_1 \left(\sum_{j=1}^n |a_j|^r \right)^{p/r},$$

for a certain constant $0 < c_1 = \mathbb{E}|\eta_1|^p$ which is finite since $p < r$. Thus letting c_2, c_3 be positive constants depending only on p, q, and r,

$$
\begin{aligned}
\mathbb{E}_\gamma \Big(\sum_{j=1}^n |\gamma_j|^r \, \|x_j\|^r \Big)^{p/r} &= c_1^{-1} \mathbb{E}_\gamma \mathbb{E}_\eta \Big| \sum_{j=1}^n \eta_j \gamma_j \|x_j\| \Big|^p \\
&= c_1^{-1} \mathbb{E}_\eta \mathbb{E}_\gamma \Big| \sum_{j=1}^n \eta_j \gamma_j \|x_j\| \Big|^p \\
&= c_2 \mathbb{E}_\eta \Big(\sum_{j=1}^n |\eta_j|^q \|x_j\|^q \Big)^{p/q} \\
&\leq c_2 \Big(\mathbb{E}_\eta \sum_{j=1}^n |\eta_j|^q \|x_j\|^q \Big)^{p/q} \\
&= c_3 \Big(\sum_{j=1}^n \|x_j\|^q \Big)^{p/q}.
\end{aligned}
$$

$\square$

The next result now follows immediately from Theorem 7.1.2:

Theorem 7.1.8. *Let X be a Banach space of type $r > 1$. Suppose that $1 \leq p < r$ and that $T : X \to L_p(\mu)$ is an operator. Then T factors through $L_q(\mu)$ for any $p < q < r$. More precisely, for each $p < q < r$ there is a strictly positive density function h on Ω so that $Sx = h^{-1/p} T x$ defines a bounded operator from $L_p(\mu)$ into $L_q(\Omega, h\, d\mu)$.*

Note here that there is a fundamental difference between the case of type $r < 2$ and type 2. In the former we only obtain a factorization through $L_q(\mu)$ when $q < r$. Can we do better and take $q = r$? The answer is no and to see why we must consider subspaces of L_p for $1 \leq p < 2$. This will be the topic of the next section, but let us mention that an improvement is possible: A later theorem of Nikishin [158] implies that T actually factors through the space "weak L_r." See [186] and the Problems.

Remark 7.1.9. An examination of the proofs of the theorems of this section shows that the main theorem (Theorem 7.1.8) will also hold if $0 < p < 1$, when L_p is no longer a Banach space; in this case we can take $r = 1$ and every Banach space has type one! Thus we conclude that if a Banach space isomorphically embeds in some L_p where $0 < p < 1$ then it embeds in every L_q for $p \le q < 1$.

The following problem, originally raised by Kwapień in 1969, is open:

Problem 7.1.10. *If X is a Banach space which embeds in L_p for some $0 < p < 1$, does X embed in L_1?*

In the isometric setting the answer is negative: a Banach space which embeds isometrically in L_p for some $0 < p < 1$ need not embed isometrically in L_1 as Koldobsky proved in 1996 [113]; see also [105]. In the isomorphic case the only known result is that X embeds in L_1 if and only if $\ell_1(X)$ embeds in some L_p when $0 < p < 1$ [104].

7.2 Subspaces of L_p for $1 \le p < 2$

We start our discussion by showing, as promised, that Theorem 7.1.7 cannot be improved to allow factorization through L_r. We will need the following simple lemma:

Lemma 7.2.1. *Suppose $f, g \in L_p$ ($1 \le p < \infty$). Then if $0 < \theta < 1$ we have $|f|^{1-\theta}|g|^\theta \in L_p$ and*
$$\||f|^{1-\theta}|g|^\theta\|_p \le \|f\|_p^{1-\theta}\|g\|_p^\theta.$$

Proof. Just note that for $s, t \ge 0$ we have $s^{1-\theta}t^\theta \le (1-\theta)s + \theta t$. Then, assuming $\|f\|_p, \|g\|_p > 0$, by convexity we have
$$\left\|\left(\frac{|f|}{\|f\|_p}\right)^\theta \left(\frac{|g|}{\|g\|_p}\right)^{1-\theta}\right\|_p \le 1,$$

and the lemma follows.

$\square$

Theorem 7.2.2. *If $1 \le p < 2$, ℓ_p cannot be strongly embedded in L_p.*

Proof. Let us suppose $(f_n)_{n=1}^\infty$ is a normalized basic sequence in L_p equivalent to the ℓ_p-basis and such that $X = [f_n]$ is strongly embedded.

Let us fix $q < p$ (in the case $p = 1$ this implies $q < 1$). Then, using Theorem 6.2.13 and Proposition 6.4.5, we can find a constant $C > 0$ such that
$$C^{-1}n^{1/p} \le \left\|\left(\sum_{j \in \mathbb{A}} |f_j|^2\right)^{\frac{1}{2}}\right\|_q \le \left\|\left(\sum_{j \in \mathbb{A}} |f_j|^2\right)^{\frac{1}{2}}\right\|_p \le Cn^{1/p},$$

for and any n and each $\mathbb{A} \subset \mathbb{N}$ with $|\mathbb{A}| = n$.

Let $N \in \mathbb{N}$ and $a > 0$. Note that, since $\|f_j\|_p = 1$, estimating $\int |f_j|^p \, dt$ gives

$$\sum_{k=1}^{\infty} \lambda\Big(|f_j| > (ak)^{\frac{1}{p}}\Big) \le a^{-1}, \qquad 1 \le j \le N,$$

where λ denotes the Lebesgue measure on $[0,1]$. Thus

$$\sum_{k=1}^{\infty} \sum_{j=1}^{N} \lambda\Big(|f_j| > (ak)^{\frac{1}{p}}\Big) \le Na^{-1}.$$

It follows that there exists at least one $m \le N$ so that

$$\sum_{j=1}^{N} \lambda\Big(|f_j| > (am)^{\frac{1}{p}}\Big) \le a^{-1}m^{-1}N\Big(\sum_{k=1}^{N}\frac{1}{k}\Big)^{-1} \le \frac{N}{am\log N}.$$

By an averaging argument over all subsets of size m we can find a subset $\mathbb{A}$ of $\{1, 2, \ldots, N\}$ with $|\mathbb{A}| = m$ such that

$$\sum_{j \in \mathbb{A}} \lambda(|f_j| > (am)^{\frac{1}{p}}) \le \frac{1}{a\log N}.$$

Let $g = \max_{j \in \mathbb{A}} |f_j|$ and $E = \{t : g(t) > (am)^{\frac{1}{p}}\}$. Then

$$\|g\chi_E\|_q \le \lambda(E)^{\frac{1}{q}-\frac{1}{p}}\|g\|_p$$

by Hölder's inequality, and

$$\|g\|_p \le \Big\|\big(\sum_{j \in \mathbb{A}}|f_j|^2\big)^{\frac{1}{2}}\Big\|_p \le Cm^{\frac{1}{p}}.$$

Thus

$$\|g\chi_E\|_q \le Cm^{\frac{1}{p}}(a\log N)^{\frac{1}{p}-\frac{1}{q}}.$$

Hence

$$\|\max_{j \in \mathbb{A}}|f_j|\|_q \le (am)^{\frac{1}{p}} + Cm^{\frac{1}{p}}(a\log N)^{\frac{1}{p}-\frac{1}{q}}.$$

It follows that given any $\delta > 0$ we can pick a and N to ensure the existence of a subset $\mathbb{A}$ of $\mathbb{N}$ of cardinality m so that

$$\|\max_{j \in \mathbb{A}}|f_j|\|_q \le \delta m^{\frac{1}{p}}.$$

On the other hand

$$\Big\|\big(\sum_{j \in \mathbb{A}}|f_j|^p\big)^{1/p}\Big\|_q \le \Big\|\big(\sum_{j \in \mathbb{A}}|f_j|^p\big)^{1/p}\Big\|_p \le m^{\frac{1}{p}}.$$

Hence

$$C^{-1}m^{\frac{1}{p}} \le \left\| \left(\sum_{j \in \mathbb{A}} |f_j|^2 \right)^{1/2} \right\|_q$$

$$\le \left\| \left(\sum_{j \in \mathbb{A}} |f_j|^p \right)^{1/p} \right\|_q^{p/2} \left\| \max_{j \in \mathbb{A}} |f_j| \right\|_q^{1-p/2}$$

$$\le \delta^{1-\frac{p}{2}} m^{\frac{1}{p}}.$$

By choosing $\delta > 0$ appropriately we reach a contradiction.

□

Remark 7.2.3. Let us observe that now it is clear that we cannot take $q = r$ in Theorem 7.1.8. Indeed, if $r < 2$ then ℓ_r is of type r and does embed into L_p for $1 \le p \le r$ by Theorem 6.4.18. However, if such a factorization of the embedding $J : \ell_r \to L_p$ were possible, we would deduce that ℓ_r strongly embeds into $L_r([0,1], h\, dt)$ for some strictly positive density function h, which contradicts Theorem 7.2.2.

We are now going to delve a little further into the structure of subspaces of L_p for $1 \le p < 2$. We need some initial observations about type in general Banach spaces; we shall establish similar results for cotype for later use.

Let X be an infinite-dimensional Banach space, and $(\varepsilon_i)_{i=1}^\infty$ a sequence of Rademachers. For each $n \in \mathbb{N}$ define $\alpha_n(X)$ to be the least constant α so that

$$\left(\mathbb{E} \left\| \sum_{i=1}^n \varepsilon_i x_i \right\|^2 \right)^{1/2} \le \alpha \left(\sum_{i=1}^n \|x_i\|^2 \right)^{1/2}, \qquad \{x_i\}_{i=1}^n \subset X;$$

and define $\beta_n(X)$ to be the least constant β such that

$$\left(\sum_{i=1}^n \|x_i\|^2 \right)^{1/2} \le \beta \left(\mathbb{E} \left\| \sum_{i=1}^n \varepsilon_i x_i \right\|^2 \right)^{1/2}, \qquad \{x_i\}_{i=1}^n \subset X.$$

Note that $1 \le \alpha_n(X), \beta_n(X) \le n^{\frac{1}{2}}$ for $n = 1, 2, \ldots$.

Lemma 7.2.4. *Both the parameters $\alpha_n(X)$ and $\beta_n(X)$ are submultiplicative, i.e.,*

$$\alpha_{mn}(X) \le \alpha_m(X)\alpha_n(X), \qquad m, n \in \mathbb{N}, \tag{7.9}$$

and

$$\beta_{mn}(X) \le \beta_m(X)\beta_n(X), \qquad m, n \in \mathbb{N}. \tag{7.10}$$

Proof. Let us take $m \times n$ vectors in the unit ball of X and consider them as a matrix $(x_{ij})_{i,j=1}^{m,n}$. Let $(\varepsilon_{ij})_{i,j=1}^{m,n}$ be a Rademacher sequence, and $(\varepsilon_i')_{i=1}^n$ be another Rademacher sequence, independent of (ε_{ij}). The independence of the Rademacher sequence $(\varepsilon_i' \varepsilon_{ij})$ yields

$$\mathbb{E} \left\| \sum_{i=1}^m \sum_{j=1}^n \varepsilon_{ij} x_{ij} \right\|^2 = \mathbb{E} \left\| \sum_{i=1}^m \varepsilon_i' \sum_{j=1}^n \varepsilon_{ij} x_{ij} \right\|^2.$$

Then,

$$\Big(\mathbb{E}\Big\|\sum_{i=1}^{m}\varepsilon_i'\sum_{j=1}^{n}\varepsilon_{ij}x_{ij}\Big\|^2\Big)^{1/2} \leq \alpha_m(X)\Big(\mathbb{E}\sum_{i=1}^{m}\Big\|\sum_{j=1}^{n}\varepsilon_{ij}x_{ij}\Big\|^2\Big)^{1/2}$$

$$\leq \alpha_m(X)\alpha_n(X)\Big(\sum_{i=1}^{m}\sum_{j=1}^{n}\|x_{ij}\|^2\Big)^{1/2}.$$

Similarly,

$$\Big(\sum_{i=1}^{m}\sum_{j=1}^{m}\|x_{ij}\|^2\Big)^{1/2} \leq \beta_n(X)\Big(\sum_{i=1}^{m}\mathbb{E}\Big\|\sum_{j=1}^{n}\varepsilon_{ij}x_{ij}\Big\|^2\Big)^{1/2}$$

$$\leq \beta_m(X)\beta_n(X)\Big(\mathbb{E}\Big\|\sum_{i=1}^{m}\sum_{j=1}^{n}\varepsilon_{ij}x_{ij}\Big\|^2\Big)^{1/2}.$$

$\square$

Proposition 7.2.5. *Suppose $p < 2 < q$.*

(a) In order that X have type r for some $p < r$ it is necessary and sufficient that for some N, $\alpha_N(X) < N^{\frac{1}{p}-\frac{1}{2}}$.

(b) In order that X have cotype s for some $s < q$ it is necessary and sufficient that for some N, $\beta_N(X) < N^{\frac{1}{2}-\frac{1}{q}}$.

Proof. One easily checks that if X has type $r > p$ [respectively, cotype $s < q$] then $\alpha_N(X) < N^{\frac{1}{p}-\frac{1}{2}}$ [respectively, $\beta_N(X) < N^{\frac{1}{2}-\frac{1}{q}}$] for some N by taking arbitrary sequences of vectors $\{x_i\}_{i=1}^n$ in X all equal to some x with $\|x\| = 1$.

Let us now complete the proof of (a). Assume N is such that $\alpha_N(X) < N^{\frac{1}{p}-\frac{1}{2}}$. Then we can write $\alpha_N(X) = N^{\theta-\frac{1}{2}}$ for some $\frac{1}{2} < \theta < \frac{1}{p}$, and by (7.9),

$$\alpha_{N^k}(X) \leq N^{k(\theta-\frac{1}{2})}, \quad k \in \mathbb{N}.$$

Given any n, if we take $k \in \mathbb{N}$ such that $N^{k-1} \leq n \leq N^k$,

$$\alpha_n(X) \leq \alpha_{N^k}(X) \leq N^{k(\theta-\frac{1}{2})} = (N^{k-1})^{\theta-\frac{1}{2}}N^{\theta-\frac{1}{2}},$$

and so we have an estimate of the form

$$\alpha_n(X) \leq Cn^{(\theta-\frac{1}{2})}, \tag{7.11}$$

for $C = N^{\theta-\frac{1}{2}}$.

Pick r such that $p < r < \frac{1}{\theta}$. Given any sequence $(x_i)_{i=1}^n$ of vectors in X, without loss of generality we will suppose that $\|x_1\| \geq \|x_2\| \geq \cdots \geq \|x_n\|$. For notational convenience let $x_i = 0$ for $i > n$. Then for $k \in \mathbb{N}$, using (7.11), we obtain

$$\left(\mathbb{E}\Big\|\sum_{i=2^{k-1}}^{2^k-1}\varepsilon_i x_i\Big\|^2\right)^{1/2} \le C2^{k(\theta-\frac{1}{2})}\left(\sum_{i=2^{k-1}}^{2^k-1}\|x_i\|^2\right)^{1/2}$$
$$\le C2^{k\theta}\|x_{2^{k-1}}\|$$
$$\le C2^{k\theta}2^{-(k-1)/r}\left(\sum_{i=1}^{\infty}\|x_i\|^r\right)^{1/r}.$$

Summing over k,

$$\left(\mathbb{E}\Big\|\sum_{i=1}^{\infty}\varepsilon_i x_i\Big\|^2\right)^{1/2} \le C2^{\frac{1}{r}}\sum_{k=1}^{\infty}2^{k(\theta-\frac{1}{r})}\left(\sum_{i=1}^{\infty}\|x_i\|^r\right)^{1/r}.$$

This implies, using the Kahane-Khintchine inequality (Theorem 6.2.5), that X has type r.

The proof of (b) is similar: Assume $\beta_N(X) < N^{\frac{1}{2}-\frac{1}{q}}$ for some N. Then in place of (7.11) we find $\theta > \frac{1}{q}$ so that, for some constant C, we have

$$\beta_n(X) \le Cn^{\frac{1}{2}-\theta}, \qquad n \in \mathbb{N}. \tag{7.12}$$

Pick s so that $\frac{1}{\theta} < s < q$. For (x_i) as in (a), for $k \in \mathbb{N}$ we have

$$\left(\sum_{i=2^{k-1}}^{2^k-1}\|x_i\|^2\right)^{1/2} \le C2^{k(\frac{1}{2}-\theta)}\left(\mathbb{E}\Big\|\sum_{i=2^{k-1}}^{2^k-1}\varepsilon_i x_i\Big\|^2\right)^{1/2}$$
$$\le C2^{k(\frac{1}{2}-\theta)}\left(\mathbb{E}\Big\|\sum_{i=1}^{\infty}\varepsilon_i x_i\Big\|^2\right)^{1/2}.$$

Now

$$\sum_{i=2^{k-1}}^{2^k-1}\|x_i\|^s \le \|x_{2^{k-1}}\|^{s-2}\sum_{i=2^{k-1}}^{2^k-1}\|x_i\|^2.$$

Thus

$$\sum_{i=1}^{\infty}\|x_i\|^s \le C^s\left(\sum_{k=1}^{\infty}2^{k(1-2\theta)}\|x_{2^{k-1}}\|^{s-2}\right)\mathbb{E}\Big\|\sum_{i=1}^{\infty}\varepsilon_i x_i\Big\|^2$$
$$\le C^s\left(\sum_{k=1}^{\infty}2^{k(1-2\theta)}2^{(1-k)(1-\frac{2}{s})}\right)\left(\sum_{i=1}^{\infty}\|x_i\|^s\right)^{1-\frac{2}{s}}\mathbb{E}\Big\|\sum_{i=1}^{\infty}\varepsilon_i x_i\Big\|^2.$$

Rearranging the last expression gives us an estimate

$$\left(\sum_{i=1}^{\infty}\|x_i\|^s\right)^{1/s} \le C'\left(\mathbb{E}\Big\|\sum_{i=1}^{\infty}\varepsilon_i x_i\Big\|^2\right)^{1/2},$$

for some constant C', and by Kahane's inequality we deduce that X has cotype s.

□

The following theorem was proved by Rosenthal in 1973 [196] using somewhat different techniques; it strongly influenced the development of factorization theory by Maurey.

Theorem 7.2.6. *Suppose X is a closed linear subspace of L_p ($1 \le p < 2$). Then the following conditions are equivalent:*

(i) X does not contain any subspace isomorphic to ℓ_p;
(ii) X does not contain any complemented subspace isomorphic to ℓ_p;
(iii) X has type r for some $r > p$;
(iv) The set $\{|f|^p : f \in B_X\} \subset L_1$ is equi-integrable;
(v) X is strongly embedded in L_p.

Moreover, if $p = 1$ these conditions are equivalent to:

(vi) X is reflexive.

Proof. Notice that in the case $p = 1$ we already have the equivalence of (i), (iv), and (vi) (see Theorem 5.2.9 and Proposition 5.6.2).

$(i) \Rightarrow (iv)$ We need only consider the case when $1 < p < 2$.

If $\{|f|^p : f \in B_X\}$ is not equi-integrable, we can find a sequence $(g_n)_{n=1}^\infty$ in B_X and a sequence of disjoint Borel sets $(A_n)_{n=1}^\infty$ so that $\|g_n \chi_{A_n}\|_p > 3\delta$ for some $\delta > 0$. Since L_p is reflexive, by passing to a subsequence we can assume that $(g_n)_{n=1}^\infty$ is weakly convergent to some $g \in L_p$ (Corollary 1.6.4). Then, by the disjointedness of the sets (A_n),

$$\sum_{n=1}^\infty \|g\chi_{A_n}\|_p^p < \infty.$$

Hence, by deleting finitely many terms, without loss of generality, we will assume that $\|g\chi_{A_n}\|_p < \delta$ for all n.

Let us consider the sequence of functions $(f_n)_{n=1}^\infty \subset B_X$ given by

$$f_n = \frac{1}{2}(g_n - g), \qquad n \in \mathbb{N}.$$

Then $\|f_n \chi_{A_n}\|_p > \delta$ for all n and $(f_n)_{n=1}^\infty$ is weakly null. We can argue that a further subsequence (which we still label $(f_n)_{n=1}^\infty$) is a basic sequence equivalent to a block basis of the Haar basis in L_p, and thus is unconditional. This uses the Bessaga-Pełczyński selection principle (Proposition 1.3.10) and the unconditionality of the Haar basis in L_p (Theorem 6.1.6). We will show that $(f_n)_{n=1}^\infty$ is equivalent to the canonical ℓ_p-basis.

For any sequence of scalars $(a_n)_{n=1}^\infty \in c_{00}$, by unconditionality there is a constant K such that

$$K^{-1}\mathbb{E}\left\|\sum_{j=1}^\infty \epsilon_j a_j f_j\right\|_p \le \left\|\sum_{j=1}^n a_j f_j\right\|_p \le K\mathbb{E}\left\|\sum_{j=1}^\infty \epsilon_j a_j f_j\right\|_p, \tag{7.13}$$

for any choice of signs (ϵ_j). Then, by the fact that L_p has type p, we obtain an upper estimate

$$\Big\| \sum_{j=1}^{\infty} a_j f_j \Big\|_p \leq C_p \Big(\sum_{j=1}^{\infty} |a_j|^p \Big)^{1/p}$$

for a suitable constant C_p.

To get a lower estimate, first we use equation (7.13) in combination with Theorem 6.2.13 and Kahane's inequality to obtain

$$\Big\| \sum_{j=1}^{n} a_j f_j \Big\|_p \geq K_p \Big\| \Big(\sum_{j=1}^{\infty} |a_j|^2 |f_j|^2 \Big)^{\frac{1}{2}} \Big\|_p,$$

for some constant K_p; and now we argue that

$$\Big\| \Big(\sum_{j=1}^{\infty} |a_j|^2 |f_j|^2 \Big)^{\frac{1}{2}} \Big\|_p \geq \| \max_j |a_j f_j| \|_p$$

$$\geq \| \max_j |a_j f_j| \chi_{A_j} \|_p$$

$$= \Big\| \sum_{j=1}^{\infty} |a_j f_j| \chi_{A_j} \Big\|_p$$

$$= \Big(\sum_{j=1}^{\infty} |a_j|^p \| f_j \chi_{A_j} \|_p^p \Big)^{1/p}$$

$$\geq \delta \Big(\sum_{j=1}^{\infty} |a_j|^p \Big)^{1/p}.$$

$(iv) \Rightarrow (iii)$ Since $\{ |f|^p : f \in B_X \}$ is equi-integrable, using Lemma 5.2.6 there is a function $\theta(M)$ with $\lim_{M \to \infty} \theta(M) = 0$ such that

$$\| f \chi_{(|f| > M)} \|_p \leq \theta(M), \qquad f \in B_X.$$

For each $N \in \mathbb{N}$ let $f_1, \ldots, f_N$ be any sequence of norm-one functions in X. Combining Theorem 6.2.13 and Kahane's inequality there is a constant C (depending only on p) so that

$$\Big(\mathbb{E} \Big\| \sum_{j=1}^{N} \varepsilon_j a_j f_j \Big\|_p^2 \Big)^{1/2} \leq C \Big\| \Big(\sum_{j=1}^{N} |a_j|^2 |f_j|^2 \Big)^{1/2} \Big\|_p,$$

for any sequence of scalars (a_j). Let us estimate the latter expression by splitting each f_j in the form $f_j = g_j + h_j$, where $|g_j| \leq M$ and $\|h_j\|_p \leq \theta(M)$:

$$\Big\| \Big(\sum_{j=1}^{N} |a_j|^2 |f_j|^2 \Big)^{\frac{1}{2}} \Big\|_p \leq \Big\| \Big(\sum_{j=1}^{N} |a_j|^2 |g_j|^2 \Big)^{\frac{1}{2}} \Big\|_p + \Big\| \Big(\sum_{j=1}^{N} |a_j|^2 |h_j|^2 \Big)^{\frac{1}{2}} \Big\|_p$$

$$\leq M\Big(\sum_{j=1}^{N}|a_j|^2\Big)^{1/2} + \Big\|\Big(\sum_{j=1}^{N}|a_j|^p|h_j|^p\Big)^{\frac{1}{p}}\Big\|_p$$

$$\leq M\Big(\sum_{j=1}^{N}|a_j|^2\Big)^{1/2} + \theta(M)\Big(\sum_{j=1}^{N}|a_j|^p\Big)^{1/p}$$

$$\leq \Big(M + \theta(M)N^{\frac{1}{p}-\frac{1}{2}}\Big)\Big(\sum_{j=1}^{N}|a_j|^2\Big)^{1/2}.$$

If we chose M so that $\theta(M) < (2C)^{-1}$ we see that for large enough N we have

$$\Big(\mathbb{E}\Big\|\sum_{j=1}^{N}\varepsilon_j a_j f_j\Big\|_p^2\Big)^{1/2} \leq \frac{1}{2}N^{\frac{1}{p}-\frac{1}{2}}\Big(\sum_{j=1}^{N}|a_j|^2\Big)^{1/2}.$$

Hence, for that N, whenever $(f_j)_{j=1}^{N} \subset X$ we have

$$\Big(\mathbb{E}\Big\|\sum_{j=1}^{N}\varepsilon_j f_j\Big\|_p^2\Big)^{1/2} \leq \frac{1}{2}N^{\frac{1}{p}-\frac{1}{2}}\Big(\sum_{j=1}^{N}\|f_j\|^2\Big)^{1/2},$$

and so X has type r for some $r > p$ (Proposition 7.2.5).

To prove $(iii) \Rightarrow (v)$ we use factorization theory. Consider the inclusion map $J : X \to L_p$. By Theorem 7.1.8, for $p < q < r$ we can find a strictly positive density function h so that $h^{-\frac{1}{p}}J$ maps X into $L_q([0,1], h\,dt)$. Since $h^{-\frac{1}{p}}J$ is also an isometry of X into $L_p([0,1], h\,dt)$ this implies that $h^{-\frac{1}{p}}J$ strongly embeds X into $L_p([0,1], h\,dt)$ by Proposition 6.4.5. But this means that convergence in measure is equivalent to norm convergence in X for the original Lebesgue measure as well.

The implication $(v) \Rightarrow (i)$ is simply Theorem 7.2.2; this completes the equivalence of (i), (iii), (iv), and (v).

Finally we note that $(i) \Rightarrow (ii)$ is trivial and that Theorem 6.4.7 shows that $(ii) \Rightarrow (v)$.

$\square$

7.3 Factoring through Hilbert spaces

In the first section of this chapter we saw that if X has type 2 and $1 \leq p < 2$ then any operator $T : X \to L_p$ factors through a Hilbert space. In this section we are giving a characterization for an operator between Banach spaces to factor through a Hilbert space.

Definition 7.3.1. Suppose that X and Y are Banach spaces. We say that an operator T from X to Y *factors through a Hilbert space* if there exist a Hilbert space H and operators $S : X \longrightarrow H$ and $R : H \longrightarrow Y$ verifying $T = RS$.

We will begin by making some remarks that will lead us to the necessary condition we are seeking. We will only consider real scalars, although at the appropriate moment we will discuss the alterations necessary to handle complex scalars. Throughout this section H will denote a generic Hilbert space with a scalar product $\langle \cdot \rangle$.

Suppose we have n arbitrary vectors $x_1, \ldots, x_n$ in H. Given a real orthogonal matrix $A = (a_{ij})_{1 \le i,j \le n}$, let us consider the new vectors in H defined from A,

$$z_i = \sum_{j=1}^n a_{ij} x_j, \quad 1 \le i \le n. \tag{7.14}$$

Then,

$$\sum_{i=1}^n \|z_i\|^2 = \sum_{i=1}^n \left\| \sum_{j=1}^n a_{ij} x_j \right\|^2$$
$$= \sum_{i=1}^n \langle \sum_{j=1}^n a_{ij} x_j, \sum_{k=1}^n a_{ik} x_k \rangle$$
$$= \sum_{i=1}^n \sum_{j=1}^n \sum_{k=1}^n a_{ij} a_{ik} \langle x_j, x_k \rangle$$
$$= \sum_{j=1}^n \langle x_j, x_j \rangle$$
$$= \sum_{j=1}^n \|x_j\|^2 .$$

Any real $n \times n$ matrix $A = (a_{ij})$ defines a linear operator (that will be denoted in the same way) $A : \ell_2^n \longrightarrow \ell_2^n$ via

$$A \begin{pmatrix} s_1 \\ s_2 \\ \vdots \\ s_n \end{pmatrix} = \begin{pmatrix} a_{11} & a_{12} & \ldots & a_{1n} \\ a_{21} & a_{22} & \ldots & a_{2n} \\ \vdots & \vdots & \ddots & \vdots \\ a_{n1} & a_{n2} & \ldots & a_{nn} \end{pmatrix} \begin{pmatrix} s_1 \\ s_2 \\ \vdots \\ s_n \end{pmatrix}.$$

The matrix $(a_{ij})_{1 \le i,j \le n}$ is orthogonal if and only if the operator A is an isometry. If $(a_{ij})_{i,j=1}^n$ is not orthogonal but $\|A\| \le 1$, it is an exercise of linear algebra to prove that (a_{ij}) can be written as a convex combination of orthogonal matrices. In fact, it is always possible to find orthonormal basis $(e_j)_{j=1}^n$ and $(f_j)_{j=1}^n$ in ℓ_2^n so that $Ae_j = \lambda_j f_j$ with $\lambda_j \ge 0$: Just find an orthonormal basis of eigenvectors $(e_j)_{j=1}^n$ for $A'A$ where A' is the transpose. Then $A = DU$ where $Df_j = \lambda_j f_j$ and $Ue_j = f_j$. U is orthogonal and since $0 \le \lambda_j \le 1$ we can write D as a convex combination of the orthogonal matrices $V_\epsilon f_j = \epsilon_j f_j$ where $\epsilon_j = \pm 1$.

Thus, if $x_1, \ldots, x_n, z_1, \ldots, z_n$ are arbitrary vectors in H satisfying equation (7.14), where $\|(a_{jk})_{j,k=1}^n\|_{\ell_2^n \to \ell_2^n} \leq 1$, we will have

$$\sum_{i=1}^n \|z_i\|^2 \leq \sum_{j=1}^n \|x_j\|^2.$$

This can easily be extended to the case of differing numbers of x_j's and z_i's by adding zeros to one of the two collections of vectors.

Theorem 7.3.2. *Let T be an operator from a Banach space X into a Banach space Y. Suppose that there exist operators $S : X \longrightarrow H$ and $R : H \longrightarrow Y$ verifying $T = RS$. If $(x_j)_{j=1}^m$ and $(z_i)_{i=1}^n$ are vectors in X related by the equation*

$$z_i = \sum_{j=1}^m a_{ij} x_j, \quad 1 \leq i \leq n, \tag{7.15}$$

where (a_{ij}) is a real $n \times m$ matrix such that $\|A\|_{\ell_2^m \to \ell_2^n} \leq 1$, then

$$\Big(\sum_{i=1}^n \|Tz_i\|^2\Big)^{1/2} \leq \|S\| \, \|R\| \, \Big(\sum_{j=1}^m \|x_j\|^2\Big)^{1/2}.$$

Proof. The proof easily follows from the comments we made. Indeed, given $x_1, \ldots, x_m$ and $z_1, \ldots, z_n$ in X satisfying (7.15), since the collections of vectors $(Sx_j)_{j=1}^m$ and $(Sz_i)_{i=1}^n$ lie inside H we have

$$\sum_{i=1}^n \|Tz_i\|^2 = \sum_{i=1}^n \|RSz_i\|^2$$

$$\leq \|R\|^2 \sum_{i=1}^n \|Sz_i\|^2$$

$$\leq \|R\|^2 \sum_{j=1}^m \|Sx_j\|^2$$

$$\leq \|R\|^2 \, \|S\|^2 \sum_{j=1}^m \|x_j\|^2.$$

$\square$

In light of the previous theorem we want to give an alternative formulation of the property that $(x_j)_{j=1}^m$ and $(z_i)_{i=1}^n$ are vectors in X related by the equation

$$z_i = \sum_{j=1}^m a_{ij} x_j, \quad 1 \leq i \leq n,$$

where $A = (a_{ij})$ is a real $n \times m$ matrix such that $\|A\|_{\ell_2^m \to \ell_2^n} \leq 1$.

Proposition 7.3.3. *Given $n, m \in \mathbb{N}$ and any two sets of vectors $(x_j)_{j=1}^m$ and $(z_i)_{i=1}^n$ in a Banach space X, the following are equivalent:*

(a) There is a real $n \times m$ matrix $A = (a_{ij})$ so that $\|A\|_{\ell_2^m \to \ell_2^n} \leq 1$ and

$$z_i = \sum_{j=1}^m a_{ij} x_j, \quad 1 \leq i \leq n;$$

(b) $\sum_{j=1}^m |x^*(z_j)|^2 \leq \sum_{i=1}^n |x^*(x_i)|^2$, for all $x^* \in X^*$.

Proof. Assume that (a) holds. Then, since $\|A\|_{\ell_2^m \to \ell_2^n} \leq 1$, it follows that

$$\sum_{i=1}^n |x^*(z_i)|^2 = \sum_{i=1}^n \left| x^*\Big(\sum_{j=1}^m a_{ij} x_j\Big) \right|^2 = \sum_{i=1}^n \left| \sum_{j=1}^m a_{ij} x^*(x_j) \right|^2 \leq \sum_{j=1}^m |x^*(x_j)|^2.$$

For the reverse implication, $(b) \Rightarrow (a)$, consider the linear operators

$$\alpha : X^* \longrightarrow \ell_2^m, \quad x^* \mapsto (x^*(x_j))_{j=1}^m$$

and

$$\beta : X^* \longrightarrow \ell_2^n, \quad x^* \mapsto (x^*(z_i))_{i=1}^n.$$

The hypothesis says that $\|\beta x^*\|_{\ell_2^n} \leq \|\alpha x^*\|_{\ell_2^m}$ for all $x^* \in X^*$. Thus we can define an operator $A_0 : \alpha(X^*) \to \beta(X^*)$ with $\|A_0\| \leq 1$ and $\beta = A_0 \circ \alpha$. Then A_0 can be extended to an operator $A : \ell_2^m \to \ell_2^n$ with $\|A\| \leq 1$. Let (a_{ij}) be the matrix associated with A.

$$x^*(z_i) = \sum_{j=1}^m a_{ij} x^*(x_j) \quad \text{for all } x^* \in X^*,$$

which implies

$$z_i = \sum_{j=1}^m a_{ij} x_j, \quad i = 1, \dots, n.$$

$\square$

The main result of this section is the following criterion:

Theorem 7.3.4. *Let X and Y be Banach spaces. Suppose E is a closed linear subspace of X and $T : E \to Y$ is an operator. In order that there exist a Hilbert space H and operators $R : X \to H$, $S : H \to Y$ with $\|R\|\|S\| \leq C$ such that $T = RS|_E$ it is necessary and sufficient that for all sets of vectors $(x_j)_{j=1}^m \subset X$ and $(z_i)_{i=1}^n \subset E$ such that*

$$\sum_{i=1}^n |x^*(z_i)|^2 \leq \sum_{j=1}^m |x^*(x_j)|^2, \quad x^* \in X^*,$$

we have

$$\Big(\sum_{i=1}^{n} \|Tz_i\|^2 \Big)^{1/2} \leq C \Big(\sum_{j=1}^{m} \|x_j\|^2 \Big)^{1/2}.$$

In the proof of this result and other ones in the next chapter we will make use of the following lemma. If $\mathcal{A}$ is a subset of real vector space we define

$$\text{cone}\,(\mathcal{A}) = \Big\{ \sum_{j=1}^{n} \alpha_j a_j \,:\, a_1, \ldots, a_n \in \mathcal{A},\ \alpha_1, \ldots, \alpha_n \geq 0,\ n = 1, 2, \ldots \Big\}.$$

Lemma 7.3.5. *Let $\mathcal{V}$ be a real vector space. Given $\mathcal{A}$, $\mathcal{B}$ two subsets of $\mathcal{V}$ such that $\mathcal{V} = \text{cone}\,(\mathcal{B}) - \text{cone}\,(\mathcal{A})$, and two functions $\phi : \mathcal{A} \to \mathbb{R}$, $\psi : \mathcal{B} \to \mathbb{R}$, the following are equivalent:*

(i) There is a linear functional $\mathcal{L}$ on $\mathcal{V}$ verifying

$$\phi(a) \leq \mathcal{L}(a), \qquad a \in \mathcal{A}$$

and

$$\psi(b) \geq \mathcal{L}(b), \qquad b \in \mathcal{B}.$$

(ii) If $(\alpha_i)_{i=1}^{m}$, $(\beta_j)_{j=1}^{n}$ are two finite sequences of nonnegative scalars such that

$$\sum_{i=1}^{m} \alpha_i a_i = \sum_{j=1}^{n} \beta_j b_j$$

for some $(a_i)_{i=1}^{m} \subset \mathcal{A}$, $(b_j)_{j=1}^{n} \subset \mathcal{B}$, then

$$\sum_{i=1}^{m} \alpha_i \phi(a_i) \leq \sum_{j=1}^{n} \beta_j \psi(b_j).$$

Proof. The implication $(i) \Rightarrow (ii)$ is immediate.

$(ii) \Rightarrow (i)$ Let us define the map $p : \mathcal{V} \to [-\infty, \infty)$ as follows:

$$p(v) = \inf \Big\{ \sum_{j=1}^{n} \beta_j \psi(b_j) - \sum_{i=1}^{m} \alpha_i \phi(a_i) \Big\},$$

the infimum being taken over all possible representations in the form $v = \sum_{j=1}^{n} \beta_j b_j - \sum_{i=1}^{m} \alpha_i a_i$, where $\alpha_1, \ldots, \alpha_m, \beta_1, \ldots, \beta_n \geq 0$, $a_1, \ldots, a_m \in \mathcal{A}$, and $b_1, \ldots, b_m \in \mathcal{B}$.

p is well-defined since $\mathcal{V} = \text{cone}\,(\mathcal{B}) - \text{cone}\,(\mathcal{A})$. Besides, one easily checks that p is positive-homogeneous and satisfies $p(v_1 + v_2) \leq p(v_1) + p(v_2)$ for any v_1, v_2 in $\mathcal{V}$. In order to prove that p is a sublinear functional we need to show that $p(v) > -\infty$ for every $v \in \mathcal{V}$. This will follow if $p(0) = 0$. Indeed, $p(v) + p(-v) \geq p(0)$, so neither $p(v)$ nor $p(-v)$ could be $-\infty$ if $p(0) = 0$.

For each representation of 0 in the form $0 = \sum_{j=1}^n \beta_j b_j - \sum_{i=1}^m \alpha_i a_i$, by the hypothesis it follows that $\sum_{j=1}^n \beta_j \psi(b_j) \geq \sum_{i=1}^m \alpha_i \phi(a_i)$. Therefore, by the definition, $p(0) \geq 0$ hence $p(0) = 0$.

As an consequence of the Hahn-Banach theorem, there is a linear functional $\mathcal{L}$ on $\mathcal{V}$ such that $\mathcal{L}(v) \leq p(v)$ for every $v \in \mathcal{V}$ and so $\phi(a) \leq \mathcal{L}(a)$ for all $a \in \mathcal{A}$ and $\mathcal{L}(b) \leq \psi(b)$ for all $b \in \mathcal{B}$.

$\square$

Proof of Theorem 7.3.4. We need only show that the condition is sufficient. Let $\mathcal{F}(X^*)$ denote the set of all functions from X^* to $\mathbb{R}$, and consider the natural map $X \to \mathcal{F}(X^*)$, $x \mapsto \hat{x}$, where

$$\hat{x}(x^*) = x^*(x), \qquad x^* \in X^*.$$

Let $\mathcal{V}$ be the linear subspace of $\mathcal{F}(X^*)$ of all finite linear combinations of functions of the form $\hat{x}\hat{z}$, with x, z in X. That is,

$$\mathcal{V} = \Big\{ \sum_{k=1}^N \lambda_k \hat{x}_k \hat{z}_k : (\lambda_k)_{k=1}^N \text{ in } \mathbb{R}, \ (x_k)_{k=1}^N \text{ and } (z_k)_{k=1}^N \text{ in } X, \text{ and } N \in \mathbb{N} \Big\}.$$

Clearly, the set $\{\hat{x}^2 : x \in X\}$ spans $\mathcal{V}$ since each product $\hat{x}\hat{z}$ with x and z in X can be written in the form

$$\hat{x}\hat{z} = \frac{1}{4} \big((\hat{x} + \hat{z})^2 - (\hat{x} - \hat{z})^2 \big).$$

We want to construct a linear functional $\mathcal{L}$ on $\mathcal{V}$ with the following properties:

$$0 \leq \mathcal{L}(\hat{x}^2) \leq C^2 \|x\|^2, \quad x \in X \tag{7.16}$$

and

$$\|Tx\|^2 \leq \mathcal{L}(\hat{x}^2), \quad x \in E. \tag{7.17}$$

To this end, let us apply Lemma 7.3.5 in the case $\mathcal{A} = \mathcal{B} = \{\hat{x}^2 : x \in X\}$ by putting

$$\phi(\hat{x}^2) = \begin{cases} 0 & \text{if } x \in X \setminus E \\ \|Tx\|^2 & \text{if } x \in E \end{cases}$$

and

$$\psi(\hat{x}^2)^2 = C^2 \|x\|^2.$$

Suppose that

$$\sum_{i=1}^n \beta_i^2 \hat{z}_i^2 = \sum_{j=1}^m \alpha_j^2 \hat{x}_j^2$$

for some $(\hat{x}_j)_{j=1}^m$, $(\hat{z}_i)_{i=1}^n$ vectors in X, and some nonnegative scalars $(\alpha_j^2)_{j=1}^m$, $(\beta_j^2)_{j=1}^n$. Let us suppose $z_1, \ldots, z_l \in E$ and $z_{l+1}, \ldots, z_n \in X \setminus E$. Then

$$\sum_{i=1}^{l} \beta_i^2 \hat{z}_i^2 \leq \sum_{j=1}^{m} \alpha_j^2 \hat{x}_j^2,$$

hence

$$\sum_{i=1}^{l} \|T(\beta_j z_i)\|^2 \leq C^2 \sum_{j=1}^{m} \|\alpha_j x_j\|^2.$$

Thus

$$\sum_{i=1}^{n} \beta_i^2 \phi(\hat{z}_i^2) \leq \sum_{j=1}^{m} \alpha_j^2 \psi(\hat{x}_j^2).$$

Lemma 7.3.5 yields a linear functional $\mathcal{L}$ on $\mathcal{V}$ with

$$\phi(\hat{x}^2) \leq \mathcal{L}(\hat{x}^2) \leq \psi(\hat{x}^2), \qquad x \in X.$$

$\mathcal{L}$, in turn, induces a symmetric bilinear form $\langle \, \cdot \, \rangle$ on X given by

$$\langle x, z \rangle = \mathcal{L}(\hat{x}\hat{z}),$$

so the map $X \longrightarrow [0, \infty)$, $x \mapsto \sqrt{\langle x, x \rangle} = \sqrt{\mathcal{L}(\hat{x}^2)}$ defines a seminorm on X.

Thus, X (modulo the subspace $\{x \; ; \; \langle x, x \rangle = 0\}$) endowed with the (now) inner product $\langle \, , \, \rangle$ is an inner product space, and $\|x\|_0 = \sqrt{\langle x, x \rangle}$ a norm on X. Let H be the completion of X_0 under this norm. H is a Hilbert space.

Take S to be the induced operator $S : X \to H$ mapping x to its equivalence class in X_0. Then we have

$$\|Sx\| \leq C\|x\|, \qquad x \in X.$$

S has norm one and dense range. By construction, if $x \in E$ we have

$$\|Tx\| \leq \|Sx\|,$$

therefore we can find an operator $R_0 : S(E) \to Y$ with $\|R_0\| \leq 1$ and $T = R_0 S|_E$. Compose R_0 with the orthogonal projection of H onto $\overline{S(E)}$ to create R.

The proof for complex scalars. In the case when X and Y are complex Banach spaces we proceed as first by "forgetting" their complex structure and treating them as real spaces. The argument creates a real symmetric bilinear form $\langle \, \cdot \, \rangle$ on X which is continuous for the original norm. We can then define a complex inner product by "recalling" the complex structure of X and setting

$$(x, z) = \frac{1}{2\pi} \int_0^{2\pi} \langle e^{i\theta} x, e^{i\theta} z \rangle - i \langle i e^{i\theta} x, e^{i\theta} z \rangle \, d\theta.$$

We leave it to the reader to check that this induces a complex inner product and that using this to define H gives the same conclusion.

$\square$

7.4 The Kwapień-Maurey theorems for type-2 spaces

We saw in Proposition 6.2.9 that if H is a Hilbert space then H has type 2 and cotype 2. More generally, since the type and cotype are isomorphic invariants, any Banach space isomorphic to a Hilbert space has type 2 and cotype 2. In 1972 Kwapień [122] showed that the converse is also true:

Theorem 7.4.1. *A Banach space X has type 2 and cotype 2 if and only if X is isomorphic to a Hilbert space.*

As Maurey noticed soon after Kwapień obtained Theorem 7.4.1, this is also a factorization theorem which follows from Theorem 7.4.2 by taking T the identity on X:

Theorem 7.4.2 (Kwapień-Maurey). *Let X and Y be Banach spaces and T an operator from X to Y. If X has type 2 and Y has cotype 2 then T factors through a Hilbert space.*

Shortly afterwards, Maurey [143] discovered a beautiful Hahn-Banach result for operators from type-2 spaces into a Hilbert space, which we now combine with Theorem 7.4.2 to give the following composite statement (that of course implies both Theorem 7.4.1 and Theorem 7.4.2 by taking $E = X$). In its proof this lemma will be needed:

Lemma 7.4.3. *Let X be a Banach space. Assume that the sets of vectors $\{z_i\}_{i=1}^n$ and $\{x_j\}_{j=1}^m$ of X satisfy the condition*

$$\sum_{i=1}^n |x^*(z_i)|^2 \le \sum_{j=1}^m |x^*(x_j)|^2, \qquad x^* \in X^*.$$

Then, if $(\gamma_i)_{i=1}^\infty$ is a sequence of independent Gaussians we have

$$\left(\mathbb{E}\left\|\sum_{i=1}^n \gamma_i z_i\right\|^2\right)^{1/2} \le \left(\mathbb{E}\left\|\sum_{j=1}^m \gamma_j x_j\right\|^2\right)^{1/2}.$$

Proof. Let F be the linear span of $\{x_1, \ldots, x_m, z_1, \ldots, z_n\}$ in X. By hypothesis, the quadratic form Q defined on F^* by

$$Q(f^*) = \sum_{j=1}^m |f^*(x_j)|^2 - \sum_{i=1}^n |f^*(z_i)|^2$$

is positive-definite. Hence we can find $z_{n+1}, \ldots, z_{n+l} \in F$ so that

$$Q(f^*) = \sum_{i=1}^l |f^*(z_{n+i})|^2, \qquad f^* \in F^*.$$

This implies that

$$\sum_{i=1}^{n+l} |x^*(z_i)|^2 = \sum_{j=1}^{m} |x^*(x_j)|^2, \qquad x^* \in X^*.$$

Then the vector-valued random variables $\sum_{i=1}^{n+l} \gamma_i z_i$ and $\sum_{j=1}^{m} \gamma_j x_j$ have the same distributions on X. As a consequence,

$$\mathbb{E}\left\|\sum_{i=1}^{n+l} \gamma_i z_i\right\|^2 = \mathbb{E}\left\|\sum_{j=1}^{m} \gamma_j x_j\right\|^2. \tag{7.18}$$

Now,

$$\left(\mathbb{E}\left\|\sum_{i=1}^{n} \gamma_i z_i\right\|^2\right)^{1/2}$$

$$\leq \frac{1}{2}\left(\mathbb{E}\left\|\sum_{i=1}^{n} \gamma_i z_i + \sum_{i=n+1}^{n+l} \gamma_i z_i\right\|^2\right)^{1/2} + \frac{1}{2}\left(\mathbb{E}\left\|\sum_{i=1}^{n} \gamma_i z_i - \sum_{i=n+1}^{n+l} \gamma_i z_i\right\|^2\right)^{1/2}$$

$$= \left(\mathbb{E}\left\|\sum_{i=1}^{n+l} \gamma_i z_i\right\|^2\right)^{1/2}$$

$$= \left(\mathbb{E}\left\|\sum_{j=1}^{m} \gamma_j x_j\right\|^2\right)^{1/2},$$

which completes the proof.

$\square$

Theorem 7.4.4. *Let X and Y be Banach spaces and E a closed subspace of X. Suppose $T : E \to Y$ is an operator. If X has type 2 and Y has cotype 2 then there is a Hilbert space H and operators $S : X \to H$, $R : H \to Y$ so that $\|R\|\|S\| \leq T_2(X)C_2(Y)\|T\|$ and $RS|_E = T$.*

Proof. We shall prove that for all sequences $(z_i)_{i=1}^{n}$ in E and $(x_j)_{j=1}^{m}$ in X such that

$$\sum_{i=1}^{n} |x^*(z_i)|^2 \leq \sum_{j=1}^{m} |x^*(x_j)|^2, \qquad x^* \in X^* \tag{7.19}$$

we have

$$\left(\sum_{i=1}^{n} \|Tz_i\|^2\right)^{1/2} \leq T_2(X)C_2(Y)\|T\|\left(\sum_{j=1}^{m} \|x_j\|^2\right)^{1/2},$$

and then we will appeal to the factorization criterion given by Theorem 7.3.4. The key to the argument is to replace the Rademacher functions in the definition of type and cotype by Gaussian random variables.

On the one hand, for any $(z_i)_{i=1}^n \subset E$, using the cotype-2 property of Y we have

$$\sum_{i=1}^n \|Tz_i\|^2 \leq C_2(Y)^2 \, \mathbb{E} \left\| \sum_{i=1}^n \varepsilon_i Tz_i \right\|^2.$$

Then, if for each $N \in \mathbb{N}$ we consider $(\varepsilon_{ki})_{1 \leq i,k \leq N}$, a sequence of $N \times N$ Rademachers,

$$\sum_{i=1}^n \|Tz_i\|^2 \leq \frac{C_2(Y)^2}{N} \, \mathbb{E} \left\| \sum_{k=1}^N \sum_{i=1}^n \varepsilon_{ki} Tz_i \right\|^2$$

$$= C_2(Y)^2 \, \mathbb{E} \left\| \sum_{i=1}^n \sum_{k=1}^N \frac{\varepsilon_{ki}}{\sqrt{N}} Tz_i \right\|^2.$$

Notice that for each $1 \leq i \leq n$, the random variables $\varepsilon_{i1}, \varepsilon_{i2}, \dots, \varepsilon_{iN}$ are independent and identically distributed, so by the Central Limit theorem, for each i the sequence $\left(\frac{\sum_{k=1}^N \varepsilon_{ik}}{\sqrt{N}} \right)_{N=1}^\infty$ converges in distribution to a Gaussian, γ_i. Thus,

$$\lim_{N \to \infty} \mathbb{E} \left\| \sum_{i=1}^n \sum_{k=1}^N \frac{\varepsilon_{ki}}{\sqrt{N}} Tz_i \right\|^2 = \mathbb{E} \left\| \sum_{i=1}^n \gamma_i Tz_i \right\|^2,$$

and, therefore,

$$\sum_{i=1}^n \|Tz_i\|^2 \leq C_2(Y)^2 \, \mathbb{E} \left\| \sum_{i=1}^n \gamma_i Tz_i \right\|^2. \qquad (7.20)$$

On the other hand, if we let $(\varepsilon_i)_{i=1}^\infty$ be a sequence of Rademachers independent of $(\gamma_i)_{i=1}^\infty$, for any sequence $(x_j)_j^m \subset X$, the symmetry of the Gaussians yields

$$\mathbb{E} \left\| \sum_{j=1}^m \gamma_i x_j \right\|^2 = \mathbb{E}\mathbb{E}_\varepsilon \left\| \sum_{j=1}^m \varepsilon_j \gamma_j x_j \right\|^2$$

$$\leq T_2(X)^2 \mathbb{E} \sum_{j=1}^m |\gamma_j|^2 \|x_j\|^2$$

$$= T_2(X)^2 \sum_{j=1}^m \|x_j\|^2 \mathbb{E}|\gamma_j|^2$$

$$= T_2(X)^2 \sum_{j=1}^m \|x_j\|^2. \qquad (7.21)$$

Suppose that the vectors $(z_i)_{i=1}^n$ in E and $(x_j)_{j=1}^m$ in X satisfy equation (7.19). Using Lemma 7.4.3 in combination with (7.18), (7.20), and (7.21) we obtain the inequality we need to apply Theorem 7.3.4:

$$\sum_{i=1}^{n} \|Tz_i\|^2 \le C_2(Y)^2 \mathbb{E} \left\| \sum_{i=1}^{n} \gamma_i z_i \right\|^2$$

$$\le C_2(Y)^2 \|T\|^2 \, \mathbb{E} \left\| \sum_{i=1}^{n} \gamma_i z_i \right\|^2$$

$$\le C_2(Y)^2 \|T\|^2 \, \mathbb{E} \left\| \sum_{j=1}^{m} \gamma_j x_j \right\|^2$$

$$\le C_2(Y)^2 T_2(X)^2 \|T\|^2 \sum_{j=1}^{m} \|x_j\|^2 \, .$$

$\square$

There is a quantitative estimate here that we would like to emphasize:

Definition 7.4.5. If X and Y are two isomorphic Banach spaces, the *Banach-Mazur distance* between X and Y, denoted $d(X, Y)$, is defined by the formula

$$d(X, Y) = \inf \left\{ \|T\| \|T^{-1}\| : \ T : X \to Y \text{ is an isomorphism} \right\}.$$

The Banach-Mazur distance is not a distance in the real sense of the term since the triangle law does not hold, but d satisfies a submultiplicative triangle inequality; that is,

$$d(X, Z) \le d(X, Y) d(Y, Z)$$

when X, Y, Z are all isomorphic. If X and Y are isometric then $d(X, Y) = 1$. The converse holds for finite-dimensional spaces but fails for infinite-dimensional spaces! (see the Problems).

In this language, Kwapień's theorem (Theorem 7.4.1) really states:

Theorem 7.4.6. *If X is a Banach space of type 2 and cotype 2 then*

$$d(X, H) \le T_2(X) C_2(X)$$

for some Hilbert space H.

We have seen (Theorem 6.4.8) that if $p > 2$ every subspace of L_p which is isomorphic to a Hilbert space is necessarily complemented. Theorem 7.4.4 shows that this phenomenon is simply a consequence of the type-2 property:

Theorem 7.4.7 (Maurey). *Let X be a Banach space of type 2. Let E be a closed subspace of X which is isomorphic to a Hilbert space. Then E is complemented in X.*

Proof. Since E has cotype 2 the identity map on E can be extended to a projection of X onto E.

$\square$

As we mentioned above, if we specialize the range space in Theorem 7.4.4 to be a Hilbert space then the assertion is a form of the Hahn-Banach theorem for Hilbert-space valued operators defined on a type-2 space. An interesting question is whether the extension property in Theorem 7.4.4 actually characterizes type-2 spaces:

Problem 7.4.8. *Suppose X is a Banach space with the property that for every closed subspace E of X and every operator $T_0 : E \to H$ (H a Hilbert space) there is a bounded extension $T : X \to H$. Must X be a space of type 2?*

For a partial positive solution of this problem we refer to [28].

Up to now the only spaces that we have considered in the context of type and cotype are the L_p-spaces (and their subspaces and quotients). It is worth pointing out that there are many other Banach spaces to which this theory can be applied. Perhaps the simplest examples are the "noncommutative" ℓ_p-spaces or *Schatten ideals*. These are ideals of operators on a separable Hilbert space which were originally introduced in 1946 by Schatten and studied in several papers by Schatten and von Neumann; an account is given in [202].

If H is a separable (complex) Hilbert space we define $\mathcal{S}_p$ to be the set of compact operators $A : H \to H$ so that the positive operator $(A^*A)^{p/2}$ has finite trace and we impose the norm

$$\|A\|_{\mathcal{S}_p} = \operatorname{tr} (A^*A)^{p/2}.$$

It is not entirely obvious, but is true, that this is a norm and that the class of such operators forms a Banach space.

In many ways the structure of $\mathcal{S}_p$ resembles that of ℓ_p. Thus if $1 \le p \le 2$, $\mathcal{S}_p$ has type p and cotype 2, while if $2 \le p < \infty$, $\mathcal{S}_p$ has cotype p and type 2 (see [215], [65]). See [5] for the structure of subspaces of $\mathcal{S}_p$.

Recently there has been considerable interest in noncommutative L_p-spaces but even to formulate the definition would take us too far afield.

Problems

7.1. For $1 \le r, p < \infty$, prove that the space $\ell_r(\ell_p)$ embeds in L_p if and only if $r = p$.

7.2. Let $p_n = 1 + \frac{1}{n}$. Consider the Banach space $X = \ell_2(\ell_{p_n}^2)$. Show that ℓ_1^2 does not embed isometrically into X but that $d(X, X \oplus_2 \ell_1^2) = 1$.

7.3. Show that any reflexive quotient of a $\mathcal{C}(K)$ space has type two.

7.4. The weak L_p-spaces, $L_{p,\infty}$.
Let (Ω, μ) be a probability measure space and $0 < p < \infty$. A measurable function f is said to belong to *weak L_p*, denoted $L_{p,\infty}$, if

$$\|f\|_{p,\infty} = \sup_{t>0} t\mu(|f| > t)^{1/p} < \infty.$$

(a) Show that $L_{p,\infty}$ is a linear space and that $\|\cdot\|_{p,\infty}$ is a *quasi-norm* on $L_{p,\infty}$, i.e., $\|\cdot\|_{p,\infty}$ satisfies the properties of a norm except the triangle law which is replaced by

$$\|f+g\|_{p,\infty} \leq C(\|f\|_{p,\infty} + \|g\|_{p,\infty}), \qquad f,g \in L_{p,\infty},$$

where $C \geq 1$ is a constant independent of f, g.

(b) Show that $L_{p,\infty}$ is complete for this quasi-norm and hence becomes a *quasi-Banach space*.

(c) Show that if $p > 1$, $\|\cdot\|_{p,\infty}$ is equivalent to the norm

$$\|f\|_{p,\infty,c} = \sup_{t>0} \sup_{\mu(A)=t} t^{1/p-1} \int_A |f| d\mu.$$

Thus $L_{p,\infty}$ can be regarded as a Banach space.

(d) Show that $L_{p,\infty} \subset L_r$ whenever $0 < r < p$.

7.5 (Nikishin [158]). (Continuation.) Suppose X is a Banach space of type p for some $1 \leq p < 2$. Suppose $1 \leq r < p$ and $T : X \rightarrow L_r(\mu)$ is a bounded linear operator.

(a) Show that for some suitable constant C we have the following estimate:

$$\mu\left(\bigcup_{j=1}^m \{|Tx_j| \geq 1\}\right)^{1/r} \leq C\left(\sum_{j=1}^m \|x_j\|^p\right)^{1/p}, \qquad x_1, \ldots, x_m \in X.$$

(b) For any constant $K > C$ consider a maximal family of disjoint sets of positive measure $(E_i)_{i \in I}$ such that we can find $x_i \in X$ with $\|x_i\| \leq 1$ and $|Tx_i| \geq K(\mu(E_i)^{-1/p})$ on E_i. Show that this collection is countable and that

$$\sum_{i \in I} \mu(E_i) \leq \left(\frac{C}{K}\right)^{\frac{rp}{p-r}}.$$

(c) Show that given $\epsilon > 0$ there is a set E with $\mu(E) > 1 - \epsilon$ so that the map $T_E f = \chi_E T f$ is a bounded operator from X into $L_{p,\infty}(\mu)$.

This gives a "factorization" through weak L_p; it is possible to obtain a more elegant "change of density" formulation (see [186]). Note that if X is an arbitrary Banach space and $r < 1$ we get boundedness of T_E into weak L_1.

7.6 (Jordan-von Neumann [96]). Show, without appealing to Kwapień's theorem, that if a Banach space X has type 2 with $T_2(X) = 1$ then X is isometrically a Hilbert space. [*Hint:* For real scalars, define an inner product by $(x,y) = \frac{1}{4}(\|x+y\|^2 - \|x-y\|^2)$.]

7.7. Let μ, ν be σ-finite measures. A linear operator $T : L_p(\mu) \rightarrow L_r(\nu)$, $0 < r,p < \infty$, is said to be a *positive operator* if $f \geq 0$ implies $Tf \geq 0$.

(a) Show that if $1 \le s \le \infty$ then for any sequence $(f_j)_{j=1}^n \in L_p(\mu)$ we have

$$\left\|(\sum_{j=1}^n |Tf_j|^s)^{1/s}\right\|_r \le \|T\| \left\|(\sum_{j=1}^n |f_j|^s)^{1/s}\right\|_p.$$

(b) Deduce that if $r < p$ and $p \ge 1$ then T factorizes through $L_p(h\nu)$ for some density function h.

7.8. Let $T : \ell_p \to L_r$, $r < p < 2$, be the linear operator defined by

$$T(\xi) = \sum_{j=1}^\infty \xi(j)\eta_j,$$

where $(\eta_j)_{j=1}^\infty$ is a sequence of independent normalized p-stable random variables.

(a) Using the boundedness of T show that the operator $S : \ell_{p/2} \to L_{r/2}$ defined by

$$S(\xi) = \sum_{j=1}^\infty \xi(j)|\eta_j|^2$$

is a bounded positive linear operator.

(b) Show that, however,

$$\left\|(n^{-1}\sum_{j=1}^n |Se_j|^{p/2})^{2/p}\right\|_{r/2} \to \infty$$

and deduce that S cannot be factored via a change of density through $L_{p/2}$. Thus the conclusion of Problem 7.7 fails when $p < 1$. [*Hint:* You need to show that

$$\lim_{n\to\infty} \left\|n^{-1}\sum_{j=1}^n |\eta_j|^p\right\|_{r/p} = \infty.$$

Consider $\min(|\eta_j|^p, M)$ for any fixed M.]

8

Absolutely Summing Operators

The theory of absolutely summing operators was one of the most profound developments in Banach space theory between 1950 and 1970. It originates in a fundamental paper of Grothendieck [76] (which actually appeared in 1956). However, some time passed before Grothendieck's remarkable work really became well-known among specialists. There are several reasons for this. One major point is that Grothendieck stopped working in the field at just about this time and moved into algebraic geometry (his work in algebraic geometry earned the Fields Medal in 1966). Thus he played no role in the dissemination of his own ideas. He also chose to publish in a relatively obscure journal that was not widely circulated; before the advent of the Internet it was much more difficult to track down copies of articles. Thus it was not until the 1968 paper of Lindenstrauss and Pełczyński [131] that Grothendieck's ideas became widely known. Since 1968, the theory of absolutely summing operators has become a cornerstone of modern Banach space theory.

In fact, most (but not all) of this chapter was known to Grothendieck although his presentation would be different. We will utilize the more modern concepts of type and cotype and use the factorization theory from Chapter 7 in our exposition. Although Grothendieck's work predates the material in Chapter 7 it can be considered as a development. In Chapter 7 we considered conditions on an operator $T : X \to Y$ that would ensure factorization through a Hilbert space; this culminated in the Kwapień-Maurey theorem (Theorem 7.4.2) which says that the conditions that X has type 2 and Y has cotype 2 are sufficient. Grothendieck inequality yields the fact that every operator $T : \mathcal{C}(K) \to L_1$ also factors through a Hilbert space even though $\mathcal{C}(K)$ is very far from type 2. This seemed quite mysterious until the work of Pisier showed that the condition X has type 2 can in certain cases be relaxed to X^* has cotype 2.

Two good references for further developments of Grothendieck theory are Pisier's CBMS conference lectures [185] and the monograph of Diestel, Jarchow, and Tonge [41].

8.1 Grothendieck's Inequality

Theorem 8.1.1 (Grothendieck Inequality). *There exists a universal constant K_G so that whenever $(a_{jk})_{j,k=1}^{m,n}$ is a real matrix such that*

$$\left| \sum_{j=1}^{m} \sum_{k=1}^{n} a_{jk} s_j t_k \right| \leq \max_j |s_j| \max_k |t_k|,$$

for any two sequences of scalars $(s_j)_{j=1}^{m}$ and $(t_k)_{k=1}^{n}$ then

$$\left| \sum_{j=1}^{m} \sum_{k=1}^{n} a_{jk} \langle u_j, v_k \rangle \right| \leq K_G \max_j \|u_j\| \max_k \|v_k\|,$$

for all sequences of vectors $(u_j)_{j=1}^{m}$ and $(v_k)_{k=1}^{n}$ in an arbitrary real Hilbert space H.

Proof. Since all Hilbert spaces are linearly isometric we can choose any Hilbert space to prove the theorem, but it is most convenient to consider the closed subspace H of L_2 spanned by a sequence of independent Gaussians $(g_k)_{k=1}^{\infty}$, equipped with the L_2-norm. Notice that if $f = \sum_{k=1}^{\infty} a_k g_k \in H$ with $\|f\|_2 = \sum_{k=1}^{\infty} |a_k|^2 = 1$ then f is also a Gaussian, and so

$$\|f\|_4^4 = \frac{1}{\sqrt{2\pi}} \int_{-\infty}^{\infty} x^4 e^{-\frac{1}{2}x^2} \, dx = 3.$$

Thus for $f \in H$ we have

$$\|f\|_2 \leq \|f\|_4 = 3^{\frac{1}{4}} \|f\|_2. \tag{8.1}$$

This shows that the subspace $(H, \|\cdot\|_2)$ is strongly embedded in L_4.

Obviously, for each matrix $A = (a_{jk})_{j,k=1}^{m,n}$ using Schwarz's inequality there is a best constant $\Gamma = \Gamma(A)$ such that for any two finite sequences of functions $(u_j)_{j=1}^{m}$ and $(v_k)_{k=1}^{n}$ in H,

$$\left| \sum_{j=1}^{m} \sum_{k=1}^{n} a_{jk} \langle u_j, v_k \rangle \right| \leq \Gamma \max_j \|u_j\|_2 \max_k \|v_k\|_2. \tag{8.2}$$

Let us assume that $\|u_j\|_2 \leq 1$ for $1 \leq j \leq m$ and $\|v_k\|_2 \leq 1$ for $1 \leq k \leq n$. For fixed M, we consider the truncations of the functions $(u_j)_{j=1}^{m}$ and $(v_k)_{k=1}^{n}$ at M:

$$u_j^M = \begin{cases} u_j & \text{if } |u_j| \leq M \\ M \operatorname{sgn} u_j & \text{if } |u_j| > M \end{cases}, \qquad v_k^M = \begin{cases} v_k & \text{if } |v_k| \leq M \\ M \operatorname{sgn} v_k & \text{if } |v_k| > M \end{cases}.$$

Taking into account that $4(x-1) \leq x^2$ for $x > 1$ we deduce that if $x > M$ then $16M^2(x-M)^2 \leq x^4$. Combining this inequality with (8.1) we obtain

$$16M^2 \int_0^1 |u_j(t) - u_j^M(t)|^2 \, dt \le \int_0^1 |u_j(t)|^4 \, dt \le 3,$$

hence

$$\left\| u_j - u_j^M \right\|_2^2 \le \frac{3}{16M^2}, \quad j = 1, \ldots, n. \tag{8.3}$$

Analogously,

$$\left\| v_k - v_k^M \right\|_2^2 \le \frac{3}{16M^2}, \quad k = 1, \ldots, n. \tag{8.4}$$

Now,

$$\left| \sum_{j=1}^m \sum_{k=1}^n a_{jk} \langle u_j, v_k \rangle \right| = \left| \sum_{j=1}^m \sum_{k=1}^n a_{jk} \int_0^1 u_j v_k \, dt \right|$$

$$\le \int_0^1 \left| \sum_{j=1}^m \sum_{k=1}^n a_{jk} u_j^M v_k^M \right| dt + \left| \sum_{j=1}^m \sum_{k=1}^n a_{jk} \int_0^1 (u_j - u_j^M) v_k^M \, dt \right|$$

$$+ \left| \sum_{j=1}^m \sum_{k=1}^n a_{jk} \int_0^1 u_j (v_k - v_k^M) \, dt \right|.$$

By the hypothesis on the matrix (a_{jk}), for each $t \in [0,1]$ we have

$$\left| \sum_{j=1}^m \sum_{k=1}^n a_{jk} u_j^M(t) v_k^M(t) \right| dt \le M^2.$$

On the other hand the equations (8.2), (8.3), and (8.4) yield

$$\left| \sum_{j=1}^m \sum_{k=1}^n a_{jk} \int_0^1 (u_j - u_j^M) v_k^M \, dt \right| = \left| \sum_{j=1}^m \sum_{k=1}^n a_{jk} \langle u_j - u_j^M, v_k^M \rangle \right| \le \Gamma \frac{\sqrt{3}}{4M}$$

and

$$\left| \sum_{j=1}^m \sum_{k=1}^n a_{jk} \int_0^1 u_j (v_k - v_k^M) \, dt \right| = \left| \sum_{j=1}^m \sum_{k=1}^n a_{jk} \langle u_j, v_k - v_k^M \rangle \right| \le \Gamma \frac{\sqrt{3}}{4M}.$$

Combining,

$$\left| \sum_{j=1}^m \sum_{k=1}^n a_{jk} \langle u_j, v_k \rangle \right| \le M^2 + \Gamma \frac{\sqrt{3}}{2M}.$$

By our assumption on Γ the following inequality must hold:

$$\Gamma \le M^2 + \Gamma \frac{\sqrt{3}}{2M}.$$

To minimize the right-hand side we take $M = \left(\frac{\sqrt{3}}{4} \Gamma \right)^{1/3}$ and thus

$$\Gamma \le 3\Big(\frac{\sqrt{3}\Gamma}{4}\Big)^{2/3},$$

which gives $\Gamma \le \frac{81}{16}$. Thus Grothendieck's inequality is proved with constant $K_G \le \frac{81}{16}$.

□

While the proof given above is, we feel, the most transparent, it is far from being effective in determining the *Grothendieck constant* K_G. Grothendieck's original argument gave $K_G \le \sinh(\pi/2)$ (see the Problems). The best estimate known is that of Krivine [120] that $K_G \le 2(\sinh^{-1} 1)^{-1} < 2$. The corresponding constant for complex scalars is known to be smaller than K_G. See [41] for a full discussion on Grothendieck's inequality.

Remark 8.1.2. Suppose (a_{jk}) is a real $m \times n$ matrix such that the bilinear form $B : \ell_\infty^m \times \ell_\infty^n \to \mathbb{R}$ given by

$$B((s_j)_{j=1}^m, (t_k)_{k=1}^n) = \sum_{j=1}^m \sum_{k=1}^n a_{jk} s_j t_k$$

has norm

$$\|B\| = \sup\left\{ \Big|\sum_{j=1}^m \sum_{k=1}^n a_{jk} s_j t_k\Big| : \max_j |s_j| \le 1, \max_k |t_k| \le 1 \right\} \le 1.$$

Suppose $(f_l)_{l=1}^N$ and $(g_l)_{l=1}^N$ are finite sequences in ℓ_∞^m and ℓ_∞^n, respectively. For each $1 \le l \le N$ let $f_l = (f_l(j))_{j=1}^m$ and $g_l = (g_l(k))_{k=1}^n$. Let us also consider the following two sets of vectors in the Hilbert space ℓ_2^N:

$$u_j = (f_l(j))_{l=1}^N, \qquad 1 \le j \le m$$

and

$$v_k = (g_l(k))_{l=1}^N, \qquad 1 \le k \le n.$$

Then Grothendieck's inequality yields

$$\Big|\sum_{l=1}^N B(f_l, g_l)\Big| = \Big|\sum_{l=1}^N \sum_{j=1}^m \sum_{k=1}^n a_{jk} f_l(j) g_l(k)\Big|$$

$$= \Big|\sum_{j=1}^m \sum_{k=1}^n a_{jk} \sum_{l=1}^N f_l(j) g_l(k)\Big|$$

$$= \Big|\sum_{j=1}^m \sum_{k=1}^n a_{jk} \langle u_j, v_k \rangle\Big|$$

$$\le K_G \max_{1 \le j \le m} \|u_j\| \max_{1 \le k \le n} \|v_k\|$$

$$= K_G \max_{1\le j\le m} \Big(\sum_{l=1}^{N} |f_l(j)|^2\Big)^{1/2} \max_{1\le k\le n} \Big(\sum_{l=1}^{N} |g_l(k)|^2\Big)^{1/2}.$$

If we put

$$\max_{1\le j\le m} \Big(\sum_{l=1}^{N} |f_l(j)|^2\Big)^{1/2} = \Big\|\Big(\sum_{l=1}^{N} |f_l|^2\Big)^{\frac12}\Big\|_\infty$$

and

$$\max_{1\le k\le n} \Big(\sum_{l=1}^{N} |g_l(k)|^2\Big)^{1/2} = \Big\|\Big(\sum_{l=1}^{N} |g_l|^2\Big)^{\frac12}\Big\|_\infty,$$

we obtain an equivalent way of stating Grothendieck's inequality: *Suppose that the bilinear form $B : \ell_\infty^m \times \ell_\infty^n \to \mathbb{R}$ has norm at most one. Then for any $(f_l)_{l=1}^N$ in ℓ_∞^m and $(g_l)_{l=1}^N$ in ℓ_∞^n,*

$$\Big|\sum_{l=1}^{N} B(f_l, g_l)\Big| \le K_G \Big\|\Big(\sum_{l=1}^{N} |f_l|^2\Big)^{\frac12}\Big\|_\infty \Big\|\Big(\sum_{l=1}^{N} |g_l|^2\Big)^{\frac12}\Big\|_\infty.$$

The space $\ell_\infty^m \times \ell_\infty^n$ can be regarded as the space of continuous functions $\mathcal{C}(K_{(m)}) \times \mathcal{C}(L_{(n)})$, where $K_{(m)}$ and $L_{(n)}$ are finite sets of cardinality m and n, respectively, equipped with the discrete topology. Our next result extends the previous remark about Grothendieck's inequality to general $\mathcal{C}(K)$-spaces.

Theorem 8.1.3. *Let K and L be two compact Hausdorff spaces and let $B : \mathcal{C}(K) \times \mathcal{C}(L) \to \mathbb{R}$ be a bounded bilinear form. Then for any $(f_k)_{k=1}^n$ in $\mathcal{C}(K)$ and $(g_k)_{k=1}^n$ in $\mathcal{C}(L)$ we have*

$$\Big|\sum_{k=1}^{n} B(f_k, g_k)\Big| \le K_G \|B\| \Big\|\Big(\sum_{k=1}^{n} |f_k|^2\Big)^{\frac12}\Big\|_\infty \Big\|\Big(\sum_{k=1}^{n} |g_k|^2\Big)^{\frac12}\Big\|_\infty,$$

where

$$\|B\| = \sup\big\{|B(f,g)| : f \in B_{\mathcal{C}(K)}, g \in B_{\mathcal{C}(L)}\big\}.$$

Proof. The proof relies on a partition of unity argument. Let $(f_k)_{k=1}^n$ be a sequence in $\mathcal{C}(K)$ and $(g_k)_{k=1}^n$ be a sequence in $\mathcal{C}(L)$. Given $\delta > 0$ one can find a finite open covering $(U_i)_{i=1}^N$ of K so that for each $1 \le k \le n$ we have $|f_k(x) - f_k(x')| < \delta$ whenever x, x' both belong to some U_i in the covering. Pick a partition of unity $(\varphi_j)_{j=1}^l$ subordinate to the covering $(U_i)_{i=1}^N$. Thus each φ_j satisfies $0 \le \varphi_j \le 1$. Furthermore, supp $\varphi_j = \overline{\{\varphi_j > 0\}}$ lies inside a set $U_{i(j)}$ in the partition, and for all $x \in K$

$$\sum_{j=1}^{l} \varphi_j(x) = 1.$$

For each $1 \le j \le l$ pick $x_j \in U_{i(j)}$ and put

$$f_k' = \sum_{j=1}^{l} f_k(x_j)\varphi_j, \qquad 1 \le k \le n.$$

Then, for any $x \in K$ with $\varphi_j(x) \ne 0$ we have $|f_k(x_j) - f_k(x)| < \delta$. Hence,

$$|f_k'(x) - f_k(x)| < \delta, \qquad x \in K, \ 1 \le k \le n.$$

That is, $\|f_k' - f_k\|_\infty < \delta$ for $1 \le k \le n$. Note also that $\|f_k'\|_\infty \le \|f_k\|_\infty$ by construction.

Similarly, for any $\delta > 0$ we may find a partition of unity $(\psi_j)_{j=1}^{m}$ on L with associated points $(y_j)_{j=1}^{m}$ so that if

$$g_k' = \sum_{j=1}^{m} g_k(y_j)\psi_j, \qquad 1 \le k \le n$$

then $\|g_k'\|_\infty \le \|g_k\|_\infty$ and

$$\|g_k' - g_k\|_\infty < \delta, \qquad 1 \le k \le n.$$

Let $(a_{jk})_{j,k=1}^{l,m}$ be the $l \times m$ matrix defined by

$$a_{jk} = B(\varphi_j, \psi_k).$$

For any $(s_j)_{j=1}^{l}$ and $(t_k)_{k=1}^{m}$ we have

$$\left| \sum_{j=1}^{l} \sum_{k=1}^{m} a_{jk} s_j t_k \right| \le \|B\| \max_j |s_j| \max_k |t_k|.$$

We select $(u_j)_{j=1}^{l}$ and $(v_k)_{k=1}^{m}$ in ℓ_2^n by

$$u_j = (f_i(x_j))_{i=1}^{n}, \qquad v_k = (g_i(y_k))_{i=1}^{n}.$$

Then

$$\sum_{i=1}^{n} B(f_i', g_i') = \sum_{i=1}^{n} \sum_{j=1}^{l} \sum_{k=1}^{m} a_{jk} f_i(x_j) g_i(y_k) = \sum_{j=1}^{l} \sum_{k=1}^{m} a_{jk} \langle u_j, v_k \rangle,$$

so by Grothendieck's inequality,

$$\left| \sum_{i=1}^{n} B(f_i', g_i') \right| \le K_G \|B\| \sup_j \left(\sum_{i=1}^{n} |f_i(x_j)|^2 \right)^{1/2} \sup_k \left(\sum_{i=1}^{n} |g_i(y_k)|^2 \right)^{1/2}.$$

Now for $1 \le i \le n$,

$$B(f_i, g_i) - B(f_i', g_i') = B(f_i - f_i', g_i) + B(f_i', g_i - g_i'),$$

and so

$$|B(f_i, g_i) - B(f_i', g_i')| \leq \delta \|B\| \left(\|f_i\|_\infty + \|g_i\|_\infty \right).$$

Putting everything together we obtain

$$\left| \sum_{i=1}^n B(f_i, g_i) \right| \leq \left| \sum_{i=1}^n B(f_i', g_i') \right| + \delta \|B\| \sum_{i=1}^n \left(\|f_i\|_\infty + \|g_i\|_\infty \right)$$

$$\leq \|B\| \left(K_G \left\| \left(\sum_{i=1}^n |f_i|^2 \right)^{\frac{1}{2}} \right\|_\infty \left\| \left(\sum_{i=1}^n |g_i|^2 \right)^{\frac{1}{2}} \right\|_\infty + \delta \sum_{i=1}^n \left(\|f_i\|_\infty + \|g_i\|_\infty \right) \right).$$

Letting $\delta \to 0$ gives the theorem.

$\square$

Theorem 8.1.3 also holds for complex scalars replacing K_G by the complex Grothendieck constant.

Remark 8.1.4 (Square-function estimates in $\mathcal{C}(K)$-spaces). In Chapter 6 we saw that in the L_p-spaces ($1 \leq p < \infty$) the following square-function estimates hold:

$$\left\| \left(\sum_{i=1}^n |f_i|^2 \right)^{\frac{1}{2}} \right\|_p \sim \left(\mathbb{E} \left\| \sum_{i=1}^n \varepsilon_i f_i \right\|_p^2 \right)^{1/2},$$

for every sequence $(f_i)_{i=1}^n$ in L_p. Now, in $\mathcal{C}(K)$-spaces, we clearly have

$$\left\| \left(\sum_{i=1}^n |f_i|^2 \right)^{1/2} \right\|_\infty \leq \left(\mathbb{E} \left\| \sum_{i=1}^n \varepsilon_i f_i \right\|_\infty^2 \right)^{1/2}$$

whenever $(f_i)_{i=1}^n \subset \mathcal{C}(K)$, but the converse estimate does not hold in general. Take for instance $\mathcal{C}(\Delta)$, the space of continuous functions on the Cantor set Δ, which we identify here with the topological product space $\{-1, 1\}^{\mathbb{N}}$. For each i, let f_i be the i-th projection from $\{-1, 1\}^{\mathbb{N}}$ onto $\{-1, 1\}$. Then for each n and any choice of signs $(\epsilon_i)_{i=1}^n$ we have

$$\left\| \sum_{i=1}^n \epsilon_i f_i \right\|_{\mathcal{C}(\Delta)} = \sup_{x \in \Delta} \left| \sum_{i=1}^n \epsilon_i f_i(x) \right| = n,$$

hence

$$\left(\mathbb{E} \left\| \sum_{i=1}^n \varepsilon_i f_i \right\|_{\mathcal{C}(K)}^2 \right)^{1/2} = n,$$

whereas, on the other hand,

$$\left\| \left(\sum_{i=1}^n |f_i|^2 \right)^{1/2} \right\|_{\mathcal{C}(\Delta)} = \sqrt{n}.$$

Theorem 8.1.5. *Suppose K is a compact Hausdorff space, that (Ω, μ) is a σ-finite measure space and that $T : \mathcal{C}(K) \to L_1(\mu)$ is a continuous operator. Then for any finite sequence $(f_k)_{k=1}^n$ in $\mathcal{C}(K)$ we have*

$$\left\| \left(\sum_{k=1}^n |Tf_k|^2 \right)^{\frac{1}{2}} \right\|_1 \leq K_G \|T\| \left\| \left(\sum_{k=1}^n |f_k|^2 \right)^{\frac{1}{2}} \right\|_\infty.$$

Proof. Let us define a bilinear form $B : \mathcal{C}(K) \times L_\infty(\mu) \to \mathbb{R}$ by

$$B(f, g) = \int_\Omega g \cdot T(f) \, d\mu.$$

Given a sequence $(f_k)_{k=1}^n$ in $\mathcal{C}(K)$, put $G = (\sum_{k=1}^n |Tf_k|^2)^{1/2}$, and then define

$$g_k(\omega) = \begin{cases} G(\omega)^{-1}(Tf_k)(\omega) & \text{if } G(\omega) \neq 0 \\ 0 & \text{if } G(\omega) = 0 \end{cases}, \quad 1 \leq k \leq n.$$

In Chapter 4 we saw that $L_\infty(\mu)$ is isometrically isomorphic to a space of continuous functions $\mathcal{C}(L)$ for some compact Hausdorff space L. With that identification we can apply Theorem 8.1.3 and obtain

$$\left\| \left(\sum_{k=1}^n |Tf_k|^2 \right)^{\frac{1}{2}} \right\|_1 = \sum_{k=1}^n \int_\Omega g_k \cdot T(f_k) \, d\mu$$

$$= \sum_{k=1}^n B(f_k, g_k)$$

$$\leq K_G \|T\| \left\| \left(\sum_{k=1}^n |f_k|^2 \right)^{\frac{1}{2}} \right\|_\infty,$$

since $\sum_{k=1}^n |g_k|^2 \leq 1$ everywhere and $\|B\| = \|T\|$. $\qquad \square$

We are now in position to apply Theorem 7.1.2.

Theorem 8.1.6. *Suppose K is a compact Hausdorff space, that (Ω, μ) is a probability measure space and that $T : \mathcal{C}(K) \to L_1(\mu)$ is a continuous operator. Then there exists a density function h on Ω such that for all $f \in \mathcal{C}(K)$,*

$$\left(\int |h^{-1} Tf|^2 h \, d\mu \right)^{1/2} \leq K_G \|T\| \|f\|.$$

In particular T factors through a Hilbert space.

Proof. It is enough to note that Theorem 8.1.5 implies that

$$\left\| \left(\sum_{i=1}^n |Tf_i|^2 \right)^{\frac{1}{2}} \right\|_1 \leq K_G \|T\| \left(\sum_{i=1}^n \|f_i\|_\infty^2 \right)^{1/2}.$$

Now Theorem 7.1.2 applies.

□

Let us recall that Kwapień's theorem (Theorem 7.4.1), or more precisely the Kwapień-Maurey theorem (Theorem 7.4.2), allows us to factorize an arbitrary operator $T : X \to Y$, where X has type 2 and Y has cotype 2, through a Hilbert space. However, in the above theorem we achieved the same result when $X = \mathcal{C}(K)$ (which fails to have any nontrivial type) and $Y = L_1(\mu)$. This is rather strange and needs explanation. If $\mathcal{C}(K)$ fails to have type 2, what is the substitute? Might the answer be that $\mathcal{C}(K)^* = \mathcal{M}(K)$ has cotype 2? Although type and cotype are not in duality, one is led to wonder if the optimal hypothesis in the Kwapień-Maurey theorem is that X^* has cotype 2. Let us prove a result in this direction:

Theorem 8.1.7. *Let X be a Banach space whose dual X^* has cotype 2. Let $T : X \to L_1$ be a bounded operator. Then T factors through a Hilbert space.*

Proof. The key here is to obtain an estimate of the form

$$\left\| \left(\sum_{j=1}^{n} |Tx_j|^2 \right)^{\frac{1}{2}} \right\|_1 \leq C \left(\sum_{j=1}^{n} \|x_j\|^2 \right)^{1/2}, \tag{8.5}$$

for some constant C and for all finite sequences $(x_j)_{j=1}^n$ in X, so that we can appeal to Theorem 7.1.2.

Assume first that T is a finite-rank operator. In this case we are guaranteed the existence of a constant so that (8.5) holds. Let the least such constant be denoted by $\Theta = \Theta(T)$. Theorem 7.1.2 yields a density function h on $[0,1]$ so that for all $x \in X$,

$$\left(\int |Tx(t)|^2 h^{-1}(t)\, dt \right)^{1/2} \leq \Theta \|x\|.$$

By Hölder's inequality,

$$\int |Tx|^{\frac{4}{3}} h^{-\frac{1}{3}}\, dt = \int |Tx|^{\frac{2}{3}} \left(|Tx|^2 h^{-1} \right)^{\frac{1}{3}}\, dt$$

$$\leq \left(\int |Tx|\, dt \right)^{2/3} \left(\int |Tx|^2 h^{-1}\, dt \right)^{1/3}$$

$$\leq \|T\|^{2/3} \Theta^{2/3} \|x\|^{4/3}.$$

Thus if we define $S : X \to L_{4/3}([0,1], h\, dt)$ by $Sx = h^{-1}Tx$, and $R : L_{4/3}([0,1], h\, dt) \to L_1$ by $Rf = hf$, we have $\|R\| = 1$, and

$$\|Sx\| \leq \|T\|^{\frac{1}{2}} \Theta^{\frac{1}{2}} \|x\|, \qquad x \in X;$$

that is, $\|S\| \leq \|T\|^{\frac{1}{2}} \Theta^{\frac{1}{2}}$.

Let us consider the adjoint $S^* : L_4([0,1], h\, dt) \to X^*$. Since L_4 has type 2 and X^* has cotype 2, we can apply Theorem 7.4.4 to deduce the existence of a Hilbert space H, and operators $U : L_4 \to H$ and $V : H \to X^*$ so that $S^* = VU$ and

$$\|V\|\|U\| \le T_2(L_4)C_2(X^*)\|S^*\| \le T_2(L_4)C_2(X^*)\|T\|^{\frac{1}{2}}\Theta^{\frac{1}{2}}.$$

It follows that we can factor $S^{**} = U^*V^* : X^{**} \to L_{4/3}([0,1], h\,dt)$ through the Hilbert space H^*. The restriction to X is a factorization of S.

For any sequence $(x_k)_{k=1}^n$ in X we have

$$\left\|\left(\sum_{k=1}^n |Tx_k|^2\right)^{1/2}\right\|_1 \le \left(\mathbb{E}\left\|\sum_{k=1}^n \varepsilon_k Tx_k\right\|_1^2\right)^{1/2}$$

$$\le \left(\mathbb{E}\left\|\sum_{k=1}^n \varepsilon_k Sx_k\right\|^2\right)^{1/2}$$

$$\le \|U\| \left(\mathbb{E}\left\|\sum_{k=1}^n \varepsilon_k V^* x_k\right\|^2\right)^{1/2}$$

$$= \|U\| \left(\sum_{k=1}^n \|V^* x_k\|^2\right)^{1/2}$$

$$\le \|V\|\|U\| \left(\sum_{k=1}^n \|x_k\|^2\right)^{1/2},$$

and so, from the definition of Θ,

$$\Theta \le \|U\|\|V\| \le T_2(L_4)C_2(X^*)\|T\|^{\frac{1}{2}}\Theta^{\frac{1}{2}},$$

which implies

$$\Theta(T) \le \left(T_2(L_4)C_2(X^*)\right)^2 \|T\|.$$

Now suppose that T is not necessarily finite-rank. Let $(S_k)_{k=1}^\infty$ be the partial-sum projections for the Haar basis in L_1. Then each $S_k T$ is finite-rank, and $\|S_k T\| \le \|T\|$ since the Haar basis is monotone. Thus

$$\Theta(S_k T) \le \left(T_2(L_4)C_2(X^*)\right)^2 \|T\|.$$

By passing to the limit in (8.5) we obtain that T satisfies such an estimate with constant $\Theta(T) \le \left(T_2(L_4)C_2(X^*)\right)^2 \|T\|$, and the result follows.

□

Notice how we needed to use finite-rank operators and a bootstrap method to obtain this result. This argument is the basis for Pisier's Abstract Grothendieck Theorem [183]:

Theorem 8.1.8 (Pisier's Abstract Grothendieck Theorem). *Let X and Y be Banach spaces so that X^* has cotype 2, Y has cotype 2, and either X or Y has the approximation property. Then any operator $T : X \to Y$ factors through a Hilbert space.*

The appearance of the approximation property here is at first unexpected, but remember we must use finite-rank approximations to our operator. Is the approximation property necessary? In a remarkable paper in 1983, Pisier [184] constructed a Banach space X so that both X and X^* have cotype 2 but X is not a Hilbert space. Applying Theorem 8.1.8 to the identity operator on this space shows that X must fail the approximation property.

8.2 Absolutely summing operators

We now introduce an important definition that goes back to the work of Grothendieck.

Definition 8.2.1. Let X, Y be Banach spaces. An operator $T : X \to Y$ is said to be *absolutely summing* if there is a constant C so that for all choices of $(x_k)_{k=1}^n$ in X,

$$\sum_{k=1}^n \|Tx_k\| \le C \sup \left\{ \sum_{k=1}^n |x^*(x_k)| : x^* \in X^*, \|x^*\| \le 1 \right\}.$$

The least such constant C is denoted $\pi_1(T)$ and is called the *absolutely summing norm* of T.

If $T : X \to Y$ is absolutely summing in particular T is bounded and $\|T\| \le \pi_1(T)$ since, by definition, for each $x \in X$

$$\|Tx\| \le \pi_1(T) \sup \{ |x^*(x)| : x^* \in B_{X^*} \} = \pi_1(T)\|x\|.$$

Notice also that for any sequence $(x_k)_{k=1}^n$ in X we have

$$\sup \left\{ \sum_{k=1}^n |x^*(x_k)| : x^* \in B_{X^*} \right\} = \sup \left\{ \left\| \sum_{k=1}^n \varepsilon_k x_k \right\| : (\varepsilon_k) \in \{-1,1\}^n \right\},$$

and so we can equivalently rewrite the definition of absolutely summing operator in terms of the right-hand side expression.

The next result identifies absolutely summing operators as exactly those operators which transform unconditionally convergent series into absolutely convergent series. We omit the routine proof (see the Problems).

Proposition 8.2.2. *An operator $T : X \longrightarrow Y$ is absolutely summing if and only if $\sum_{n=1}^\infty \|Tx_n\| < \infty$ whenever $\sum_{n=1}^\infty x_n$ is unconditionally convergent (or simply a (WUC) series).*

Recall that a classical theorem of Riemann asserts that if $\sum x_n$ is a series of real numbers then $\sum |x_n| < \infty$ if and only if $\sum x_n$ converges unconditionally. This easily extends to any finite-dimensional Banach space. During the late 1940s there was a flurry of interest in the problem of whether the same phenomenon could occur in any *infinite-dimensional* Banach space. In our language this asks whether the identity operator I_X can ever be absolutely summing if X is infinite-dimensional. Note for example that if X is a Hilbert space and $(e_n)_{n=1}^\infty$ is an orthonormal sequence then $\sum \frac{1}{n} e_n$ converges unconditionally but $\sum \frac{1}{n} = \infty$. Before addressing this problem let us introduce a more general definition:

Definition 8.2.3. Let X, Y be Banach spaces. An operator $T : X \to Y$ is called *p-absolutely summing* $(1 \le p < \infty)$ if there exists a constant C such that for all choices of $(x_k)_{k=1}^n$ in X we have

$$\left(\sum_{k=1}^n \|Tx_k\|^p \right)^{1/p} \le C \sup \left\{ \left(\sum_{k=1}^n |x^*(x_k)|^p \right)^{1/p} : x^* \in B_{X^*} \right\}. \qquad (8.6)$$

The least such constant C is denoted $\pi_p(T)$ and is called the *p-absolutely summing norm* of T.

Let us point out that, in practice, we will only use the most important cases, when $p = 1$ or $p = 2$. In fact, 2-absolutely summing operators play a very important role in further developments.

Theorem 8.2.4. *Let T be an operator between the Banach spaces X and Y. If $1 \le r < p < \infty$ and T is r-absolutely summing then T is p-absolutely summing with $\pi_p(T) \le \pi_r(T)$.*

Proof. Given $p > r$ let us pick q such that $1/p + 1/q = 1/r$. Suppose $(x_i)_{i=1}^n$ in X satisfy $(\sum_{i=1}^n |x^*(x_i)|^p)^{1/p} \le 1$ for all $x^* \in B_{X^*}$. Then for any $(c_i)_{i=1}^n$ scalars so that $(\sum_{i=1}^n |c_i|^q)^{1/q} \le 1$, using Hölder's inequality with the conjugate indices q/r and p/r we have

$$\left(\sum_{i=1}^n |c_i|^r |x^*(x_i)|^r \right)^{1/r} \le \left(\sum_{i=1}^n |x^*(x_i)|^p \right)^{1/p} \le 1, \qquad x^* \in B_{X^*}.$$

Hence

$$\left(\sum_{i=1}^n |c_i|^r \|Tx_i\|^r \right)^{1/r} \le \pi_r(T),$$

and by Hölder's inequality,

$$\left(\sum_{i=1}^n \|Tx_i\|^p \right)^{1/p} \le \pi_r(T).$$

Finally, a standard homogeneity argument immediately yields that

$$\left(\sum_{i=1}^n \|Tx_i\|^p \right)^{1/p} \le \pi_r(T) \sup_{\|x^*\| \le 1} \left(\sum_{i=1}^n |x^*(x_i)|^p \right)^{1/p},$$

for any choice of vectors $(x_i)_{i=1}^n$ in X. That is, T is p-absolutely summing and $\pi_p(T) \le \pi_r(T)$.

$\square$

Before proceeding, let us note the obvious *ideal* properties of the absolutely summing norms whose proof we leave for the Problems.

Proposition 8.2.5. *Suppose* $1 \le p < \infty$.

(i) If $S, T : X \to Y$ *are p-absolutely summing operators then* $S + T$ *is also p-absolutely summing and* $\pi_p(S + T) \le \pi_p(S) + \pi_p(T)$.

(ii) Suppose $T : X \to Y$, $S : Y \to Z$, *and* $R : Z \to W$ *are operators. If* S *is p-absolutely summing then so is* RST *and* $\pi_p(RST) \le \|R\|\pi_p(S)\|T\|$.

There is an extensive theory of operator ideals primarily developed by Pietsch and his school; we refer the reader to the recent survey [40].

Next we will recast the results of the previous section in the language of absolutely summing operators, but first let us make the following useful remark:

Remark 8.2.6. Suppose X is a subspace of $\mathcal{C}(K)$, where K is a compact Hausdorff topological space. Using Jensen's inequality, and the fact that $\nu \in \mathcal{C}(K)^* = \mathcal{M}(K)$ is an extreme point of the unit ball of $\mathcal{C}(K)^*$ if and only if $\nu = \pm \delta_s$, where $\delta_s(f) = f(s)$ for $f \in \mathcal{C}(K)$, given any $(f_j)_{j=1}^n$ in X we have

$$\sup_{x^* \in B_{X^*}} \sum_{j=1}^n |x^*(f_j)|^p = \sup \left\{ \sum_{j=1}^n \left| \int_K f_j \, d\nu \right|^p : \nu \in B_{\mathcal{C}(K)^*} \right\}$$

$$\le \sup \left\{ \sum_{j=1}^n \int_K |f_j|^p \, d|\nu| : \nu \in B_{\mathcal{M}(K)} \right\}$$

$$= \max_{s \in K} \sum_{j=1}^n |f_j(s)|^p.$$

Theorem 8.2.7. *Let* K *be a compact Hausdorff space and let* μ *be a* σ-*finite measure. Then every bounded operator* $T : \mathcal{C}(K) \to L_1(\mu)$ *is 2-absolutely summing with* $\pi_2(T) \le K_G \|T\|$.

Proof. Using Lemma 6.2.16 in combination with Theorem 8.1.5, given any $(f_i)_{i=1}^n$ in $\mathcal{C}(K)$ we obtain

$$\left(\sum_{i=1}^n \|Tf_i\|_1^2 \right)^{1/2} = \left(\sum_{i=1}^n \left\| |Tf_i|^2 \right\|_{1/2} \right)^{1/2}$$

$$\le \left\| \sum_{i=1}^n |Tf_i|^2 \right\|_{1/2}^{1/2}$$

$$= \left\| \left(\sum_{i=1}^n |Tf_i|^2 \right)^{\frac{1}{2}} \right\|_1$$

$$\leq K_G \|T\| \left\| \left(\sum_{i=1}^{n} |f_i|^2 \right)^{\frac{1}{2}} \right\|_{\infty}.$$

To complete the proof we need only observe that

$$\left\| \left(\sum_{i=1}^{n} |f_i|^2 \right)^{\frac{1}{2}} \right\|_{\infty} = \max_{s \in K} \left(\sum_{i=1}^{n} |f_i(s)|^2 \right)^{1/2}$$

$$= \max_{s \in K} \left(\sum_{i=1}^{n} |\delta_s(f_i)|^2 \right)^{1/2}$$

$$= \sup \left\{ \left(\sum_{i=1}^{n} |x^*(f_i)|^2 \right)^{1/2} : x^* \in B_{\mathcal{C}(K)^*} \right\}.$$

$\square$

The next theorem is a fundamental link with factorization theory. It is due to Pietsch (1966) [181].

Theorem 8.2.8. *Suppose X is a closed subspace of $\mathcal{C}(K)$ (K compact Hausdorff). An operator T from X into a Banach space Y is p-absolutely summing for some $1 \leq p < \infty$ with $\pi_p(T) \leq C$ if and only if there is a regular Borel probability measure ν on K so that for all $f \in X$,*

$$\|Tf\| \leq C \left(\int_K |f|^p \, d\nu \right)^{1/p}. \tag{8.7}$$

Proof. Assume first that $0 \neq T$ is a p-absolutely summing operator. We will use Lemma 7.3.5 to find a linear functional $\mathcal{L}$ on $\mathcal{C}(K)$ satisfying:

$$\mathcal{L}(f) \leq \max_{s \in K} f(s), \qquad f \in \mathcal{C}(K) \tag{8.8}$$

and

$$\pi_p(T)^{-p} \|Tf\|^p \leq \mathcal{L}(|f|^p), \qquad f \in X. \tag{8.9}$$

To this end, suppose we have functions $f_1, \ldots, f_n \in \mathcal{C}(K)$, $g_1, \ldots, g_m \in X$, and nonnegative scalars $\alpha_1, \ldots, \alpha_n, \beta_1, \ldots, \beta_m$ such that

$$\sum_{i=1}^{n} \alpha_i f_i = \sum_{j=1}^{m} \beta_j |g_j|^p.$$

Then

$$\pi_p(T)^{-p} \sum_{j=1}^{m} \beta_j \|Tg_j\|^p \leq \max_{s \in K} \sum_{j=1}^{m} \beta_j |g_j(s)|^p$$

$$= \max_{s \in K} \sum_{i=1}^{n} \alpha_i f_i(s)$$

$$\leq \sum_{i=1}^{n} \alpha_i \max_{s \in K} f_i(s).$$

This guarantees the existence of a linear functional $\mathcal{L}$ on $\mathcal{C}(K)$ verifying both (8.8) and (8.9). In particular, $\mathcal{L}$ is a positive functional since $\mathcal{L}(f) \leq 0$ whenever $f < 0$, and $\mathcal{L}(-1) \leq -1$. By the Riesz representation theorem there is a regular Borel probability measure ν on K so that $\mathcal{L}f = \int_K f \, d\nu$ for all $f \in \mathcal{C}(K)$. It is then clear that ν solves our problem.

Suppose conversely that there is a regular Borel probability measure ν on K so that for all $f \in X$,

$$\|Tf\|^p \leq C^p \int_K |f|^p \, d\nu.$$

Then for any $f_1, \ldots, f_n \in X$ we have

$$\sum_{j=1}^{n} \|Tf_j\|^p \leq C^p \sum_{j=1}^{n} \int_K |f_j|^p \, d\nu$$

$$\leq C^p \max_{s \in K} \sum_{j=1}^{n} |f_j(s)|^p$$

$$= C^p \sup \left\{ \sum_{j=1}^{n} \left| \int_K f_j \, d\nu \right|^p : \nu \in \mathcal{M}(K), \|\nu\| = 1 \right\}.$$

$\square$

Remark 8.2.9. Notice that we just showed that, if ν is a probability measure on some compact Hausdorff space K, then the inclusion maps $j_p : \mathcal{C}(K) \to L_p(K, \nu)$ and $\iota_p : L_\infty(K, \nu) \to L_p(K, \nu)$ are canonical examples of p-absolutely summing operators $(1 \leq p < \infty)$.

Since every Banach space X can be considered as a closed subspace of $\mathcal{C}(B_{X^*})$ (where B_{X^*} has the weak* topology), one usually states Theorem 8.2.8 in the following form:

Theorem 8.2.10 (Pietsch Factorization Theorem). *An operator $T : X \to Y$ is p-absolutely summing if and only if there is a regular Borel probability measure ν on B_{X^*} (in its weak* topology) so that for each $x \in X$*

$$\|Tx\| \leq \pi_p(T) \left(\int_{B_{X^*}} |x^*(x)|^p \, d\nu(x^*) \right)^{1/p}. \tag{8.10}$$

Interpretation. Let us denote by $j_p : \mathcal{C}(B_{X^*}) \to L_p(B_{X^*}, \nu)$ the canonical inclusion map and by X_p the closure in $L_p(B_{X^*}, \nu)$ of the natural copy of X in $\mathcal{C}(B_{X^*})$. Then we can induce an operator $S : X_p \to Y$ with $\|S\| = \pi_p(T)$ and so that $T = S \circ j_p$. We thus have the following picture:

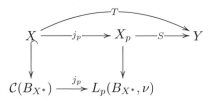

Remark 8.2.11. The case $p = 2$ is special. Suppose $T : X \to Y$ is 2-absolutely summing. Then, since there is an orthogonal projection from $L_2(B_{X^*}, \nu)$ onto the subspace X_2, we can factor T in the following manner:

$$
\begin{array}{ccc}
X & \xrightarrow{\quad T \quad} & Y \\
\downarrow & & \uparrow {\scriptstyle \tilde{S}} \\
\mathcal{C}(B_{X^*}) & \xrightarrow{\;\; j_2 \;\;} & L_2(B_{X^*}, \nu)
\end{array}
$$

An immediate consequence is

Theorem 8.2.12. *If an operator $T : X \to Y$ is 2-absolutely summing then it factors through a Hilbert space.*

Theorem 8.2.13. *Suppose that X, Y are Banach spaces and that E is a closed subspace of X. Suppose the operator $T : E \to Y$ is 2-absolutely summing. Then there exists a 2-absolutely summing operator $\tilde{T} : X \to Y$ such that $\tilde{T}|_E = T$ and $\pi_2(\tilde{T}) = \pi_2(T)$.*

Proof. We can factor the operator $T : E \to Y$ using Remark 8.2.11:

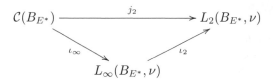

On the other hand, the natural inclusion $j_2 : \mathcal{C}(B_{E^*}) \to L_2(B_{E^*}, \nu)$ admits a factorization through $L_\infty(B_{E^*}, \nu)$:

$$
\begin{array}{ccc}
\mathcal{C}(B_{E^*}) & \xrightarrow{\qquad\quad j_2 \qquad\quad} & L_2(B_{E^*}, \nu) \\
{\scriptstyle \iota_\infty} \searrow & & \nearrow {\scriptstyle \iota_2} \\
& L_\infty(B_{E^*}, \nu) &
\end{array}
$$

If we combine these two diagrams we see that the operator $\iota_\infty \circ \iota_E$ maps continuously E into $L_\infty(B_{E^*}, \nu)$, which is an isometrically injective space. Thus $\iota_\infty \circ \iota_E$ can be extended with preservation of norm to an operator R defined on X:

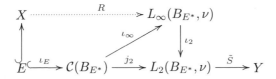

Clearly, the operator $\tilde{T} = \tilde{S}\iota_2 R$ is an extension of T to X. From Proposition 8.2.5 (ii) and Remark 8.2.9 we deduce that $\tilde{T}$ is 2-absolutely summing and, since $\pi_2(\iota_2) = 1$, $\|R\| = 1$, and $\|\tilde{S}\| = \pi_2(T)$, it follows that $\pi_2(\tilde{T}) = \pi_2(T)$.

$\square$

We can now answer the question we raised on the converse of the Riemann theorem. This result was proved by Dvoretzky and Rogers [50] in 1950, which predates the entire theory of absolutely summing operators. In fact the proof of Dvoretzky and Rogers that we will touch on later (see Proposition 12.3.4 and Problem 12.8) is quite different and relies on geometrical ideas. With the passage of time the theorem looks a lot easier today than it did in 1950!

Theorem 8.2.14 (Dvoretzky-Rogers Theorem). *Let X be a Banach space such that every unconditionally convergent series in X is absolutely convergent. Then X is finite-dimensional.*

Proof. By Proposition 8.2.2, our hypothesis is equivalent to saying that the identity operator $I_X : X \to X$ is absolutely summing; hence it is also 2-absolutely summing by Theorem 8.2.4. Now by Theorem 8.2.12 we deduce that X is isomorphic to a Hilbert space. But we have already seen that any infinite-dimensional Hilbert space contains an unconditionally convergent series which is not absolutely convergent, namely, $\sum_{n=1}^{\infty} \frac{1}{n}e_n$, where $(e_n)_{n=1}^{\infty}$ is an orthonormal sequence.

$\square$

If we combine Theorem 8.2.7 and Theorem 8.2.8 we obtain an alternative way to see that every operator $T : \mathcal{C}(K) \to L_1(\mu)$ factors through a Hilbert space. This approach is dual to the methods of the previous section, like for example in Theorem 8.1.6 . We are in effect introducing a "density" on K rather than on Ω.

Corollary 8.2.15. *If X and Y are Banach spaces and $T : X \to Y$ is a p-absolutely summing operator for some $1 \le p < \infty$ then T is Dunford-Pettis and weakly compact. In particular, if $T : X \to Y$ is p-absolutely summing and X is reflexive then T is compact.*

Proof. Without loss of generality we can assume $p > 1$. The Pietsch factorization theorem tells us that T factors through a subspace X_p of $L_p(B_{X^*}, \nu)$ for some probability measure ν, hence T must be weakly compact by the reflexivity of X_p.

Assume now that (x_n) is a weakly null sequence in X. By equation (8.10), for each x_n we have

$$\|Tx_n\| \leq \pi_p(T) \left(\int_{B_{X^*}} |x^*(x_n)|^p \, d\nu(x^*) \right)^{1/p}.$$

The Lebesgue Dominated Convergence theorem easily yields that $\lim_{n\to\infty} \|Tx_n\| = 0$, so T is Dunford-Pettis.

$\square$

We conclude this section by identifying the 2-absolutely summing operators on a Hilbert space with the well-known class of Hilbert-Schmidt operators. In a certain sense we can regard the class of 2-absolutely summing operators as the natural generalization to arbitrary Banach spaces of this class.

Definition 8.2.16. Suppose H_1, H_2 are separable Hilbert spaces. We assume H_1, H_2 infinite-dimensional for notational convenience. An operator $T : H_1 \to H_2$ is said to be *Hilbert-Schmidt* if $\sum_{n=1}^{\infty} \|Te_n\|^2 < \infty$ for some orthonormal basis $(e_n)_{n=1}^{\infty}$ of H_1.

Let $(e_n)_{n=1}^{\infty}$ be an orthonormal basis of H_1 and $(f_n)_{n=1}^{\infty}$ be an orthonormal basis of H_2. Then, by Parseval's identity,

$$\sum_{n=1}^{\infty} \|Te_n\|^2 = \sum_{n=1}^{\infty}\sum_{k=1}^{\infty} |\langle Te_n, f_k\rangle|^2 = \sum_{k=1}^{\infty}\sum_{n=1}^{\infty} |\langle e_n, T^*f_k\rangle|^2 = \sum_{k=1}^{\infty} \|T^*f_k\|^2.$$
$$(8.11)$$

This implies that the expression $\sum_{n=1}^{\infty} \|Te_n\|^2$ is independent of the choice of orthonormal basis in H_1. The quantity

$$\|T\|_{HS} = \left(\sum_{n=1}^{\infty} \|Te_n\|^2 \right)^{1/2}$$

is called the *Hilbert-Schmidt norm of T*. Notice that equation (8.11) also shows that $\|T\|_{HS} = \|T^*\|_{HS}$, so $T : H_1 \to H_2$ is Hilbert-Schmidt if and only if $T^* : H_2 \to H_1$ is.

Remark 8.2.17. (a) If $T : H \to H$ is Hilbert-Schmidt then $\|T\| \leq \|T\|_{HS}$.

(b) If $T : H \to H$ is Hilbert-Schmidt then T is compact. Indeed, take $(P_m)_{m=1}^{\infty}$ the partial sum projections associated to an orthonormal basis (e_n) of H and let I_H be the identity operator on H. Then,

$$\|T - TP_m\|_{HS} = \|T(I_H - P_m)\|_{HS} = \|T|_{[e_j; j>m+1]}\|_{HS} \to 0.$$

Therefore $\|T - TP_m\| \to 0$. Since $(TP_m)_{m=1}^{\infty}$ are finite-rank operators, it follows that T is compact.

Theorem 8.2.18. *An operator $T : H_1 \to H_2$ is Hilbert-Schmidt if and only if T is 2-absolutely summing. Furthermore, $\|T\|_{HS} = \pi_2(T)$.*

Proof. Suppose first that T is 2-absolutely summing. If $(e_j)_{j=1}^\infty$ is an orthonormal basis of H_1 then for each $n \in N$ we have

$$\sup\left\{\left(\sum_{j=1}^n |\langle e_j, x\rangle|^2\right)^{1/2} : x \in H_1, \|x\| \le 1\right\} = 1,$$

and so

$$\left(\sum_{j=1}^n \|Te_j\|^2\right)^{1/2} \le \pi_2(T).$$

Hence T is Hilbert-Schmidt and $\|T\|_{HS} \le \pi_2(T)$.

Suppose conversely that T is Hilbert-Schmidt. Let $(x_j)_{j=1}^n$ in H_1 have the property that

$$\sup\left\{\left(\sum_{j=1}^n |\langle x_j, x\rangle|^2\right)^{1/2} : x \in H_1, \|x\| \le 1\right\} \le 1.$$

Then the operator $S : H_1 \to H_1$ defined by $Se_j = x_j$ for $1 \le j \le n$ and $Se_j = 0$ for $j > n$ satisfies $\|S\| \le 1$. Hence

$$\|TS\|_{HS} = \|S^*T^*\|_{HS} \le \|T^*\|_{HS} = \|T\|_{HS}.$$

Thus

$$\sum_{j=1}^n \|Tx_j\|^2 = \sum_{j=1}^n \|TSe_j\|^2 \le \|T\|_{HS}^2,$$

which implies that T is 2-absolutely summing with $\pi_2(T) \le \|T\|_{HS}$.   □

8.3 Absolutely summing operators on $L_1(\mu)$-spaces and an application to uniqueness of unconditional bases

We now revisit Grothendieck's inequality to obtain another rather startling application from Grothendieck's Résumé [76].

Theorem 8.3.1. *Suppose* $T : L_1(\mu) \to \ell_2$ *is a bounded operator. Then* T *is absolutely summing and* $\pi_1(T) \le K_G\|T\|$.

Proof. Suppose $(f_i)_{i=1}^n$ in $L_1(\mu)$ are such that

$$\sup\left\{\sum_{i=1}^n \left|\int_\Omega f_i g\, d\mu\right| : g \in L_\infty(\mu), \|g\|_\infty \le 1\right\} \le 1.$$

We must show that $\sum_{i=1}^{n} \|Tf_i\| \leq K_G\|T\|$. Notice that it is enough to prove the latter inequality when $(f_i)_{i=1}^{n}$ are simple functions so that there is decomposition of Ω into finitely many measurable sets $A_1, \ldots, A_m$ of positive measure so that each f_i is a linear combination of $\{\chi_{A_j}\}_{j=1}^{m}$. Thus it suffices to prove the result for an operator $T : \ell_1^m \to \ell_2^m$.

Let $T : \ell_1^m \to \ell_2^m$ with $\|T\| \leq 1$. Suppose $(x_i)_{i=1}^{n}$ in ℓ_1^m satisfy

$$\sup \left\{ \sum_{i=1}^{n} |\langle x_i, \eta \rangle| : \eta \in \ell_\infty^n, \|\eta\|_\infty \leq 1 \right\} \leq 1.$$

If for each $1 \leq i \leq n$ we let $x_i = (x_{ik})_{k=1}^{m}$, it is easy to see that

$$\left| \sum_{i=1}^{n} \sum_{k=1}^{n} x_{ik} s_i t_k \right| \leq 1$$

whenever $\max |s_i|, \max |t_k| \leq 1$.

Let $(e_k)_{k=1}^{m}$ denote the canonical basis of ℓ_1^m and put $u_k = Te_k \in \ell_2^m$. By our assumption on T, $\|u_j\|_2 \leq 1$. For $1 \leq j \leq n$ pick v_j so that $\langle T\xi_j, v_j \rangle = \|T\xi_j\|_2$ and $\|v_j\|_2 = 1$. Then

$$\sum_{j=1}^{n} \|T\xi_j\|_2 = \sum_{j=1}^{n} \langle T\xi_j, v_j \rangle$$

$$= \sum_{j=1}^{n} \sum_{k=1}^{m} \xi_{jk} \langle u_k, v_j \rangle$$

$$\leq K_G$$

by Grothendieck's inequality 8.1.1. This establishes the result.

$\square$

Remark 8.3.2. (a) Since ℓ_1 is an $L_1(\mu)$-space for a suitable μ, Theorem 8.3.1 holds for operators $T : \ell_1 \to \ell_2$. In particular it also holds for a quotient map of ℓ_1 onto ℓ_2. This is in sharp contrast to the fact that every absolutely p-summing operator (for any p) on a reflexive space is compact.

(b) Theorem 8.3.1 is actually equivalent to Grothendieck's inequality in the sense that Grothendieck's inequality could equally be derived from this theorem. It is also equivalent to either Theorem 8.1.5 or Theorem 8.2.7.

Lindenstrauss and Pełczyński [131] discovered a very neat application of Theorem 8.3.1 to the isomorphic theory of Banach spaces by showing that the spaces c_0 and ℓ_1 have essentially (i.e., up to equivalence) only one unconditional basis, namely, the unit vector basis. It is almost unfortunate that, later, Johnson (cf. [91]) found an "elementary" proof which completely circumvents the use of Grothendieck's inequality!

Theorem 8.3.3. *Every normalized unconditional basis in ℓ_1 [respectively, c_0] is equivalent to the canonical basis of ℓ_1 [respectively, c_0].*

Proof. Assume that $(u_n)_{n=1}^{\infty}$ is a normalized K-unconditional basis in ℓ_1. For any sequence of scalars (a_i) we have

$$\Big(\sum_{i=1}^{n}|a_i|^2\Big)^{1/2} \le C_2(\ell_1)\Big(\mathbb{E}\Big\|\sum_{i=1}^{n}\varepsilon_i a_i u_i\Big\|^2\Big)^{1/2} \le C_2(\ell_1)K\Big\|\sum_{i=1}^{n}a_i u_i\Big\|,$$

where $C_2(\ell_1)$ is the cotype-2 constant of ℓ_1. From here it follows that the operator $T : \ell_1 \to \ell_2$ defined by

$$T\Big(\sum_{i=1}^{\infty}a_i u_i\Big) = (u_i^*(x))_{i=1}^{\infty} = (a_1, a_2, \ldots, a_i, \ldots)$$

is bounded with $\|T\| \le C_2(\ell_1)K$. Therefore, by Theorem 8.3.1, T is absolutely summing and $\pi_1(T) \le K_G C_2(\ell_1)K$. Thus

$$\sum_{i=1}^{n}|a_i| = \sum_{i=1}^{n}\|T(a_i u_i)\|$$

$$\le K_G C_2(\ell_1)K \sup_{\varepsilon_i=\pm 1}\Big\|\sum_{i=1}^{n}\varepsilon_i a_i u_i\Big\|$$

$$\le K_G C_2(\ell_1)K^2\Big\|\sum_{i=1}^{n}a_i u_i\Big\|,$$

which shows that $(u_n)_{n=1}^{\infty}$ is equivalent to the canonical basis of ℓ_1.

Suppose now that $(u_n)_{n=1}^{\infty}$ is a normalized K-unconditional basis of c_0. We know that every unconditional basis of c_0 is shrinking by James's theorem (Theorem 3.3.1), hence the biorthogonal functionals $(u_n^*)_{n=1}^{\infty}$ form an unconditional basis of ℓ_1. By the first part of the proof, $(u_n^*/\|u_n^*\|)_{n=1}^{\infty}$ is equivalent to the canonical basis of ℓ_1 and, since $1 \le \|u_n^*\| \le K$, $(u_n^*)_{n=1}^{\infty}$ is equivalent to the canonical ℓ_1-basis. Hence there exists a constant M (depending only on the basis (u_n)) so that for each $x^* \in \ell_1 = c_0^*$ we have

$$\sum_{k=1}^{n}|x^*(u_k)| \le M\|x^*\|.$$

Then for any scalars (a_i) and each $x^* \in B_{\ell_1}$,

$$\Big|x^*\Big(\sum_{i=1}^{n}a_i u_i\Big)\Big| \le M \max_{1\le i\le n}|a_i|,$$

that is,

$$\Big\|\sum_{i=1}^{n}a_i u_i\Big\| \le M \max_{1\le i\le n}|a_i|.$$

The other estimate follows immediately from the unconditionality of (u_n).

□

Remark 8.3.4. Notice that the argument of Theorem 8.3.3 could be applied to an unconditional basis of L_1; the conclusions would be that every normalized unconditional basis of L_1 is equivalent to the canonical basis of ℓ_1. Since L_1 is not isomorphic to ℓ_1 this provides yet another proof that L_1 has no unconditional basis. Similarly any $\mathcal{C}(K)$-space with unconditional basis must be isomorphic to c_0 (we have already seen this for quite different reasons in Chapter 4, Theorem 4.5.2).

Let us observe that unconditional bases of Hilbert spaces also share this uniqueness property:

Theorem 8.3.5. *If $(u_n)_{n=1}^{\infty}$ is a normalized unconditional basis of a Hilbert space then $(u_n)_{n=1}^{\infty}$ is equivalent to the canonical basis of ℓ_2.*

Proof. Let K be the unconditional basis constant of $(u_n)_{n=1}^{\infty}$. The unconditionality of the basis and the generalized parallelogram law yield

$$\left\| \sum_{i=1}^{n} a_i u_i \right\| \leq K \left(\mathbb{E} \left\| \sum_{i=1}^{n} \varepsilon_i a_i u_i \right\|^2 \right)^{1/2} = K \left(\sum_{i=1}^{n} |a_i|^2 \right)^{1/2},$$

for any scalars (a_i). The other estimate we need to show the equivalence of bases follows in the same way.

□

Definition 8.3.6. If X is a Banach space with a normalized unconditional basis $(e_n)_{n=1}^{\infty}$ we say that X has *unique unconditional basis* if whenever $(u_n)_{n=1}^{\infty}$ is another normalized unconditional basis of X, then $(u_n)_{n=1}^{\infty}$ is equivalent to $(e_n)_{n=1}^{\infty}$. That is, there is a constant D so that

$$D^{-1} \left\| \sum_{n=1}^{\infty} a_n u_n \right\| \leq \left\| \sum_{n=1}^{\infty} a_i e_i \right\| \leq D \left\| \sum_{n=1}^{\infty} a_n u_n \right\|,$$

for any $(a_n)_{n=1}^{\infty} \in c_{00}$.

The fact that the three spaces ℓ_1, ℓ_2, and c_0 have the property of uniqueness of unconditional basis leads us to consider what other spaces might have the same property. We will resolve this problem later, but let us first show how to construct essentially different unconditional bases in ℓ_p when $1 < p < \infty$ and $p \neq 2$. This is due to Pełczyński [169] and it beautifully illustrates the usage of "L_p-methods" to deduce properties about their relatives, the spaces ℓ_p.

Proposition 8.3.7. *If $1 < p < \infty$, $p \neq 2$, then ℓ_p has at least two nonequivalent unconditional bases.*

Proof. Let $1 < p < \infty$, $p \neq 2$. We saw in Proposition 6.4.2 that the operator P defined in L_p by

$$P(f) = \sum_{k=1}^{\infty} \left(\int_0^1 f(t) r_k(t)\, dt \right) r_k$$

is a projection onto R_p, the closed subspace spanned in L_p by the Rademacher functions. For each n let $F_p^{(n)}$ denote the subspace of L_p spanned by the characteristic functions on the dyadic intervals of the family $[\frac{k}{2^n}, \frac{k+1}{2^n}]$, $k = 0, 1, \ldots, 2^{n-1}$, and let $R_p^{(n)} = [r_k]_{k=1}^n$. Clearly the space $F_p^{(n)}$ is isometric to $\ell_p^{2^n}$ and $R_p^{(n)}$ is isometric to ℓ_2^n. Moreover, $P|_{F_p^{(n)}}$ is a projection from $F_p^{(n)}$ onto its subspace $R_p^{(n)}$ (with projection constant independent of n). It is easy to see that this defines (coordinatewise) a projection from $\ell_p(F_p^{(n)})$ onto $\ell_p(R_p^{(n)})$. Obviously $\ell_p(F_p^{(n)})$ is isometric to $\ell_p(\ell_p^{2^n}) = \ell_p$ and $\ell_p(R_p^{(n)})$ is isometric to $\ell_p(\ell_2^n)$. Since ℓ_p is prime and $\ell_p(\ell_2^n)$ is complemented in ℓ_p, it follows that $\ell_p(\ell_2^n)$ is isomorphic to ℓ_p.

Then, if ℓ_p had a unique unconditional basis, in particular the canonical basis of ℓ_p and the canonical basis of $\ell_p(\ell_2^n)$ would be equivalent, which is not true.

$\square$

Problems

8.1. Grothendieck's original proof of the Grothendieck inequality.
(a) Let g_1, g_2 be (normalized) Gaussians. Show that

$$\mathbb{E}(\operatorname{sgn} g_1)(\operatorname{sgn} (g_1 \cos \theta + g_2 \sin \theta)) = 1 - \frac{2}{\pi}\theta, \qquad 0 \leq \theta \leq \pi.$$

Now let X be the space of $m \times n$ real matrices with the norm

$$\|A\|_X = \sup_{|s_i| \leq 1} \sup_{|t_j| \leq 1} \left| \sum_{i=1}^m \sum_{j=1}^n a_{ij} s_i t_j \right|$$

and define the *multiplier norm* of an $m \times n$ matrix B by

$$\|B\|_{\mathcal{M}} = \sup_{\|A\|_X \leq 1} \|B \cdot A\|_X,$$

where $B \cdot A$ is the matrix $(b_{ij} a_{ij})_{i,j=1}^{m,n}$.

(b) Let $u_i, v_j \in \ell_2^N$ for $i = 1, 2 \ldots, m$ and $j = 1, 2, \ldots, n$. Suppose $\|u_i\|_2 = \|v_j\|_2 = 1$ for all i, j. By considering $\sum_{k=1}^N u_i(k) g_k$ and $\sum_{k=1}^N v_j(k) g_k$ where $g_1, \ldots, g_N$ are normalized Gaussians show that

$$\left\| \left(1 - \frac{2}{\pi}\theta_{ij}\right)_{i,j=1}^{m,n} \right\|_{\mathcal{M}} \leq 1,$$

where θ_{ij} is the unique solution of $0 \leq \theta_{ij} \leq \pi$ and $\cos\theta_{ij} = \langle u_i, v_j \rangle$.

(c) Using the fact that $\cos\theta_{ij} = \sin(\pi/2 - \theta_{ij})$ show that

$$\|(\cos\theta_{ij})_{i,j=1}^{m,n}\|_m \leq \sinh\frac{\pi}{2}.$$

(d) Deduce Grothendieck's inequality with $K_G \leq \sinh\frac{\pi}{2}$.

8.2. (a) Show that Grothendieck's inequality is equivalent to the statement that every bounded operator $T : \ell_1 \to \ell_2$ is absolutely summing (Theorem 8.3.1).

(b) Deduce that Grothendieck's inequality is equivalent to the statement there is a quotient map $Q : X \to \ell_2$ which is absolutely summing for some separable Banach space X.

Using (a) and (b), Pełczyński and Wojtaszczyk proved that Grothendieck's inequality follows from a classical inequality of Paley (if rather indirectly) [176].

8.3. Prove Proposition 8.2.2.

8.4. Prove Proposition 8.2.5.

8.5. Prove that the identity operator I_X on an infinite-dimensional Banach space X is never p-absolutely summing for any $p < \infty$.

8.6. Prove the dual form of Theorem 8.1.7: Suppose X is a Banach space that has cotype 2. Then every operator $T : \mathcal{C}(K) \to X$ factors through a Hilbert space and hence T is 2-absolutely summing.
Deduce that if $T : c_0 \to X$, then there exist $a_n \geq 0$ with $\sum_{n=1}^{\infty} a_n = 1$ and

$$\|T(\xi)\| \leq C\left(\sum_{j=1}^{\infty} |\xi(j)|^2 a_j\right)^{1/2}.$$

8.7. (a) Show if $T : c_0 \to \ell_2$ is a bounded operator and $S : \ell_2 \to \ell_2$ is Hilbert-Schmidt then (if $(e_n)_{n=1}^{\infty}$ is the canonical basis),

$$\sum_{n=1}^{\infty} \|STe_n\| < \infty.$$

(b) Deduce (using Problem 8.6) that if X has cotype 2 then any 2-absolutely summing operator $R : X \to \ell_2$ is absolutely summing.

8.8. (a) Let $T : \ell_2 \to \ell_2$ be a p-absolutely summing operator where $p > 2$. Show that T is Hilbert-Schmidt.

(b) Conversely if T is Hilbert-Schmidt show that T is absolutely summing.

These results are due to Pietsch [181] and Pełczyński [173]. The best constants involved were found by Garling [64].

8.9. (a) Let X be a Banach space. Show that an operator $T : X \to \ell_2$ is 2-absolutely summing if and only if for every operator $S : \ell_2 \to X$ the composition TS is Hilbert-Schmidt.

(b) Show that if every operator $T : X^* \to \ell_2$ is 2-absolutely summing then every operator $T : X \to \ell_2$ is also 2-absolutely summing.

9

Perfectly Homogeneous Bases and Their Applications

In this chapter we first prove a characterization of the canonical bases of the spaces ℓ_p $(1 \leq p < \infty)$ and c_0 due to Zippin [223]. In the remainder of the chapter we show how this is used in several different contexts to prove general theorems by reduction to the ℓ_p case. For example, we show that the Lindenstrauss-Pełczyński theorem on the uniqueness of the unconditional basis in c_0, ℓ_1, and ℓ_2 (Theorem 8.3.3) has a converse due to Lindenstrauss and Zippin; these are the only three such spaces. We also deduce a characterization of c_0 and ℓ_p in terms of complementation of block basic sequences due to Lindenstrauss and Tzafriri [135] and apply it to prove a result of Pełczyński and Singer [177] on the existence of conditional bases in any Banach space with a basis.

9.1 Perfectly homogeneous bases

The canonical bases of ℓ_p and c_0 have a very special property in that *every* normalized block basic sequence is equivalent to the original basis (Lemma 2.1.1). This property was given the name *perfect homogeneity*.

In the 1960s several papers appeared which proved results for a Banach space with a perfectly homogeneous basis mimicking known result for the ℓ_p-spaces. However, it turns out that this property actually characterizes the canonical bases of the ℓ_p-spaces! This is a very useful result proved in 1966 by Zippin [223]. Thus the concept is quite redundant.

We shall define perfectly homogeneous bases in a slightly different way, which is, with hindsight, equivalent.

Definition 9.1.1. A block basis sequence $(u_n)_{n=1}^{\infty}$ of a basis $(e_n)_{n=1}^{\infty}$,

$$u_n = \sum_{p_{n-1}+1}^{p_n} a_i e_i,$$

is a *constant coefficient block basic sequence* if for each n there is a constant c_n so that $a_i = c_n$ or $a_i = 0$ for $p_{n-1} + 1 \leq i \leq p_n$; that is,

$$u_n = c_n \sum_{i \in A_n} e_i,$$

where A_n is a subset of integers contained in $(p_{n-1}, p_n]$.

Definition 9.1.2. A basis $(e_n)_{n=1}^{\infty}$ of a Banach space X is *perfectly homogeneous* if every normalized constant coefficient block basic sequence of $(e_n)_{n=1}^{\infty}$ is equivalent to $(e_n)_{n=1}^{\infty}$.

This definition is enough to force any perfectly homogeneous basis to be unconditional since $(e_n)_{n=1}^{\infty}$ must be equivalent to $(\epsilon_n e_n)_{n=1}^{\infty}$ for every choice of signs $\epsilon_n = \pm 1$.

Lemma 9.1.3. *Let $(e_n)_{n=1}^{\infty}$ be a normalized perfectly homogeneous basis of a Banach space X. Then $(e_n)_{n=1}^{\infty}$ is uniformly equivalent to all its normalized constant coefficient block basic sequences. That is, there is a constant $M \geq 1$ such that for any normalized constant coefficient block basic sequences $(u_n)_{n=1}^{\infty}$ and $(v_n)_{n=1}^{\infty}$ of $(e_n)_{n=1}^{\infty}$ we have*

$$M^{-1} \left\| \sum_{k=1}^{n} a_k u_k \right\| \leq \left\| \sum_{k=1}^{n} a_k v_k \right\| \leq M \left\| \sum_{k=1}^{n} a_k u_k \right\|,$$

for any choice of scalars $(a_i)_{i=1}^{n}$ and every $n \in \mathbb{N}$.

Proof. It suffices to prove such an inequality for the basic sequence $(e_n)_{n=n_0+1}^{\infty}$ for some n_0. If the lemma fails, we can inductively build constant coefficient block basic sequences $(u_n)_{n=1}^{\infty}$ and $(v_n)_{n=1}^{\infty}$ of $(e_n)_{n=1}^{\infty}$ so that for some increasing sequence of integers $(p_n)_{n=0}^{\infty}$ with $p_0 = 0$ and some scalars $(a_i)_{i=1}^{\infty}$ we have

$$\left\| \sum_{i=p_{n-1}+1}^{p_n} a_i u_i \right\| < 2^{-n},$$

but

$$\left\| \sum_{i=p_{n-1}+1}^{p_n} a_i v_i \right\| > 2^{-n},$$

which contradicts the assumption of perfect homogeneity.

$\square$

Let us suppose that $(e_n)_{n=1}^{\infty}$ is a normalized basis for a Banach space X. For each $n \in \mathbb{N}$ put

$$\lambda(n) = \left\| \sum_{k=1}^{n} e_k \right\|.$$

Obviously,

$$K^{-1} \leq \lambda(n) \leq n, \qquad n \in \mathbb{N}, \tag{9.1}$$

where $K \geq 1$ is the basis constant. Notice that if $(e_n)_{n=1}^{\infty}$ is 1-unconditional then the sequence $(\lambda(n))_{n=1}^{\infty}$ is nondecreasing.

Lemma 9.1.4. *Suppose that $(e_n)_{n=1}^{\infty}$ is a normalized, unconditional basis of a Banach space X. If $\sup_n \lambda(n) < \infty$ then $(e_n)_{n=1}^{\infty}$ is equivalent to the canonical basis of c_0.*

Proof. For any n and any choice of signs $(\epsilon_i)_{i=1}^{n}$ we have

$$\left\| \sum_{j=1}^{n} \epsilon_j e_j \right\| \leq C,$$

where C depends on $\sup_n \lambda(n)$ and the unconditional basis constant of $(e_n)_{n=1}^{\infty}$. Hence, by Lemma 2.4.6, $\sum e_j$ is a WUC series and so $\sum a_j e_j$ converges for all $(a_n)_{n=1}^{\infty} \in c_0$. This shows that $(e_n)_{n=1}^{\infty}$ is equivalent to the canonical c_0-basis.

$\square$

Lemma 9.1.5. *Let $(e_i)_{i=1}^{\infty}$ be a normalized perfectly homogeneous basis of a Banach space X. Then, if M is the constant given by Lemma 9.1.3, we have*

$$\frac{1}{M^3} \lambda(n)\lambda(m) \leq \lambda(nm) \leq M^3 \lambda(n)\lambda(m) \tag{9.2}$$

for all m, n in $\mathbb{N}$.

Proof. Note that M can also serve as an unconditional constant (of course, not necessarily the optimal) for $(e_n)_{n=1}^{\infty}$.

Let us consider a family

$$f_j = \sum_{i=(j-1)n+1}^{jn} e_i, \qquad j = 1, \dots, m,$$

of m disjoint blocks of length n of the basis $(e_i)_{i=1}^{\infty}$. Let $c_j = \|f_j\|$ for $j = 1, \dots, m$. By hypothesis,

$$M^{-1}\lambda(n) \leq c_j \leq M\lambda(n), \qquad j = 1, 2, \dots, m,$$

and so

$$\frac{1}{M^2\lambda(n)} \left\| \sum_{j=1}^{m} f_j \right\| \leq \left\| \sum_{j=1}^{m} c_j^{-1} f_j \right\| \leq \frac{M^2}{\lambda(n)} \left\| \sum_{j=1}^{m} f_j \right\|.$$

Now, again by Lemma 9.1.3,

$$M^{-1}\lambda(m) \leq \left\| \sum_{j=1}^{m} c_j^{-1} f_j \right\| \leq M\lambda(m).$$

Hence,

$$\frac{\lambda(mn)}{M^3\lambda(n)} \le \lambda(m) \le \frac{M^3\lambda(mn)}{\lambda(n)}.$$

□

Before continuing we need the following lemma, which is very useful in many different contexts:

Lemma 9.1.6.

(i) Suppose that $(s_n)_{n=1}^{\infty}$ is a sequence of real numbers such that

$$s_{m+n} \le s_m + s_n, \quad m, n \in \mathbb{N}.$$

Then $\lim_{n\to\infty} \frac{s_n}{n}$ exists (possibly equal to $-\infty$) and

$$\lim_{n\to\infty} \frac{s_n}{n} = \inf_n \frac{s_n}{n}.$$

(ii) Suppose that $(s_n)_{n=1}^{\infty}$ is a sequence of real numbers such that

$$|s_{m+n} - s_m - s_n| \le 1$$

for all $m, n \in \mathbb{N}$. Then there is a constant c so that

$$|s_n - cn| \le 1, \qquad n = 1, 2, \ldots.$$

Proof. (i) Fix $n \in \mathbb{N}$. Then, each $m \in \mathbb{N}$ can be written as $m = ln + r$ for some $0 \le l$ and $0 \le r < n$. The hypothesis implies that

$$s_{ln} \le ls_n, \qquad s_{ln+r} \le ls_n + s_r.$$

Thus

$$\frac{s_m}{m} = \frac{s_{ln+r}}{ln+r} \le \frac{l}{ln+r}s_n + \frac{s_r}{ln+r} \le \frac{s_n}{n} + \frac{\max_{0 \le r < n} s_r}{m}$$

and so

$$\limsup_{m\to\infty} \frac{s_m}{m} \le \frac{s_n}{n}, \qquad n \in \mathbb{N}. \tag{9.3}$$

Hence,

$$\limsup_{m\to\infty} \frac{s_m}{m} \le \inf_n \frac{s_n}{n}.$$

(ii) Let $t_n = s_n + 1$ and $u_n = s_n - 1$. Then $(t_n)_{n=1}^{\infty}$ and $(-u_n)_{n=1}^{\infty}$ both obey the conditions of (i). Hence $\lim_{n\to\infty} t_n/n = \lim_{n\to\infty} u_n/n$ both exist and are finite; let c be their common value. By (i) we have

$$\frac{u_n}{n} \le c \le \frac{t_n}{n}, \qquad n = 1, 2, \ldots$$

and the conclusion follows.

□

Lemma 9.1.7. *Let $(e_n)_{n=1}^\infty$ be a normalized, perfectly homogeneous basis of a Banach space X. Then, either $(e_n)_{n=1}^\infty$ is equivalent to the canonical basis of c_0 or there exist a constant C and $1 \le p < \infty$ such that*

$$C^{-1}|A|^{\frac{1}{p}} \le \left\| \sum_{k \in A} e_k \right\| \le C|A|^{\frac{1}{p}},$$

for any finite subset A of $\mathbb{N}$.

Proof. If we use equation (9.2) with $m = 2^k$ and $n = 2^j$ we obtain

$$\frac{1}{M^3}\lambda(2^k)\lambda(2^j) \le \lambda(2^{j+k}) \le M^3\lambda(2^k)\lambda(2^j). \qquad (9.4)$$

For $k = 0, 1, 2, \ldots$ let $h(k) = \log_2 \lambda(2^k)$. From (9.4) we get

$$|h(j) + h(k) - h(j + k)| \le 3 \log_2 M.$$

By (ii) of Lemma 9.1.6 there is a constant c so that

$$|h(j) - cj| \le 3 \log_2 M, \qquad j = 1, 2, \ldots.$$

By equation (9.1), $K^{-1} \le \lambda(2^k) \le 2^k$ for each $k = 0, 1, 2, \ldots$, which implies $\log_2 K^{-1} \le h(k) \le k$, and so $0 \le c \le 1$.

If $c = 0$ we would have $\lambda(2^j) \le M^3$ for all $j \in \mathbb{N}$ hence $(\lambda(n))_{n=1}^\infty$ would be bounded and so $(e_n)_{n=1}^\infty$ would be equivalent to the canonical basis of c_0 by Lemma 9.1.4.

Otherwise, if $0 < c \le 1$, there is $p \in [1, \infty)$ such that $c = \frac{1}{p}$. Thus we can rewrite equation (9.4) in the form

$$\frac{1}{M^3}2^{\frac{j}{p}} \le \lambda(2^j) \le M^3 2^{\frac{j}{p}}, \quad j \in \mathbb{N}. \qquad (9.5)$$

Since for any n with $2^{j-1} \le n \le 2^j$ we have

$$M^{-1}\lambda(2^{j-1}) \le \lambda(n) \le M\lambda(2^j)$$

we conclude that

$$M^{-4}n^{\frac{1}{p}} \le \lambda(n) \le M^4 n^{\frac{1}{p}}.$$

Finally, if A is any finite subset of $\mathbb{N}$ we have

$$M^{-1}\lambda(|A|) \le \left\| \sum_{j \in A} e_j \right\| \le M\lambda(|A|)$$

and so the lemma follows with $C = M^5$. $\qquad \qquad \square$

We now come to Zippin's theorem [223].

Theorem 9.1.8 (Zippin). *Let X be a Banach space with normalized basis $(e_n)_{n=1}^{\infty}$. Suppose that $(e_n)_{n=1}^{\infty}$ is perfectly homogeneous. Then $(e_n)_{n=1}^{\infty}$ is equivalent either to the canonical basis of c_0 or the canonical basis of ℓ_p for some $1 \le p < \infty$.*

Proof. If the sequence $(\lambda(n))_{n=1}^{\infty}$ is bounded above then $(e_n)_{n=1}^{\infty}$ is equivalent to the standard unit vector basis of c_0. If $(\lambda(n))_{n=1}^{\infty}$ is unbounded we can use the preceding lemma to deduce the existence of $1 \le p < \infty$ so that

$$C^{-1}|A|^{\frac{1}{p}} \le \left\| \sum_{k \in A} e_k \right\| \le C|A|^{\frac{1}{p}},$$

for any finite subset A of $\mathbb{N}$.

Suppose $(a_i)_{i=1}^{n}$ is any finite sequence of scalars such that $\sum_{i=1}^{n} a_i^p = 1$. We will suppose that $(a_i)_{i=1}^{n}$ are such that $|a_i|^p \in \mathbb{Q}$ for all $i = 1, \ldots, n$. Hence each a_i^p can be written in the form $a_i^p = m_i/m$, where $m_i \in \mathbb{N}$, m is the common denominator of the a_i's, and $\sum_{i=1}^{n} m_i = m$.

Let E_1 be the interval of natural numbers $[1, m_1]$ and for $i = 2, \ldots, n$, let $E_i = [m_1 + \cdots + m_{i-1} + 1, m_1 + \cdots + m_i]$. $E_1, \ldots, E_n$ are disjoint intervals of $\mathbb{N}$ such that $|E_i| = m_i$ for each $i = 1, \ldots, n$. Consider the normalized constant coefficient block basic sequence defined for every $i = 1, \ldots, n$ as

$$u_i = c_i^{-1} \sum_{k \in E_i} e_k,$$

where $c_i = \left\| \sum_{k \in E_i} e_k \right\|$. Since $(e_n)_{n=1}^{\infty}$ is perfectly homogeneous, Lemma 9.1.3 yields

$$M^{-1}\lambda(m_i) \le c_i \le M\lambda(m_i)$$

for all $1 \le i \le n$, and so by Lemma 9.1.7,

$$C^{-1}M^{-1}m_i^{\frac{1}{p}} \le c_i \le CMm_i^{\frac{1}{p}}.$$

Therefore,

$$\frac{1}{CM^2 m^{\frac{1}{p}}} \left\| \sum_{i=1}^{n} \sum_{j \in E_i} e_j \right\| \le \left\| \sum_{i=1}^{n} a_i u_i \right\| \le \frac{CM^2}{m^{\frac{1}{p}}} \left\| \sum_{i=1}^{n} \sum_{j \in E_i} e_j \right\|.$$

This reduces to

$$\frac{\lambda(m)}{CM^2 m^{\frac{1}{p}}} \le \left\| \sum_{i=1}^{n} a_i u_i \right\| \le \frac{CM^2 \lambda(m)}{m^{\frac{1}{p}}},$$

hence

$$\frac{1}{C^2 M^2} \le \left\| \sum_{i=1}^{n} a_i u_i \right\| \le C^2 M^2.$$

Using perfect homogeneity again, we have

$$\frac{1}{C^2 M^3} \le \left\| \sum_{i=1}^{n} a_i e_i \right\| \le C^2 M^3. \tag{9.6}$$

To finish the proof we note that a simple density argument shows that equation (9.6) holds whenever $\sum_{i=1}^{n} |a_i|^p = 1$ (i.e., without the assumption that $|a_i|^p$ is rational).

$\square$

9.2 Symmetric bases

We next study a special class of bases which include the canonical bases of the spaces ℓ_p and c_0.

Definition 9.2.1. An unconditional basis $(e_n)_{n=1}^{\infty}$ of a Banach space X is *symmetric* if $(e_n)_{n=1}^{\infty}$ is equivalent to $(e_{\pi(n)})_{n=1}^{\infty}$ for any permutation π of $\mathbb{N}$.

A symmetric basis of a Banach space has the property of being equivalent to all its (infinite) subsequences, as the next lemma states. The converse need not be true. In fact, the summing basis of c_0 is equivalent to all its subsequences and is not even unconditional.

Lemma 9.2.2. *Suppose $(e_n)_{n=1}^{\infty}$ is a symmetric basis of a Banach space X. Then there exists a constant D such that*

$$D^{-1} \left\| \sum_{i=1}^{N} a_i e_{j_i} \right\| \le \left\| \sum_{i=1}^{N} a_i e_{k_i} \right\| \le D \left\| \sum_{i=1}^{N} a_i e_{j_i} \right\|$$

for any choice of scalars $(a_i)_{i=1}^{N}$, any $N \in \mathbb{N}$, and any two families of distinct natural numbers $\{j_1, \ldots, j_N\}$ and $\{k_1, \ldots, k_N\}$.

Proof. It is enough to prove the lemma for the basic sequence $(e_n)_{n \ge n_0}$ for some n_0. If it is false, then for every n_0 we can build a strictly increasing sequence of natural numbers $(p_n)_{n=0}^{\infty}$ with $p_0 = 0$, natural numbers $m_n \le p_n - p_{n-1}$, scalars $(a_{n,i})_{n=1,i=1}^{\infty,m_n}$, and families $\{j_{n,1}, \ldots, j_{n,m_n}\}, \{k_{n,1}, \ldots, k_{n,m_n}\}$ such that for all $n = 1, 2, \ldots$ we have

$$p_{n-1} + 1 \le j_{n,i}, k_{n,i} \le p_n, \qquad 1 \le i \le m_n,$$

$$\left\| \sum_{i=1}^{m_n} a_{n,i} e_{j_{n,i}} \right\| < 2^{-n},$$

and

$$\left\| \sum_{i=1}^{m_n} a_{n,i} e_{k_{n,i}} \right\| > 2^n.$$

Now one can make a permutation π of $\mathbb{N}$ so that $\pi[p_{n-1}+1, p_n] = [p_{n-1}+1, p_n]$ and $\pi(j_{n,i}) = k_{n,i}$ and this will contradict the equivalence of $(e_n)_{n=1}^{\infty}$ with $(e_{\pi(n)})_{n=1}^{\infty}$.

□

Definition 9.2.3. If $(e_n)_{n=1}^\infty$ is a symmetric basis of a Banach space X then the best constant K such that for all $x = \sum_{n=1}^\infty a_n e_n \in X$ the inequality

$$\left\| \sum_{n=1}^\infty \epsilon_n a_n e_{\pi(n)} \right\| \leq K \left\| \sum_{n=1}^\infty a_n e_n \right\|$$

holds for all choices of signs (ϵ_n) and all permutations π, is called the *symmetric constant* of $(e_n)_{n=1}^\infty$. In this case we also say that $(e_n)_{n=1}^\infty$ is *K-symmetric*.

For every $x = \sum_{n=1}^\infty a_n e_n \in X$, put

$$|||x||| = \sup \left\| \sum_{n=1}^\infty t_n a_n e_{\pi(n)} \right\|, \qquad (9.7)$$

the supremum being taken over all choices of scalars (ϵ_n) of signs and all permutations of the natural numbers. Equation (9.7) defines a new norm on X equivalent to $\| \cdot \|$ since $\|x\| \leq |||x||| \leq K \|x\|$ for all $x \in X$. With respect to this norm, $(e_n)_{n=1}^\infty$ is a 1-symmetric basis of X.

Definition 9.2.4. A basis (e_n) of a Banach space X is *subsymmetric* provided it is unconditional and for every increasing sequence of integers $\{n_i\}_{i=1}^\infty$, the subbasis $(e_{n_i})_{i=1}^\infty$ is equivalent to (e_n). The *subsymmetric constant* of (e_n) is the smallest constant $C \geq 1$ such that given any scalars $(a_i) \in c_{00}$, we have

$$\left\| \sum_{i=1}^\infty \epsilon_i a_i e_{n_i} \right\| \leq C \left\| \sum_{i=1}^\infty a_i e_i \right\|$$

for all sequences of signs (ϵ_i) and all increasing sequences of integers $\{n_i\}_{i=1}^\infty$. In this case we say that (e_n) is *C-subsymmetric*.

Remark 9.2.5. The concepts of symmetric and subsymmetric basis do not coincide, as shown by the following example due to Garling [63]. Let X be the space of all sequences of scalars $\xi = (\xi_n)_{n=1}^\infty$ for which

$$\|\xi\| = \sup \sum_{k=1}^\infty \frac{|\xi_{n_k}|}{\sqrt{k}} < \infty,$$

the supremum being taken over all increasing sequences of integers $(n_k)_{k=1}^\infty$. We leave for the reader the task to check that X, endowed with the norm defined above, is a Banach space whose unit vectors $(e_n)_{n=1}^\infty$ form a subsymmetric basis which is not symmetric.

Theorem 9.2.6. *Let X be a Banach space with normalized, 1-symmetric basis $(e_n)_{n=1}^\infty$. Suppose that $(u_n)_{n=1}^\infty$ is a normalized constant coefficient block basic sequence. Then the subspace $[u_n]$ is complemented in X by a norm-one projection.*

Proof. For each $k = 1, 2, \ldots$, let $u_k = c_k \sum_{j \in A_k} e_k$, where $(A_k)_{k=1}^{\infty}$ is a sequence of mutually disjoint subsets of $\mathbb{N}$ (notice that, since $(e_n)_{n=1}^{\infty}$ is 1-symmetric, the blocks of the basis need not be in increasing order). For every fixed $n \in \mathbb{N}$, let Π_n denote the set of all permutations π of $\mathbb{N}$ such that for each $1 \le k \le n$, π restricted to A_k acts as a cyclic permutation of the elements of A_k (in particular $\pi(A_k) = A_k$), and $\pi(j) = j$ for all $j \notin \cup_{k=1}^{n} A_k$. Every $\pi \in \Pi_n$ has associated an operator on X defined for $x = \sum_{j=1}^{\infty} a_j e_j$ as

$$T_{n,\pi}(\sum_{j=1}^{\infty} a_j e_j) = \sum_{j=1}^{\infty} a_j e_{\pi(j)}.$$

Notice that, due to the 1-symmetry of $(e_n)_{n=1}^{\infty}$, we have $\|T_{n,\pi}(x)\| = \|x\|$.

Let us define an operator on X by averaging over all possible choices of permutations $\pi \in \Pi_n$: given $x = \sum_{j=1}^{\infty} a_j e_j$,

$$T_n(x) = \frac{1}{|\Pi_n|} \sum_{\pi \in \Pi_n} T_{n,\pi}(x) = \sum_{k=1}^{n} \left(\frac{1}{|A_k|} \sum_{j \in A_k} a_j\right) \sum_{j \in A_k} e_j + \sum_{j \notin \cup_{k=1}^{n} A_k} a_j e_j.$$

Then,

$$\|T_n(x)\| = \left\|\frac{1}{|\Pi_n|} \sum_{\pi \in \Pi_n} T_{n,\pi}(x)\right\| \le \frac{1}{|\Pi_n|} \sum_{\pi \in \Pi_n} \|T_{n,\pi}(x)\| = \|x\|.$$

Therefore, for each $n \in \mathbb{N}$ the operator

$$P_n(x) = \sum_{k=1}^{n} \left(\frac{1}{|A_k|} \sum_{j \in A_k} a_j\right) \sum_{j \in A_k} e_j, \qquad x \in X$$

is a norm-one projection onto $[u_k]_{k=1}^{n}$. Now it readily follows that

$$P(x) = \sum_{k=1}^{\infty} \left(\frac{1}{|A_k|} \sum_{j \in A_k} a_j\right) \underbrace{\sum_{j \in A_k} e_j}_{c_k^{-1} u_k}$$

is a well defined projection from X onto $[u_k]$ with $\|P\| = 1$. $\qquad \square$

9.3 Uniqueness of unconditional basis

Zippin's theorem (Theorem 9.1.8) has a number of very elegant applications. We give a couple in this section. The first relates to the theorem of Lindenstrauss and Pełczyński proved in Section 8.3. There we saw that the normalized unconditional bases of the three spaces c_0, ℓ_1, and ℓ_2 are unique (up to equivalence); we also saw that, in contrast, the spaces ℓ_p for $p \neq 1, 2$ have at least two nonequivalent normalized unconditional bases.

In 1969, Lindenstrauss and Zippin [140] completed the story by showing that the list ends with these three spaces!

Theorem 9.3.1 (Lindenstrauss, Zippin). *A Banach space X has a unique unconditional basis (up to equivalence) if and only if X is isomorphic to one of the following three spaces: c_0, ℓ_1, or ℓ_2.*

Proof. Suppose that X has a unique normalized unconditional basis, $(e_n)_{n=1}^\infty$. Then, in particular, the basis $(e_{\pi(n)})_{n=1}^\infty$ is equivalent to $(e_n)_{n=1}^\infty$ for each permutation π of $\mathbb{N}$. That is, $(e_n)_{n=1}^\infty$ is a symmetric basis of X. Without loss of generality we can assume that its symmetric constant is 1.

Let $(u_n)_{n=1}^\infty$ be a normalized constant coefficient block basic sequence with respect to $(e_n)_{n=1}^\infty$ such that there are infinitely many blocks of size k for all $k \in \mathbb{N}$. That is,

$$\left| \{u_n \ : \ |\text{supp } u_n| = k\} \right| = \infty$$

for each $k \in \mathbb{N}$. Let us call Y the closed linear span of the sequence $(u_n)_{n=1}^\infty$.

The subspace Y is complemented in X by Theorem 9.2.6.

On the other hand, the subsequence of $(u_n)_{n=1}^\infty$ consisting of the blocks whose supports have size 1 spans a subspace isometrically isomorphic to X, which is complemented in Y because of the unconditionality of $(u_n)_{n=1}^\infty$.

By the symmetry of the basis $(e_n)_{n=1}^\infty$, X is isomorphic to X^2.

Analogously, if we split the natural numbers in two subsets S_1, S_2 such that

$$\left| \{n \in S_1 \ ; \ |\text{supp} u_n| = k\} \right| = \left| \{n \in S_2 \ ; \ |\text{supp} u_n| = k\} \right| = \infty$$

for all $k \in \mathbb{N}$, we see that

$$[u_n]_{n=1}^\infty \approx [u_n]_{n \in S_1} \oplus [u_n]_{n \in S_2} \approx [u_n]_{n=1}^\infty \oplus [u_n]_{n=1}^\infty.$$

Hence $Y \approx Y^2$.

Using Pełczyński's decomposition technique (Theorem 2.2.3) we deduce that $X \approx Y$.

Since $(u_n)_{n=1}^\infty$ is an unconditional basis of Y, by the hypothesis it must be equivalent to $(e_n)_{n=1}^\infty$. In particular $(u_n)_{n=1}^\infty$ is symmetric and, therefore, equivalent to all of its subsequences. Hence $(e_n)_{n=1}^\infty$ is perfectly homogeneous. Theorem 9.1.8 implies that $(e_n)_{n=1}^\infty$ is equivalent either to the canonical basis of c_0 or ℓ_p for some $1 \leq p < \infty$. But we saw in the previous chapter (Proposition 8.3.7) that if $p \in (1, \infty) \setminus \{2\}$ then ℓ_p has an unconditional basis which is not equivalent to the standard unit vector basis. The only remaining possibilities for the space X are c_0, ℓ_1, or ℓ_2.

$\square$

The Lindenstrauss-Zippin theorem thus completes the classification of those Banach spaces with a unique unconditional basis. The elegance of this result encouraged further work in this direction. One obvious modification is to require uniqueness of unconditional basis up to a permutation, (UTAP). In many ways this is a more natural concept for unconditional bases, whose order is irrelevant.

Definition 9.3.2. Two unconditional bases $(e_n)_{n=1}^{\infty}$ and $(f_n)_{n=1}^{\infty}$ of a Banach space X are said to be *permutatively equivalent* if there is a permutation π of $\mathbb{N}$ so that $(e_{\pi(n)})_{n=1}^{\infty}$ and $(f_n)_{n=1}^{\infty}$ are equivalent. Then we say that a Banach space X has a (UTAP) unconditional basis $(e_n)_{n=1}^{\infty}$ if every normalized unconditional basis in X is permutatively equivalent to $(e_n)_{n=1}^{\infty}$.

Classifying spaces with (UTAP) bases is more difficult because the initial step (reduction to symmetric bases) is no longer available.

The first step toward this classification was taken in 1976 by Edelstein and Wojtaszczyk [52], who showed that the finite direct sums of the spaces c_0, ℓ_1, and ℓ_2 have (UTAP) bases (thus adding four new spaces to the already known ones). After their work, Bourgain, Casazza, Lindenstrauss, and Tzafriri embarked on a comprehensive study completed in 1985 [15]. They added the spaces $c_0(\ell_1), \ell_1(c_0)$ and $\ell_1(\ell_2)$ to the list, but showed, remarkably, that $\ell_2(\ell_1)$ fails to have a (UTAP) basis! However, all hopes of a really satisfactory classification of Banach spaces having a (UTAP) basis were dashed when they also found a *nonclassical* Banach space which also has (UTAP). This space was a modification of Tsirelson space, to be constructed in the next chapter, which contains no copy of any space isomorphic to an ℓ_p $(1 \le p < \infty)$ or c_0. The subject was revisited in [26] and [27], and several other "pathological" spaces with (UTAP) bases have been discovered, including the original Tsirelson space. For an account of this topic see [218].

For the classification of symmetric basic sequences in L_p spaces we refer to [18], [93], and [194].

9.4 Complementation of block basic sequences

We now turn our attention to the study of complementation of subspaces of a Banach space. Starting with the example of c_0 in ℓ_∞ we saw that a subspace of a Banach space need not be complemented. Using Zippin's theorem we will now study the complementation in a Banach space of the span of block basic sequences of unconditional bases.

Lemma 9.4.1. *Let $(e_n)_{n=1}^{\infty}$ be an unconditional basis of a Banach space X. Suppose that $(u_k)_{k=1}^{\infty}$ is a normalized block basic sequence of $(e_n)_{n=1}^{\infty}$ such that the subspace $[u_k]$ is complemented in X. Then there is a projection Q from X onto $[u_k]$ of the form*

$$Q(x) = \sum_{k=1}^{\infty} u_k^*(x)u_k,$$

where supp $u_k^* \subseteq$ supp u_k for all $k \in \mathbb{N}$.

Proof. Suppose

$$u_k = \sum_{j \in A_k} a_j e_j,$$

where $A_k = \text{supp } u_k$, and that P is a bounded projection onto $[u_k]$. For each k let Q_k be the projection onto $[e_j]_{j \in A_k}$ given by

$$Q_k x = \sum_{j \in A_k} e_j^*(x) e_j.$$

We will show that the formula

$$Qx = \sum_{k=1}^{\infty} Q_k P Q_k x, \qquad x \in X$$

defines a bounded projection onto $[u_k]$ (and it is clearly of the prescribed form).

Suppose $x = \sum_{j=1}^{m} e_j^*(x) e_j$ for some m. Then for a suitable N so that supp $x \subset A_1 \cup \cdots \cup A_N$ we have

$$Qx = \sum_{k=1}^{N} Q_k P Q_k x$$

$$= \underset{\epsilon_k = \pm 1}{\text{Average}} \sum_{j=1}^{N} \sum_{k=1}^{N} \epsilon_j \epsilon_k Q_j P Q_k x$$

$$= \underset{\epsilon_k = \pm 1}{\text{Average}} \left(\sum_{j=1}^{N} \epsilon_j Q_j \right) P \left(\sum_{k=1}^{N} \epsilon_k Q_k \right) x.$$

By the unconditionality of the original basis,

$$\|Qx\| \leq K^2 \|P\| \|x\|,$$

where K is the unconditional basis constant. It is now easy to check that Q extends to a bounded operator and has the required properties.

$\square$

The following characterization of the canonical bases of the ℓ_p-spaces and c_0 is due to Lindenstrauss and Tzafriri [135].

Theorem 9.4.2. *Let $(e_n)_{n=1}^{\infty}$ be an unconditional basis of a Banach space X. Suppose that for every block basic sequence $(u_n)_{n=1}^{\infty}$ of a permutation of $(e_n)_{n=1}^{\infty}$, the subspace $[u_n]$ is complemented in X. Then $(e_n)_{n=1}^{\infty}$ is equivalent to the canonical basis of c_0 or ℓ_p for some $1 \leq p < \infty$.*

Proof. Without loss of generality we may assume that the constant of unconditionality of the basis $(e_n)_{n=1}^{\infty}$ is 1. Our first goal is to show that whenever we have

$$u_n = \sum_{k \in A_n} \alpha_k e_k, \quad v_n = \sum_{k \in B_n} \beta_k e_k, \quad n \in \mathbb{N}$$

any two normalized block basic sequences of $(e_n)_{n=1}^\infty$ such that $A_n \cap B_m = \emptyset$ for all n, m, then $(u_n)_{n=1}^\infty \sim (v_n)_{n=1}^\infty$.

First we will prove that if $(a_n)_{n=1}^\infty$ is a sequence of scalars for which $\sum_{n=1}^\infty a_n u_n$ converges, then the series $\sum_{n=1}^\infty s_n a_n v_n$ converges for every sequence of scalars $(s_n)_{n=1}^\infty$ tending to 0. For each $n \in \mathbb{N}$ consider

$$w_n = u_n + s_n v_n, \quad n \in \mathbb{N}.$$

$(w_n)_{n=1}^\infty$ is a seminormalized block basic sequence with respect to a permutation of $(e_n)_{n=1}^\infty$. To be precise, supp $w_n = A_n \cup B_n$ for each n and $1 \leq \|w_n\| \leq 2$ (for n big enough so that $|s_n| \leq 1$). By the hypothesis, the subspace $[w_n]$ is complemented in X. Lemma 9.4.1 yields a projection $Q : X \to X$ of the form

$$Q(x) = \sum_{n=1}^\infty w_n^*(x) w_n,$$

where the elements of the sequence $(w_n^*)_{n=1}^\infty \subset X^*$ satisfy supp $w_n^* \subseteq A_n \cup B_n$. Moreover, it is easy to see that $\|w_n^*\| \leq \|Q\|$ for all n.

The series

$$\sum_{n=1}^\infty a_n Q(u_n) = \sum_{n=1}^\infty a_n w_n^*(u_n) w_n = \sum_{n=1}^\infty a_n w_n^*(u_n)(u_n + s_n v_n)$$

converges because $\sum_{n=1}^\infty a_n u_n$ does. Therefore, by unconditionality, it follows that $\sum_{n=1}^\infty a_n w_n^*(u_n) s_n v_n$ converges as well. From here we deduce the convergence of the series $\sum_{n=1}^\infty a_n s_n v_n$ by noticing that $w_n^*(u_n) \to 1$ since

$$w_n^*(u_n) = 1 - s_n w_n^*(v_n)$$

and

$$0 \leq |s_n w_n^*(v_n)| \leq |s_n| \, \|w_n^*\| \leq \|Q\| \, |s_n| \to 0.$$

Now, if $(a_n)_{n=1}^\infty$ is a sequence of scalars for which $\sum_{n=1}^\infty a_n u_n$ converges, we can find a sequence of scalars $(t_n)_{n=1}^\infty$ tending to ∞ such that $\sum_{n=1}^\infty t_n a_n u_n$ converges. Since $(1/t_n)_{n=1}^\infty$ tends to 0, the previous argument applies so $\sum_{n=1}^\infty a_n v_n$ converges.

Reversing the roles of (u_n) and (v_n) we get the equivalence of these two block basic sequences.

This argument applies not only to block basic sequences of $(e_n)_{n=1}^\infty$ but to block basic sequences of a permutation of $(e_n)_{n=1}^\infty$. Thus $(u_n)_{n=1}^\infty$ is equivalent to every permutation of $(v_n)_{n=1}^\infty$. This implies that $(e_{2n})_{n=1}^\infty$ and $(e_{2n-1})_{n=1}^\infty$ are both perfectly homogeneous and equivalent to each other. We conclude the proof by applying Zippin's theorem (Theorem 9.1.8).

□

Remark 9.4.3. In the above theorem, it is necessary to allow complementation of the span of block basic sequences with respect to a permutation of $(e_n)_{n=1}^{\infty}$. One may show that the canonical basis of $\ell_p(\ell_r^n)$ where $r \neq p$ has the property that every block basic sequence spans a complemented subspace, but obviously it is not equivalent to the canonical basis of ℓ_p or c_0 (see the Problems).

In [135], Lindenstrauss and Tzafriri solved the *Complemented Subspace Problem* discussed in Chapter 2. We cannot quite prove this yet in full generality as it requires more machinery, but in this section we will see the proof in the case of spaces with unconditional basis.

Theorem 9.4.4. *Let X be a Banach space with unconditional basis. If every closed subspace of X is complemented in X then X is isomorphic to ℓ_2.*

Proof. Let $(x_n)_{n=1}^{\infty}$ be an unconditional basis of such an X. By Theorem 9.4.2, (x_n) is equivalent either to the canonical basis of c_0 or to the canonical basis of ℓ_p for some $1 \leq p < \infty$.

Suppose that (x_n) is equivalent to the canonical basis of ℓ_p for some $1 < p < \infty$, $p \neq 2$. We know that, in this case, ℓ_p is isomorphic to $\ell_p(\ell_2^n)$ and that the canonical basis of $\ell_p(\ell_2^n)$ is not equivalent to the standard basis of ℓ_p. Therefore X contains an unconditional basis (u_n) equivalent to the canonical basis of $\ell_p(\ell_2^n)$. Repeating the argument at the beginning of the proof with (u_n) would lead to a contradiction.

Thus the possibilities for X are reduced to three spaces: X is either c_0, ℓ_1, or ℓ_2. To complete the proof we need only show that c_0 and ℓ_1 have uncomplemented subspaces. In fact, in the case of ℓ_1 we have already seen examples (Corollary 2.3.3).

Let us consider first the case of c_0. For each n, ℓ_1^n embeds isometrically in $\ell_{\infty}^{2^n}$. This follows from the fact that the norm of each element $(a_i)_{i=1}^n$ in ℓ_1^n can be written, using duality, as

$$\|(a_i)_{i=1}^n\| = \max \left| \sum_{k=1}^n \varepsilon_k a_k \right|,$$

the maximum being taken over the 2^n possible choices for the sequence of signs $(\varepsilon_k)_{k=1}^n$. Thus the embedding of ℓ_1^n into $\ell_{\infty}^{2^n}$ is given by the map

$$(a_i)_{i=1}^n \mapsto \left(\sum_{i=1}^n \varepsilon_i a_i \right)_{(\varepsilon_i)_{i=1}^n \in \{-1,1\}^n} \in \ell_{\infty}^{2^n}.$$

Hence, $c_0(\ell_1^n)$ embeds in $c_0(\ell_{\infty}^{2^n})$, which is isometrically isomorphic to c_0. As before, the subspace $c_0(\ell_1^n)$ cannot be complemented in c_0 because the canonical basis of $c_0(\ell_1)$ is not equivalent to the standard c_0-basis.   □

Remark 9.4.5. In this proof we could have also shown that ℓ_1 has an uncomplemented subspace using an argument similar to that for c_0: For each n, the space $L_1([0,1], \Sigma_n)$ is isometric to $\ell_1^{2^n}$ and, by Khintchine's inequality, it contains an isomorphic copy of ℓ_2^n (namely, the space spanned by $\{r_1, r_2, \ldots, r_n\}$) with isomorphism constants uniform on n. Then $\ell_1(\ell_2^n)$ embeds in $\ell_1(\ell_1^{2^n})$, which is isometrically isomorphic to ℓ_1. If the subspace $\ell_1(\ell_2^n)$ were complemented in ℓ_1 then it would be isomorphic to ℓ_1 and so, as a consequence, ℓ_1 would have an unconditional basis equivalent to the canonical basis of $\ell_1(\ell_2^n)$, which is not true.

9.5 The existence of conditional bases

In this section we prove an earlier result of Pełczyński and Singer from 1964 [177] to the effect that every Banach space with a basis has a basis which is not unconditional. The original argument was more involved and does not use Zippin's theorem (Theorem 9.1.8) which it predates.

Definition 9.5.1. A normalized basis $(x_n)_{n=1}^{\infty}$ of a Banach space X is called *conditional* if it is not unconditional.

In Chapter 3 we saw that c_0 has, at least, one conditional basis, the summing basis. On the other hand, the vectors $e_1, e_1 - e_2, e_2 - e_3, e_3 - e_4, \ldots$, form a conditional basis of ℓ_1, where, as usual, $(e_n)_{n=1}^{\infty}$ denotes the standard ℓ_1-basis basis of ℓ_1. As for ℓ_2 the existence of conditional basis requires a bit of elaboration. This was originally proved by Babenko [7] as a consequence of harmonic analysis methods. Our proof is based on a later argument by McCarthy and Schwartz [148]. However, the McCarthy-Schwartz argument is in a certain sense a very close relative of the Babenko approach.

Theorem 9.5.2. ℓ_2 *has a conditional basis.*

Proof. Let $(e_n)_{n=1}^{\infty}$ be the canonical orthonormal basis of ℓ_2. We pick a sequence of nonnegative real numbers $(a_n)_{n=1}^{\infty}$ such that

$$\sum_{n=1}^{\infty} a_n = \infty, \quad \sum_{n=1}^{\infty} n a_n^2 < \infty.$$

One may suppose that $a_n \sim 1/(n \log n)$ for n large to get such a sequence. We now define a sequence $(f_n)_{n=1}^{\infty}$ by

$$f_{2n-1} = e_{2n-1}$$

and

$$f_{2n} = e_{2n} + \sum_{j=1}^{n} a_j e_{2n+1-2j}.$$

We will investigate conditions under which $(f_n)_{n=1}^{\infty}$ is (a) a basis and (b) an unconditional basis.

Let us define an infinite matrix $B = (b_{ij})$ by

$$b_{ij} = \begin{cases} a_k & j - i = 2k - 1 \\ 0 & \text{otherwise.} \end{cases}$$

Thus

$$B = \begin{pmatrix} 0 & a_1 & 0 & a_2 & 0 & a_3 & 0 & \dots \\ 0 & 0 & 0 & 0 & 0 & 0 & 0 & \dots \\ 0 & 0 & 0 & a_1 & 0 & a_2 & 0 & \dots \\ 0 & 0 & 0 & 0 & 0 & 0 & 0 & \dots \\ \cdot & \cdot & \cdot & \cdot & \cdot & \cdot & \cdot & \dots \\ \cdot & \cdot & \cdot & \cdot & \cdot & \cdot & \cdot & \dots \end{pmatrix}.$$

Now B as a matrix acts on c_{00} (when we regard each entry as an infinite column vector). Furthermore $f_j = (I + B)e_j$.

Notice that B^2 can be computed (since every column has most finitely many entries) and in fact $B^2 = 0$. Consider the partial sum operators with respect to the basis P_n say. In matrix terms we have

$$P_n = \begin{pmatrix} I_n & 0 \\ 0 & 0 \end{pmatrix}$$

as a partitioned matrix. We also have $BP_nB = 0$.

The matrix $I + B$ is invertible (as a linear endomorphism of c_{00}) with inverse $I - B$. It follows that $(f_j)_{j=1}^{\infty}$ is always a Hamel basis of the countable dimensional space c_{00}. The partial sum operators with respect to this Hamel basis are given by $(I + B)P_n(I - B) = I + BP_n - P_nB$. For $(f_n)_{n=1}^{\infty}$ to be a basis of ℓ_2 simply requires that the operators $BP_n - P_nB$ extend to a uniformly bounded sequence of operators on ℓ_2. Now $BP_n - P_nB$ is just the restriction of the matrix B to the set of (i, j) so that $i \le n < j$ (i.e., to the top right-hand corner). We claim that this operator is actually the restriction of a Hilbert-Schmidt operator since

$$\sum_{i=1}^{n} \sum_{j=n+1}^{\infty} |b_{ij}|^2 \le \sum_{k=1}^{\infty} k a_k^2.$$

It follows that we have a uniform bound

$$\|BP_n - P_nB\| \le \left(\sum_{k=1}^{\infty} k a_k^2\right)^{1/2}.$$

The uniform bound establishes that $(f_n)_{n=1}^{\infty}$ is a basis of ℓ_2.

Assume that $(f_n)_{n=1}^{\infty}$ is unconditional. Then, since $1 \le \|f_n\| \le M$ for some M, $(f_n)_{n=1}^{\infty}$ must be equivalent to the canonical ℓ_2-basis, and the operator

$I + B$ must define a bounded operator on ℓ_2; thus so does B. On the other hand, summing over the top left-hand corner square, we obtain

$$\left\langle B(\sum_{j=1}^{2n} e_j), \sum_{j=1}^{2n} e_j \right\rangle = \sum_{i=1}^{2n}\sum_{j=1}^{2n} b_{ij} = \sum_{k=1}^{n}(n-k+1)a_k.$$

Thus, if B defines a bounded operator,

$$\sum_{k=1}^{n}(n-k+1)a_k \le 2n\|B\|,$$

i.e.,

$$\sum_{k=1}^{n}(1 - \frac{k-1}{n})a_k \le 2\|B\|.$$

Letting $n \to \infty$ we would conclude that $\sum_{k=1}^{\infty} a_k < \infty$, which would contradict our initial choice.

□

Babenko's argument is based on considering weighted L_2-spaces. We consider complex Hilbert spaces. Let w be a density function on $\mathbb{T}$ and consider the space $L_2(w(\theta)d\theta)$. Then it may be shown that the sequence $\{1, e^{i\theta}, e^{-i\theta}, e^{2i\theta}, \dots\}$ is a basis of $L_2(w\,d\theta)$ if and only if the Riesz projection $f \to \sum_{n\ge 0} \hat{f}(n)e^{in\theta}$ (or the Hilbert transform) acts boundedly on $L_2(w\,d\theta)$. This happens if and only if w is an A_2-weight (e.g., see [73]). On the other hand unconditionality implies

$$\|f\|_{L_2(w\,d\theta)} \approx \left(\sum_{n\in\mathbb{Z}} |\hat{f}(n)|^2\right)^{1/2} \approx \|f\|_{L_2(d\theta)}$$

so that $w, w^{-1} \in L_\infty$. So, to give an example one needs an A_2-weight w with w or w^{-1} unbounded. Babenko used the weight $|\theta|^\alpha$ where $0 < \alpha < 1$. However, the argument given in Theorem 9.5.2 can also be rephrased as a proof of the existence of unbounded A_2-weights.

We are headed to show the result of Pełczyński and Singer [177] that every Banach space with a basis has a conditional basis. To this end, first we need a few lemmas. Our next lemma gives us a criterion for the construction of a new basis of a Banach space with a given basis.

Lemma 9.5.3. *Suppose that $(e_n)_{n=1}^\infty$ is a basis of a Banach space X and that $(r_n)_{n=0}^\infty$ is an increasing sequence of integers with $r_0 = 0$. For each n let E_n be the closed subspace spanned by the basis elements $\{e_{r_{n-1}+1}, \dots, e_{r_n}\}$. Further assume that $(f_n)_{n=1}^\infty$ is a sequence in X such that:*

(i) $(f_{r_{n-1}+1}, \dots, f_{r_n})$ is a basis of E_n for all n;
(ii) $\sup_n K_n = M < \infty$, where K_n is basis constant of $(f_{r_{n-1}+1}, \dots, f_{r_n})$

Then $(f_n)_{n=1}^{\infty}$ is a basis of X.

Proof. Let K be the basis constant of $(e_n)_{n=1}^{\infty}$ and let (S_N) be the sequence of natural projections associated with this basis. Since $[f_n] = [e_n] = X$, it suffices to show that there is a constant $C > 0$ such that given m and p in $\mathbb{N}$ with $m \leq p$, the inequality

$$\Big\| \sum_{k=1}^{m} \alpha_k f_k \Big\| \leq C \Big\| \sum_{k=1}^{p} \alpha_k f_k \Big\|$$

holds for any scalars $(\alpha_k)_{k=1}^{p}$.

Given any two integers m, p with $m \leq p$, there are integers n, q such that $r_{n-1} < m \leq r_n$ and $r_{q-1} < p \leq r_q$. We have two possibilities: either $n < q$ or $n = q$. Assume first that $n < q$. Then,

$$\Big\| \sum_{k=1}^{m} \alpha_k f_k \Big\| \leq \Big\| \sum_{k=1}^{r_{n-1}} \alpha_k f_k \Big\| + \Big\| \sum_{k=r_{n-1}+1}^{m} \alpha_k f_k \Big\|$$

$$\leq \Big\| S_{r_{n-1}} \Big(\sum_{k=1}^{p} \alpha_k f_k \Big) \Big\| + M \Big\| \sum_{k=r_{n-1}+1}^{r_n} \alpha_k f_k \Big\|$$

$$\leq K \Big\| \sum_{k=1}^{p} \alpha_k f_k \Big\| + M \Big\| S_{r_n} \Big(\sum_{k=1}^{p} \alpha_k f_k \Big) - S_{r_{n-1}} \Big(\sum_{k=1}^{p} \alpha_k f_k \Big) \Big\|$$

$$\leq (K + 2KM) \Big\| \sum_{k=1}^{p} \alpha_k f_k \Big\|.$$

If $n = q$, analogously we have

$$\Big\| \sum_{k=1}^{m} \alpha_k f_k \Big\| \leq \Big\| \sum_{k=1}^{r_{n-1}} \alpha_k f_k \Big\| + \Big\| \sum_{k=r_{n-1}+1}^{m} \alpha_k f_k \Big\|$$

$$\leq \Big\| S_{r_{n-1}} \Big(\sum_{k=1}^{p} \alpha_k f_k \Big) \Big\| + M \Big\| \sum_{k=r_{n-1}+1}^{p} \alpha_k f_k \Big\|$$

$$\leq K \Big\| \sum_{k=1}^{p} \alpha_k f_k \Big\| + M \Big\| S_{r_n} \Big(\sum_{k=1}^{p} \alpha_k f_k \Big) - S_{r_{n-1}} \Big(\sum_{k=1}^{p} \alpha_k f_k \Big) \Big\|$$

$$\leq (K + 2KM) \Big\| \sum_{k=1}^{p} \alpha_k f_k \Big\|.$$

$\square$

The following two lemmas are due to Zippin [224].

Lemma 9.5.4. *Let E, F be two closed subspaces of codimension 1 of a Banach space X. Then there exists an isomorphism $T : E \to F$ so that $\|T\|\|T^{-1}\| \leq 25$.*

Proof. Unless $E = F$, $E \cap F$ is a subspace of X of codimension 2. Let us pick $x_0 \in E \setminus (E \cap F)$ such that $1 = \|x_0\| d(x_0, E \cap F) \leq 2$. Analogously, pick $x_1 \in F$ such that $1 = \|x_1\| d(x_1, E \cap F) \leq 2$.

Each element of E can be written in a unique way in the form $\lambda x_0 + y$ for some scalar λ and some $y \in E \cap F$. Analogously, the elements of F admit a unique representation in the fashion $\lambda x_1 + y$, where $\lambda \in \mathbb{R}$ and $y \in E \cap F$. Define $T : E \to F$ as $T(\lambda x_0 + y) = \lambda x_1 + y$. On the one hand we have

$$\|\lambda x_1 + y\| \leq |\lambda| \, \|x_1\| + \|y\| \leq 2|\lambda| + \|y\| \leq 2 \max \big\{ |\lambda|, \|y\| \big\}. \tag{9.8}$$

On the other,

$$\|\lambda x_0 + y\| = |\lambda| \Big\| x_0 + \frac{y}{|\lambda|} \Big\| = |\lambda| \Big\| x_0 - \big(- \frac{y}{|\lambda|} \big) \Big\| \geq |\lambda| d(x_0, E \cap F) = |\lambda|$$

and

$$\|y + \lambda x_0\| \geq \|y\| - 2|\lambda|.$$

Hence,

$$\|y + \lambda x_0\| \geq \max \big\{ |\lambda|, \|y\| - 2|\lambda| \big\} \geq \max \big\{ |\lambda|, \frac{1}{3} \|y\| \big\}. \tag{9.9}$$

Combining (9.8) and (9.9) we obtain

$$\|T(\lambda x_0 + y)\| \leq 5 \|\lambda x_0 + y\|,$$

so $\|T\| \leq 5$. We would follow exactly the same steps to find a bound for $\|T^{-1}\|$, which would yield $\|T\| \|T^{-1}\| \leq 25$.

$\square$

Lemma 9.5.5. *Suppose that $(e_n)_{n=1}^{\infty}$ is a basis of a Banach space X and that $(u_n)_{n=1}^{\infty}$ is a block basic sequence of $(e_n)_{n=1}^{\infty}$. Then there exists a basis $(f_n)_{n=1}^{\infty}$ of X such that $(u_n)_{n=1}^{\infty}$ is a subbasis of $(f_n)_{n=1}^{\infty}$.*

Proof. For each $n \in \mathbb{N}$ suppose that u_n is normalized and supported on the basis elements $\{e_{r_{n-1}+1}, \ldots, e_{r_n}\}$, where $(r_n)_{n=1}^{\infty}$ is an increasing sequence of positive integers with $r_1 = 1$. Let $E_n = [e_{r_{n-1}+1}, \ldots, e_{r_n}]$. By the Hahn-Banach theorem there exists u_n^* in the dual space of the finite-dimensional normed space E_n such that $u_n^*(u_n) = \|u_n\| = 1$. Let $F_n = \ker u_n^*$. F_n is a subspace of codimension 1 of E_n. By Lemma 9.5.4 there is an isomorphism

$$T_n : [e_{r_{n-1}+1}, \ldots, e_{r_n-1}] \longrightarrow F_n$$

with $\|T_n\| \|T_n\|^{-1} \leq 25$. Pick $f_i = T_n(e_i)$ for $i = r_{n-1} + 1, \ldots, r_n - 1$. Then $\{f_{r_{n-1}+1}, \ldots, f_{r_n-1}\}$ is a basis of F_n with basis constant bounded by $25K$, K being the basis constant of $(e_n)_{n=1}^{\infty}$. Thus, if we take $f_{r_n} = u_n$ for each n, by Lemma 9.5.3 the sequence $(f_n)_{n=1}^{\infty}$ is a basis of X that satisfies the lemma.

$\square$

Theorem 9.5.6 (Pełczyński-Singer). *Let X be any Banach space with a basis. Then X has a conditional basis.*

Proof. Assume that every basis of X is unconditional and let $(e_n)_{n=1}^\infty$ be one of them. Suppose $(u_k)_{k=1}^\infty$ is a block basic sequence of $(e_n)_{n=1}^\infty$. Then, using Lemma 9.5.5, X has a basis $(f_n)_{n=1}^\infty$ of which $(u_k)_{k=1}^\infty$ is subsequence. Moreover, $(f_n)_{n=1}^\infty$ is unconditional by our assumption, hence $[u_k]$ is a complemented subspace in X. This argument will also apply to every permutation of $(e_n)_{n=1}^\infty$. Hence every block basic sequence of every permutation of $(e_n)_{n=1}^\infty$ spans a complemented subspace. By Theorem 9.4.2, $(e_n)_{n=1}^\infty$ must be equivalent to the canonical basis of c_0 or ℓ_p for some $1 \leq p < \infty$. This is a contradiction because, on the one hand, ℓ_p has an unconditional basis which is not equivalent to the canonical basis of the space if $1 < p < \infty$, $p \neq 2$, as we saw in Proposition 8.3.7, and, on the other hand, c_0, ℓ_1, and ℓ_2 have conditional bases.

$\square$

9.6 Greedy bases

This section deals with nonlinear approximation in (separable) Banach spaces with respect to a given basis of the space. This is a recent development which was spurred by problems in approximation theory related to data compression. As will be seen the idea is closely related to the theory of symmetric bases.

Let $(e_n)_{n=1}^\infty$ be a seminormalized basis of a Banach space X (i.e., $1/c \leq \|e_n\| \leq c$ for some c) with biorthogonal functionals $(e_n^*)_{n=1}^\infty$. For each $m = 0, 1, 2, \ldots$ we let Σ_m denote the collection of all elements of X which can be expressed as a linear combination of m elements of $(e_n)_{n=1}^\infty$:

$$\Sigma_m = \Big\{ y = \sum_{j \in B} \alpha_j e_j : B \subset \mathbb{N}, \, |B| = m, \, \alpha_j \in \mathbb{R} \Big\}.$$

Let us notice that, in some cases, it may be possible to write an element from Σ_m in more than one way, and that the space Σ_m is not linear: the sum of two elements from Σ_m is generally not in Σ_m, it is in Σ_{2m}.

For $x \in X$, we define its *best m-term approximation error* (with respect to the given basis) as

$$\sigma_m(x) = \inf_{y \in \Sigma_m} \|x - y\|.$$

The fundamental question here is to study how to construct an algorithm which for each $x \in X$ and each $m = 0, 1, 2, \ldots$ provides an element $y_m \in \Sigma_m$ so that the error of the approximation of x by y_m is (uniformly) comparable with $\sigma_m(x)$, i.e.,

$$\|x - y_m\| \leq C\sigma_m(x),$$

where C is an absolute constant.

The answer to this question in some particular cases is simple. For instance, if $X = H$ is a Hilbert space and $(e_n)_{n=1}^\infty$ is an orthonormal basis, any element $x \in H$ has an expansion in the form

$$x = \sum_{n=1}^\infty \langle x, e_n \rangle e_n$$

and

$$\|x\|^2 = \sum_{n=1}^\infty |\langle x, e_n \rangle|^2.$$

One easily realizes that a best approximation s_m to x from Σ_m is obtained as follows. We order the Fourier coefficients $(\langle x, e_j \rangle)_{j=1}^\infty$ of x according to the absolute value of their size and we choose Λ_m as the set of indices j for which $|\langle x, e_j \rangle|$ is largest. Then

$$s_m = \sum_{j \in \Lambda_m} \langle x, e_j \rangle e_j$$

is a best approximation to x from Σ_m and

$$\sigma_m(x)^2 = \|x - s_m\|^2 = \sum_{j \notin \Lambda_m} |\langle x, e_j \rangle|^2.$$

This is an example of a *Greedy Algorithm*. The most obvious and natural form to generalize such an algorithm is to consider $(\mathcal{G}_m)_{m=1}^\infty$, a sequence of maps from X to X where, for each x, $\mathcal{G}_m(x)$ is obtained by taking the largest m coefficients of x. To be precise, for $x \in X$ put

$$\mathcal{G}_m(x) = \sum_{j \in B} e_j^*(x) e_j,$$

where the set $B \subset \mathbb{N}$ is chosen in such a way that $|B| = m$ and $|e_j^*(x)| \geq |e_k^*(x)|$ whenever $j \in B$ and $k \notin B$.

A few comments about the maps $(\mathcal{G}_m)_{m=1}^\infty$ are in order. First, it may happen that for some x and m the set B, hence the element $\mathcal{G}_m(x)$, is not uniquely determined by the previous conditions. In such a case, we pick either of them. Besides, the maps $(\mathcal{G}_m)_{m=1}^\infty$ are neither linear (even when the sets B are uniquely determined) nor continuous.

Definition 9.6.1. A basis (e_n) is *greedy* if there is a constant $C \geq 1$ such that for any $x \in X$ and $m \in \mathbb{N}$ we have

$$\|x - \mathcal{G}_m(x)\| \leq C \sigma_m(x).$$

The smallest such constant C will be called the *greedy constant* of (e_n).

This means that the Greedy Algorithm $(\mathcal{G}_m)_{m=1}^\infty$ realizes near best m-term approximation. Now we will provide a characterization of greedy bases. To state it we need the following concept.

Definition 9.6.2. A basis (e_n) is called *democratic* if there is a constant $D \geq 1$ such that for any two finite subsets A, B of $\mathbb{N}$ with $|A| = |B|$ we have

$$\left\| \sum_{k \in A} e_k \right\| \leq D \left\| \sum_{k \in B} e_k \right\|.$$

Note that a democratic basis is automatically seminormalized.

The following characterization of greedy bases was proved by Konyagin and Temlyakov in 1999 [116].

Theorem 9.6.3. *A basis (e_n) is greedy if and only if it is unconditional and democratic.*

Proof. Let us assume, first, that (e_n) is greedy with greedy constant C. For any finite set $S \subset \mathbb{N}$ we denote P_S the projection

$$P_S(x) = \sum_{n \in S} e_n^*(x) e_n.$$

We will prove the unconditionality of (e_n) by showing that for each $x \in X$ and any finite set $S \subset \mathbb{N}$ we have

$$\|P_S(x)\| \leq (C + 1)\|x\|. \tag{9.10}$$

Let us fix a finite set $S \subset \mathbb{N}$ of cardinality m, $x \in X$ and a number $\alpha > \sup_{n \notin S} |e_n^*(x)|$. Consider the vector

$$y = x - P_S(x) + \alpha \sum_{n \in S} e_n.$$

Clearly $\sigma_m(y) \leq \|x\|$ and $\mathcal{G}_m(y) = \alpha \sum_{n \in S} e_n$. Thus, by our assumption that (e_n) is greedy, we get

$$\|x - P_S(x)\| = \|y - \mathcal{G}_m(y)\| \leq C \sigma_m(y) \leq C\|x\|.$$

This implies (9.10).

To show that (e_n) is democratic, let us pick two finite sets P, Q of the same cardinality m. Take a third subset S such that $|S| = m$ and $P \cap S = \emptyset = Q \cap S$. Fix any $\epsilon > 0$ and consider

$$x = (1 + \epsilon) \sum_{n \in P} e_n + \sum_{n \in S} e_n.$$

We have

$$\sigma_m(x) \leq (1 + \epsilon) \left\| \sum_{n \in P} e_n \right\|$$

and

$$\left\|\sum_{n\in S} e_n\right\| = \|x - \mathcal{G}_m(x)\| \le C\sigma_m(x) \le C(1+\epsilon)\left\|\sum_{n\in P} e_n\right\|. \qquad (9.11)$$

Analogously we get

$$\left\|\sum_{n\in Q} e_n\right\| \le C(1+\epsilon)\left\|\sum_{n\in S} e_n\right\|. \qquad (9.12)$$

Combining (9.11) and (9.12) and taking into account that ϵ is arbitrarily small, we obtain

$$\left\|\sum_{n\in Q} e_n\right\| \le C^2\left\|\sum_{n\in P} e_n\right\|.$$

Now we will prove the converse part of the theorem. Assume that (e_n) is K-unconditional and D-democratic. Fix $x \in X$ and $m = 1, 2, \ldots$. Given any $\epsilon > 0$ we pick

$$p_m = \sum_{n\in B} \alpha_n e_n \in \Sigma_m$$

such that

$$\|x - p_m\| \le \sigma_m(x) + \epsilon.$$

Clearly, we can write

$$\mathcal{G}_m(x) = \sum_{n\in S} e_n^*(x) e_n = P_S(x),$$

for some $S \subset \mathbb{N}$ with $|S| = m$. Then,

$$\|x - \mathcal{G}_m(x)\| = \|x - P_S x + P_B x - P_B x\| = \|x - P_B x + P_{B\setminus S} x - P_{S\setminus B} x\|. \qquad (9.13)$$

The assumption that (e_n) is K-unconditional implies that

$$\begin{aligned}
\|x - P_B x - P_{S\setminus B} x\| &= \|x - P_{B\cup S} x\| \\
&= \|P_{\mathbb{N}\setminus(B\cup S)}(x - p_m)\| \\
&\le K\|x - p_m\| \\
&\le K(\sigma_m(x) + \epsilon),
\end{aligned} \qquad (9.14)$$

and that

$$\|P_{S\setminus B} x\| \le K\|x - p_m\| \le K(\sigma_m(x) + \epsilon).$$

From the definition of $\mathcal{G}_m$ it is immediate to see that

$$\gamma := \min_{j\in S\setminus B} |e_j^*(x)| \ge \max_{j\in B\setminus S} |e_j^*(x)| := \beta,$$

so, from the unconditionality of (e_n), we obtain

$$\gamma\left\|\sum_{j\in S\setminus B} e_j\right\| \le K\|P_{S\setminus B} x\| \qquad (9.15)$$

and

$$\|P_{B\setminus S}x\| \leq K\beta \left\| \sum_{j\in B\setminus S} e_j \right\|. \tag{9.16}$$

Since $|B \setminus S| = |S \setminus B|$, using the D-democracy of the basis and (9.15) and (9.16) we get

$$\|P_{B\setminus S}x\| \leq K^2 D \|P_{S\setminus B}x\|. \tag{9.17}$$

Combining (9.13), (9.14), and (9.17), and taking into account that ϵ was arbitrarily small, the inequality

$$\|x - \mathcal{G}_m(x)\| \leq (K + K^3 D)\sigma_m(x)$$

holds.

□

There has been quite a bit of recent research on greedy bases in concrete spaces. It is clear and quite trivial that symmetric bases are greedy, but there are nonsymmetric greedy bases. An important result due to Temlyakov [213] is that the normalized Haar system in L_p is a greedy basis when $1 < p < \infty$. Note this basis cannot be symmetric, since it is easy to find a subsequence of the basis equivalent to the canonical ℓ_p-basis. A good reference for a survey of applications is to be found in [214].

Problems

9.1. Suppose $(x_n)_{n=1}^\infty$ is a basis for a Banach space X. Suppose there is a constant $C \geq 1$ such that whenever $p_0 = 0 < p_1 < \dots$ and $(u_n)_{n=1}^\infty$ and $(v_n)_{n=1}^\infty$ are two normalized block basic sequences of $(x_n)_{n=1}^\infty$ of the form

$$u_n = \sum_{i=p_{n-1}+1}^{p_n} a_i x_i,$$

$$v_n = \sum_{i=p_{n-1}+1}^{p_n} b_i x_i,$$

then $(u_n)_{n=1}^\infty$ and $(v_n)_{n=1}^\infty$ are C-equivalent. Show that the closed linear span of a block basic sequence of $(x_n)_{n=1}^\infty$ is always complemented.

9.2. Show that every block basic sequence of $\ell_p(\ell_r^n)$ where $1 \leq r \neq p < \infty$ spans a complemented subspace.

9.3. Show that ℓ_p for $1 \leq p < \infty$ has a unique (up to equivalence) symmetric basis.

9.4. Lorentz sequence spaces.

For every $1 \leq p < \infty$ and every nonincreasing sequence of positive numbers $w = (w_n)_{n=1}^{\infty}$ we consider the *Lorentz sequence space* $d(w, p)$ of all sequences of scalars $x = (a_n)_{n=1}^{\infty}$ for which

$$\|x\| = \sup \left(\sum_{n=1}^{\infty} |a_{\pi(n)}|^p w_n \right)^{1/p} < \infty, \tag{9.18}$$

where π ranges over all permutations of $\mathbb{N}$. One easily checks that $d(w, p)$ equipped with the norm defined by (9.18) is a Banach space.

(a) Show that if $\inf_n w_n > 0$ then $d(w, p) \approx \ell_p$.

(b) Show that if $\sum_{n=1}^{\infty} w_n < \infty$ then $d(w, p) \approx \ell_\infty$.

Therefore to avoid trivial cases we shall assume that $w_1 = 1$, $\lim_{n\to\infty} w_n = 0$, and $\sum_{n=1}^{\infty} w_n = \infty$.

(c) Show that no nontrivial Lorentz sequence space is isomorphic to an ℓ_p-space.

(d) Show that the unit vectors $(e_n)_{n=1}^{\infty}$ form a normalized symmetric basis for $d(w, p)$.

The reader interested in knowing more about Lorentz sequence spaces will find these properties and other, deeper ones in [138].

9.5 (Lindenstrauss-Tzafriri [136]).

Let F be an Orlicz function satisfying the additional condition that for some $q < \infty$ the function $F(x)/x^q$ is decreasing.

(a) Let E_F be the subset of $\mathcal{C}[0, 1]$ defined as the closure of the set of all functions of the form $F_t(x) = F(tx)/F(t)$ for $0 < t \leq 1$. Show that E_F is compact.

(b) Let C_F be the closed convex hull of E_F. Show that every normalized block basic sequence has a subsequence equivalent to the canonical basis of ℓ_G for some $G \in C_F$. Conversely show that for every $G \in C_F$ there is a normalized block basic sequence equivalent to the canonical ℓ_G-basis.

(c) Show that every symmetric basic sequence in ℓ_F is equivalent to the canonical basis of some ℓ_G where $G \in C_F$.

(d) Show that if $G \in E_F$ then ℓ_G is isomorphic to a complemented subspace of ℓ_F.

9.6 (Lindenstrauss-Tzafriri [136]).

(Continuation of 9.5) For $0 < s < 1$ define $T_s(F) \in \mathcal{C}[0, 1]$ by $T_s F(x) = F(sx)/F(s)$.

(a) Show that $T_s : C_F \to C_F$ is continuous.

(b) Show that there is a common fixed point for $\{T_s : 0 < s < 1\}$ and hence that $x^p \in C_F$ for some $1 \leq p < \infty$. (This uses the Schauder Fixed Point theorem, Theorem E.4). Deduce that every ℓ_F has a closed subspace isomorphic to some ℓ_p.

For a more precise result see [137].

9.7 (Zippin [224]). (Compare with Problem 3.8)

(a) Let X be a Banach space with a basis which is not boundedly complete. Show that X has a normalized basis $(x_n)_{n=1}^\infty$ so that for some subsequence $(x_{p_n})_{n=1}^\infty$ we have $\sup_n \|\sum_{j=1}^n x_{p_j}\| < \infty$. Deduce that X has a basis which is not shrinking.

(b) Show that X is reflexive whenever (i) every basis is shrinking, or (ii) every basis is boundedly complete

9.8 ([102]). Let X be a Banach space with a basis and suppose X has the following property: whenever $(x_n)_{n=1}^\infty$ is a basis of X and $(\sum_{j=1}^n a_j x_j)_{n=1}^\infty$ is a weakly Cauchy sequence then $\sum_{j=1}^\infty a_j x_j$ converges.

(a) Show that every weakly Cauchy block basic sequence of a basis $(x_n)_{n=1}^\infty$ is weakly null. [*Hint*: Use Zippin's lemma (Lemma 9.5.5).]

(b) Show that if $(y_n)_{n=1}^\infty$ is a weakly Cauchy sequence then there is a subsequence $(y_{n_k})_{k=1}^\infty$ and a sequence $(z_k)_{k=1}^\infty$ of the form

$$z_k = \sum_{j=1}^{p_k} a_j x_j + \sum_{j=p_k+1}^{p_{k+1}-1} b_j x_j$$

so that $\lim_{k\to\infty} \|y_{n_k} - z_k\| = 0$.

(c) Show that X is weakly sequentially complete.

9.9. Show that every unconditional basis of L_p $(1 < p < \infty)$ has a subsequence equivalent to the canonical basis of ℓ_p. Deduce that:

(a) If $p \neq 2$, L_p has no symmetric basis.

(b) If $(f_n)_{n=1}^\infty$ is a greedy basis of L_p then there exist $0 < c < C < \infty$ so that

$$cn^{1/p} \leq \left\| \sum_{k=1}^n f_n \right\|_p \leq Cn^{1/p}.$$

9.10 (Wojtaszczyk [222]). A basis $(e_n)_{n=1}^\infty$ of a Banach space X is called *quasi-greedy* if $\mathcal{G}_m(x) \to x$ for every $x \in X$. Show that $(e_n)_{n=1}^\infty$ is quasi-greedy if and only if there is a constant K such that

$$\|\mathcal{G}_m(x)\| \leq K\|x\|, \qquad x \in X.$$

(*Caution*: The maps $(\mathcal{G}_m)$ are highly nonlinear and hence you cannot use the Uniform Boundedness principle!)

9.11 (Edelstein-Wojtaszczyk [52]). Let $(x_n)_{n=1}^\infty$ be a normalized unconditional basis of $\ell_1 \oplus \ell_2$. Show that one can partition $\mathbb{N}$ into two infinite sets $\mathbb{A}$ and $\mathbb{B}$ so that $(x_n)_{n\in\mathbb{A}}$ is equivalent to the canonical basis of ℓ_1 and $(x_n)_{n\in\mathbb{B}}$ is equivalent to the canonical basis of ℓ_2. [*Hint*: Suppose $x_n = (y_n, z_n)$ with $y_n \in \ell_1$ and $z_n \in \ell_2$. Let $x_n^* = (y_n^*, z_n^*) \in \ell_\infty \oplus \ell_2$. Let $\mathbb{A} = \{n : y_n^*(y_n) \geq \frac{1}{2}\}$.]

ℓ_p-Subspaces of Banach Spaces

In the previous chapters the spaces ℓ_p ($1 \le p < \infty$) and c_0 have played a pivotal role in the development of the theory. This suggests that we should ask when we can embed one of these spaces in an arbitrary Banach space. For c_0 we have a complete answer: c_0 embeds into X if and only if X contains a WUC series which is not unconditionally convergent (Theorem 2.4.11).

In this chapter we present a remarkable theorem of Rosenthal from 1974 [197] which gives a precise necessary and sufficient condition for ℓ_1 to be isomorphic to a subspace of a Banach space X; this is analogous to, but much more difficult than, the characterization of Banach spaces containing c_0. It requires us to develop so-called Ramsey theory, which has proved a very productive contributor to infinite-dimensional Banach space theory. Rosenthal's theorem asserts that either a Banach space contains ℓ_1 or every bounded sequence has a weakly Cauchy subsequence.

The rest of the chapter is devoted to the construction of an important example, *Tsirelson space*. During the 1960s a potential picture of the structure of Banach spaces emerged in which the ℓ_p-spaces and c_0 were considered as potential building blocks. A question then arose as to whether every Banach space must contain a copy of one of these spaces. This was solved by Tsirelson [217], who constructed an elegant counterexample. Tsirelson's space has had a very profound influence on the further development of the subject.

10.1 Ramsey theory

Let $\mathcal{P}\mathbb{N}$ denote the power set $2^{\mathbb{N}}$ of the natural numbers, i.e., the collection of all subsets of $\mathbb{N}$. $\mathcal{P}\mathbb{N}$ can be identified with the Cantor set $\Delta = \{0,1\}^{\mathbb{N}}$ via the mapping $A \to \chi_A$ where $\chi_A(n) = 1$ if $n \in A$ and 0 otherwise. Let $\mathcal{P}_\infty \mathbb{N}$ be the subset of $\mathcal{P}\mathbb{N}$ of all infinite subsets of $\mathbb{N}$. The complementary set of $\mathcal{P}_\infty \mathbb{N}$ in $\mathcal{P}\mathbb{N}$ of all finite subsets of $\mathbb{N}$ is denoted $\mathcal{F}\mathbb{N}$.

Given any $M \in \mathcal{P}\mathbb{N}$, $\mathcal{F}_r(M)$ will be the collection of all finite subsets of M of cardinality r.

If $M \in \mathcal{P}_\infty M$ and $f : \mathcal{F}_r(\mathbb{N}) \to \mathbb{R}$ is any function, we will write

$$\lim_{A \in \mathcal{F}_r(M)} f(A) = \alpha$$

to mean that given $\epsilon > 0$ there exists $N \in \mathbb{N}$ so that if $A \in \mathcal{F}_r(\mathbb{N})$ and $A \subset [N, \infty)$ then $|f(A) - \alpha| < \epsilon$.

We shall start by proving a generalization of the original **Ramsey theorem** [192]. This is far too simple for our purposes and we will need to go much further. The original Ramsey theorem corresponds to the case $r = 2$ of (ii) of the following theorem. We will use Theorem 10.1.1 (i) in the next chapter.

Theorem 10.1.1.

(i) *Suppose $r \in \mathbb{N}$ and $f : \mathcal{F}_r(\mathbb{N}) \to \mathbb{R}$ is a bounded function. Then there exists $M \in \mathcal{P}_\infty(\mathbb{N})$ so that $\lim_{A \in \mathcal{F}_r(M)} f(A)$ exists.*

(ii) *If $\mathcal{A} \subset \mathcal{F}_r(\mathbb{N})$ then there exists $M \in \mathcal{P}_\infty(\mathbb{N})$ so that either $\mathcal{F}_r(M) \subset \mathcal{A}$ or $\mathcal{F}_r(M) \cap \mathcal{A} = \emptyset$.*

Proof. (ii) follows directly from (i) if we define $f(A) = \chi_\mathcal{A}(A)$.

The proof of (i) is done by induction on r. For $r = 1$ it is trivially true. Assume that $r \geq 2$ and that (i) holds for $r - 1$; we must deduce that (i) is also true for r.

For distinct integers $m_1, \ldots, m_r$, put

$$f(m_1, m_2, \ldots, m_r) = f(\{m_1, \ldots, m_r\}).$$

We first use a diagonal procedure to obtain a subsequence (or subset) M_1 of $\mathbb{N}$ so that for every distinct $m_1, \ldots, m_{r-1}$,

$$\lim_{m_r \in M_1} f(m_1, m_2, \ldots, m_{r-1}, m_r) = g(m_1, m_2, \ldots, m_{r-1})$$

exists. g is independent of the order of $m_1, \ldots, m_{r-1}$ so we may write it as a bounded map $g : \mathcal{F}_{r-1}(\mathbb{N}) \to \mathbb{R}$. It follows from the inductive hypothesis that M_1 has an infinite subset M_2 so that

$$\lim_{A \in \mathcal{F}_{r-1}(M_2)} g(A) = \alpha$$

for some real α.

If $A \in \mathcal{F}_{r-1}(M_2)$ and $\epsilon > 0$, we can find an integer $N = N(A, \epsilon)$ so that if $n \geq N(A, \epsilon)$ and $n \in M_2$ then $n \notin A$, and

$$|f(A \cup \{n\} - g(A))| < \epsilon.$$

We next choose an infinite subset of M_2. Pick $r - 1$ initial points. Then if $m_1 < m_2 < \cdots < m_n$ have been chosen with $n \geq r - 1$, pick $m_{n+1} > m_n$ so that

$$m_{n+1} > \max_{A \in \mathcal{F}_{r-1}\{m_1,\ldots,m_n\}} N(A, 2^{-n}).$$

Finally let $M = \{m_j\}_{j=1}^{\infty}$.

Given $\epsilon > 0$ we may take $n \in \mathbb{N}$ so that, on the one hand, if $A \subset [m_n, \infty)$ with $A \in \mathcal{F}_{r-1}(M)$ then $|g(A) - \alpha| < \frac{1}{2}\epsilon$, and, on the other hand, n is large enough so that $2^{-n} < \frac{1}{2}\epsilon$. Suppose $A \in \mathcal{F}_r(M)$ with $A \subset [m_n, \infty)$. Let m_k be its largest member and let $B = A \setminus \{m_k\}$. Then

$$|f(A) - g(B)| < 2^{-(k-1)} \le 2^{-n} \le \epsilon/2$$

and

$$|g(B) - \alpha| < \epsilon/2,$$

which shows that

$$|f(A) - \alpha| < \epsilon.$$

Hence

$$\lim_{A \in \mathcal{F}_r(M)} f(A) = \alpha.$$

$\square$

We will need an infinite version of Theorem 10.1.1 (*ii*) when $\mathcal{A}$ becomes a subset of $\mathcal{P}_\infty \mathbb{N}$. This requires some topological restrictions.

$\mathcal{P}_\infty \mathbb{N}$ inherits a metric topology from the Cantor set which we call the *Cantor topology*. Since $\mathcal{P}_\infty \mathbb{N}$ is a G_δ-set in $\mathcal{P}\mathbb{N}$, and the Cantor set is compact, this topology can be given by a complete metric.

We shall also be interested in a second stronger topology which is known as the *Ellentuck topology*. If $A \in \mathcal{F}\mathbb{N}$ and $E \in \mathcal{P}_\infty \mathbb{N}$, we define $\mathcal{P}_\infty(A, E)$ to be the collection of all infinite subsets of $A \cup E$ which contain A. In the special case $A = \emptyset$ we write $\mathcal{P}_\infty(\emptyset, E) = \mathcal{P}_\infty(E)$.

Let us say that a set $\mathcal{U} \subset \mathcal{P}_\infty \mathbb{N}$ is *open for the Ellentuck topology* or *Ellentuck-open* if whenever $E \in \mathcal{U}$ there exists a finite set $A \subset E$ so that $\mathcal{P}_\infty(A, E) \subset \mathcal{U}$. This is easily seen to define a topology (the *Ellentuck topology*) on $\mathcal{P}_\infty \mathbb{N}$.

Our aim is to study a dichotomy result. We want to put conditions on a subset $\mathcal{V}$ of $\mathcal{P}_\infty \mathbb{N}$ so that either there is an $M \in \mathcal{P}_\infty \mathbb{N}$ with $\mathcal{P}_\infty(M) \subset \mathcal{V}$ or there is an $M \in \mathcal{P}_\infty \mathbb{N}$ with $\mathcal{P}_\infty(M) \cap \mathcal{V} = \emptyset$. If such a dichotomy holds we say that $\mathcal{V}$ has the *Ramsey property* (or that $\mathcal{V}$ is a *Ramsey set*). However, it turns out to be easier to study a stronger property.

We say that $\mathcal{V}$ is *completely Ramsey* if for finite A and infinite E either there exists an $M \in \mathcal{P}_\infty(E)$ with $\mathcal{P}_\infty(A, M) \subset \mathcal{V}$ or there exists $M \in \mathcal{P}_\infty(E)$ with $\mathcal{P}_\infty(A, M) \cap \mathcal{V} = \emptyset$.

The main result in this section is a theorem of Galvin and Prikry [62] which says that a set which is Borel for the Ellentuck topology is completely Ramsey. In particular this implies that a set which is Borel for the Cantor topology is completely Ramsey. Loosely speaking, this means that if we have a subset of $\mathcal{P}_\infty \mathbb{N}$ which may be defined by countably many conditions then we expect it to be completely Ramsey. This is very useful as we shall see because most sets which arise in analysis are of this type. In fact we will only use the special case of open sets for the Cantor topology, and this follows from the next result.

Theorem 10.1.2. *Suppose $\mathcal{U}$ is an Ellentuck-open set in $\mathcal{P}_\infty\mathbb{N}$. Then $\mathcal{U}$ is completely Ramsey.*

Proof. Let us introduce some notation. If A is finite and E is infinite we shall say that (A, E) is a *pair*. The pair (A, E) is *good* (for $\mathcal{U}$) if there is an infinite subset M of E with $P_\infty(A, M) \subset \mathcal{U}$. Otherwise we shall say that (A, E) is *bad*. Of course, if (A, E) is bad and $F \in \mathcal{P}_\infty(E)$ then (A, F) is also bad. Notice also that if the symmetric difference $E \Delta F$ is finite then (A, E) and (A, F) are either both good or both bad. We will show that if (A, E) is bad then there exists $M \in \mathcal{P}_\infty(E)$ with the property that $\mathcal{P}_\infty(A, M) \cap \mathcal{U} = \emptyset$. To achieve this we do not use the fact that $\mathcal{U}$ is Ellentuck open until the very last step.

Step 1. Suppose $(A_j)_{j=1}^m$ are finite sets and E is an infinite set such that the pair (A_j, E) is bad for $1 \le j \le m$. Then we claim that we can find $n \in E \setminus \bigcup_{j=1}^m A_j$ and $F \in \mathcal{P}_\infty(E)$ so that the pair $(A_j \cup \{n\}, F)$ is also bad for $1 \le j \le m$.

Suppose this is false. Then we may inductively pick an increasing sequence $(n_k)_{k=1}^\infty$, a decreasing sequence of infinite sets $(E_k)_{k=0}^\infty$ with $E_0 = E$, and a sequence $(p(k))_{k=1}^\infty$ of integers with $1 \le p(k) \le m$ so that $n_k \in E_{k-1} \setminus \bigcup_{j=1}^m A_j$ and $\mathcal{P}_\infty(A_{p(k)} \cup \{n_k\}, E_k) \subset \mathcal{U}$.

Now, there exists $1 \le p \le n$ so that the set $\{k \in \mathbb{N} : p(k) = p\}$ is infinite. Let $M = \{n_k : p(k) = p\}$. Suppose $G \in \mathcal{P}_\infty(A_p, M)$. Let k be the least integer such that $n_k \in G$. Then $G \in \mathcal{P}_\infty(A_{p(k)} \cup \{n_k\}, E_k) \subset \mathcal{U}$. Hence $\mathcal{P}_\infty(A_p, M) \subset \mathcal{U}$, contradicting our hypothesis.

Step 2. We show that if a pair (A, E) is bad we can find $M \in \mathcal{P}_\infty(E)$ so that the pair (B, M) is bad for *every* finite set B with $A \subset B \subset A \cup M$.

This is achieved again by an inductive construction. To start the induction we use Step 1. Set $E_0 = E$; there exists $n_1 \in E_0$ and an infinite set $E_1 \in \mathcal{P}_\infty(E_0)$ for which the pair (B, E_1) is bad if $A \subset B \subset A \cup \{n_1\}$. Suppose we have chosen sets $E_0, E_1, \ldots, E_k$ with $E_j \subset E_{j-1}$ for $1 \le j \le n$, and integers $n_1, n_2, \ldots, n_k$ with $n_j \in E_{j-1}$ for $1 \le j \le n$, such that (B, E_j) is bad if $A \subset B \subset A \cup \{n_1, \ldots, n_j\}$ for $1 \le j \le n$. Then, according to Step 1, we can find $n_{k+1} \in E_k$ with $n_{k+1} > n_k$ and $E_{k+1} \subset E_k$ so that $(B \cup \{n_{k+1}\}, E_{k+1})$ is bad for every $A \subset B \subset A \cup \{n_1, \ldots, n_k\}$.

It remains to show that $M = \{n_1, n_2, \ldots\}$ has the desired property. If B is a finite subset of $A \cup M$, let k be the largest natural number so that $n_k \in B$. Then $B \subset A \cup \{n_1, \ldots, n_k\}$ so that (B, E_k) is bad. However, $M \subset E_k \cup \{n_1, \ldots, n_k\}$ so (B, M) is also bad.

Step 3. Let us complete the proof, recalling finally that $\mathcal{U}$ is supposed Ellentuck open. If a pair (A, E) is bad, we determine $M \subset E$ according to Step 2 so that (B, M) is bad whenever B is finite and $A \subset B \subset A \cup M$. Suppose $\mathcal{P}_\infty(A, M)$ meets $\mathcal{U}$, so there exists $G \in \mathcal{P}_\infty(A, M) \cap \mathcal{U}$. Since $\mathcal{U}$ is open there exists a finite set B, which can be assumed to contain A so that

$\mathcal{P}_\infty(B, G) \subset \mathcal{U}$. This implies that (B, M) is good, and we have reached a contradiction. Hence the only possible conclusion is that $\mathcal{P}_\infty(A, M) \cap \mathcal{U} = \emptyset$.

□

Now we come to the theorem of Galvin and Prikry [62] mentioned before.

Theorem 10.1.3. *Let $\mathcal{V}$ be a subset of $\mathcal{P}_\infty \mathbb{N}$ which is Borel for the Ellentuck topology. Then $\mathcal{V}$ is completely Ramsey.*

Proof. We first remark that if $\mathcal{U}$ is dense and open for the Ellentuck topology, then Theorem 10.1.2 yields that for every pair (A, E) there exists $M \in \mathcal{P}_\infty(E)$ with $\mathcal{P}_\infty(A, M) \subset \mathcal{U}$. This is because there is no pair (A, M) with $\mathcal{P}_\infty(A, M) \cap \mathcal{U} = \emptyset$.

Step 1. We claim that for any pair (A, E), if $B \subset E$ is finite then there exists $M \in \mathcal{P}_\infty(B, E)$ so that $\mathcal{P}_\infty(A, M) \subset \mathcal{U}$.

Indeed, we list all subsets $(B_j)_{j=1}^N$ of B. Find $H_1 \in \mathcal{P}_\infty(E)$ so that $\mathcal{P}_\infty(A \cup B_1, H_1) \subset \mathcal{U}$ and then inductively $H_j \in \mathcal{P}_\infty(H_{j-1})$ so that $\mathcal{P}_\infty(A \cup B_j, H_j) \subset \mathcal{U}$. Finally let $M = H_N$. If $G \in \mathcal{P}_\infty(A, M)$ let $G \cap B = B_j$. Then $G \in \mathcal{P}_\infty(A \cup B_j, M) \subset \mathcal{P}_\infty(A \cup B_j, H_j) \subset \mathcal{U}$.

Step 2. Suppose $\mathcal{G}$ is an intersection of a countable family of open dense sets for the Ellentuck topology. Then we can find a descending sequence of dense open sets $(\mathcal{U}_n)_{n=1}^\infty$ with $\mathcal{G} = \cap_{n=1}^\infty \mathcal{U}_n$. We will show that if (A, E) is any pair we can find $M \in \mathcal{P}_\infty(E)$ so that $\mathcal{P}_\infty(A, M) \subset \mathcal{G}$.

As usual we inductively pick an increasing sequence of integers $(n_k)_{k=1}^\infty$ and a descending sequence of infinite sets $(E_k)_{k=0}^\infty$ with $E_0 = E$, such that $n_k \in E_j$ for all j and $\mathcal{P}_\infty(A, E_k) \subset \mathcal{U}_k$. We pick $n_1 \in E_0$ arbitrarily and let $E_1 \subset E_0$ be so that $n_1 \in E_1$ and $\mathcal{P}_\infty(A, E_1) \subset \mathcal{U}_1$. If $n_1, \ldots, n_{k-1}, E_1, \ldots, E_{k-1}$ have been picked we choose $n_k \in E_{k-1}$ with $n_k > n_{k-1}$ and then use Step 1 to pick $E_k \subset E_{k-1}$ so that $\{n_1, \ldots, n_k\} \subset E_k$ and $\mathcal{P}_\infty(A, E_k) \subset \mathcal{U}_k$.

Finally let $M = \{n_1, n_2, \ldots\}$. If $G \in \mathcal{P}_\infty(A, M)$ then for every k, $G \in \mathcal{P}_\infty(A, E_k)$ which implies $G \in \mathcal{U}_k$. Hence $G \in \mathcal{G}$.

Step 3. Let us complete the proof supposing that $\mathcal{V}$ is a Borel set for the Ellentuck topology. Then there is a set $\mathcal{G}$ which is the intersection of a sequence of dense open sets $(\mathcal{U}_n)_{n=1}^\infty$, so that $\mathcal{G} \cap \mathcal{V} = \mathcal{G} \cap \mathcal{U}$ for some Ellentuck open set $\mathcal{U}$ (see the Problems). If (A, E) is any pair, we may first find $G \in \mathcal{P}_\infty(E)$ so that $\mathcal{P}_\infty(A, G) \subset \mathcal{G}$ by Step 2. Now, there exists $M \in \mathcal{P}_\infty(G)$ so that either $\mathcal{P}_\infty(A, M) \subset \mathcal{U}$ or $\mathcal{P}_\infty(A, M) \cap \mathcal{U} = \emptyset$. But then either $\mathcal{P}_\infty(A, M) \subset \mathcal{V}$ or $\mathcal{P}_\infty(A, M) \cap \mathcal{V} = \emptyset$.

□

10.2 Rosenthal's ℓ_1 theorem

The motivation for the main result in this section comes from the problem of finding a criterion to be able to extract a weakly Cauchy subsequence from any bounded sequence in a Banach space X. If X is reflexive, this follows from the Eberlein-Šmulian theorem. What if X is not reflexive? It was known to Banach that if X^* is separable, then every bounded sequence in X has a weakly Cauchy subsequence. But in other spaces this is not possible.

For instance, the canonical basis $(e_n)_{n=1}^\infty$ of ℓ_1 has no weakly Cauchy subsequences. Rosenthal's ℓ_1 theorem says that, in some sense, this is the only possible example. Rosenthal proved this for real Banach spaces, and the necessary modifications for complex Banach spaces were given shortly after by Dor [44]. Our proof will work for both real and complex scalars.

Theorem 10.2.1 (Rosenthal's ℓ_1 Theorem [197]). *Let $(x_n)_{n=1}^\infty$ be a bounded sequence in an infinite-dimensional Banach space X. Then either:*

(a) $(x_n)_{n=1}^\infty$ has a subsequence which is weakly Cauchy, or
(b) $(x_n)_{n=1}^\infty$ has a subsequence which is basic and equivalent to the canonical basis of ℓ_1.

Proof. Let $(x_n)_{n=1}^\infty$ be a bounded sequence in a Banach space X which has no weakly Cauchy subsequence. We will suppose that $\|x_n\| \leq 1$ for all n. We begin by passing to a subsequence which is basic. This is achieved by Theorem 1.5.6 since, obviously, the set $\{x_n\}_{n=1}^\infty$ does not have any weakly convergent subsequences. Thus we can assume that the sequence $(x_n)_{n=1}^\infty$ is already basic.

If M is any infinite subset of $\mathbb{N}$, in order to measure how far the sequence of elements in M is from being weakly Cauchy we define

$$\text{osc}\,(M) = \sup_{\|x^*\|\leq 1} \lim_{k\to\infty} \sup_{\substack{m,n>k \\ m,n\in M}} |x^*(x_m) - x^*(x_n)|.$$

We claim that there exists $M \in \mathcal{P}_\infty\mathbb{N}$ so that if $M' \in \mathcal{P}_\infty(M)$ then $\text{osc}\,(M') = \text{osc}\,(M) > 0$.

Indeed, let us inductively define infinite sets $\mathbb{N} = M_0 \supset M_1 \supset M_2 \supset M_3 \ldots$ so that

$$\text{osc}\,(M_k) < \inf_{M'\in\mathcal{P}_\infty(M_{k-1})} \text{osc}\,(M') + k^{-1}, \qquad k = 1,2,\ldots.$$

Let M be chosen by a diagonal procedure so that $M \subset M_k \cup F_k$ where each F_k is finite. M has the desired property that $\text{osc}\,(M') = \text{osc}\,(M)$ if $M' \in \mathcal{P}_\infty(M)$. Then, $\text{osc}\,(M) > 0$ follows from the fact there is no weakly Cauchy subsequence.

We may make one further reduction by finding $u^* \in B_{X^*}$ and $M' \subset M$ so that $\lim_{n\in M'} u^*(x_n) = \theta$ where $|\theta| \geq \frac{1}{2}\text{osc}\,(M)$.

Again for convenience of notation we may suppose that the original sequence has these properties, i.e., $\text{osc}\,(M) = 4\delta > 0$ is constant for every infinite set M and $\lim_{n\to\infty} u^*(x_n) = \theta$ for some $u^* \in B_{X^*}$ and $|\theta| > \delta$.

Since $(x_n)_{n=1}^\infty$ is basic and bounded away from zero, there exist biorthogonal functionals $(x_n^*)_{n=1}^\infty$ in X^* and we have a bound $\|x_n^*\| \leq B$ for some constant B.

Let $C = 1 + \delta^{-1} + \delta^{-2}$. Let us consider the subset $\mathcal{V}$ of $\mathcal{P}_\infty \mathbb{N}$ of all $M = \{m_j\}_{j=1}^\infty$ where $(m_j)_{j=1}^\infty$ is strictly increasing such that there exists $x^* \in X^*$ with $\|x^*\| \le C$ and $x^*(x_{m_j}) = (-1)^j$ for all j.

It follows immediately from the weak* compactness of $\{x^* : \|x^*\| \le C\}$ that the set $\mathcal{V}$ is closed for the Cantor topology, and hence closed for the Ellentuck topology. Thus, $\mathcal{V}$ has the Ramsey property (note here we only use Theorem 10.1.2).

Suppose M is any infinite subset of $\mathbb{N}$. Since osc$(M) = \delta$ we can find a subsequence $(m_j)_{j=1}^\infty$ of M so that for some $y^* \in B_{X^*}$ we have $\lim_{j \to \infty} y^*(x_{m_{2j}}) = \alpha$ and $\lim_{j \to \infty} y^*(x_{m_{2j-1}}) = \beta$ where $|\alpha - \beta| \ge 2\delta$. Next let

$$v^* = \frac{2}{(\alpha - \beta)} y^* - \frac{\alpha + \beta}{\theta(\alpha - \beta)} u^*.$$

Then

$$\|v^*\| \le (1 + \theta^{-1})\delta^{-1} \le \delta^{-1} + \delta^{-2}$$

and

$$\lim_{j \to \infty} v^*(x_{m_{2j}}) = 1, \quad \lim_{j \to \infty} v^*(x_{m_{2j-1}}) = -1.$$

By passing to a further subsequence we can suppose that if $c_j = v^*(x_{m_j}) - (-1)^j$ then $|c_j| \le 2^{-j} B^{-1}$. Then consider

$$x^* = v^* + \sum_{j=1}^\infty c_j x_{m_j}^*.$$

We have

$$\|x^*\| \le 1 + \delta^{-1} + \delta^{-2} = C.$$

Further $x^*(x_{m_j}) = (-1)^j$.

It follows that $M' \in \mathcal{V}$ and thus there is no M so that $\mathcal{P}_\infty(M) \cap \mathcal{V} = \emptyset$. Hence there is an infinite subset M so that every $M' \in \mathcal{P}_\infty(M)$ is in $\mathcal{V}$.

Let $M = \{m_j\}_{j=1}^\infty$ where (m_j) is increasing. Then the sequence $(m_{2j})_{j=1}^\infty$ has the property that for every sequence of signs (ϵ_j) we can find x^* with $\|x^*\| \le C$ and $x^*(x_{m_{2j}}) = \epsilon_j$.

If X is real, it is clear that for any sequence of scalars $(a_j)_{j=1}^n$, we can pick $\epsilon_j = \pm 1$ with $\epsilon_j a_j = |a_j|$ and then find $x^* \in X^*$ with $\|x^*\| \le C$ so that $x^*(x_{m_{2j}}) = \epsilon_j$. Thus,

$$\left\| \sum_{j=1}^n a_j x_{m_{2j}} \right\| \ge \frac{1}{C} \sum_{j=1}^n |a_j|,$$

and so $(x_{m_{2j}})$ is equivalent to the canonical ℓ_1-basis.

If X is complex, the same reasoning shows that

$$\left\| \sum_{j=1}^n a_j x_{m_{2j}} \right\| \ge \frac{1}{C} \sum_{j=1}^n |\Re a_j|,$$

and, similarly,

$$\left\| \sum_{j=1}^{n} a_j x_{m_{2j}} \right\| \geq \frac{1}{C} \sum_{j=1}^{n} |\Im a_j|.$$

Thus,

$$\left\| \sum_{j=1}^{n} a_j x_{m_{2j}} \right\| \geq \frac{1}{2C} \sum_{j=1}^{n} |a_j|.$$

$\square$

Corollary 10.2.2. *A Banach space X contains no copy of ℓ_1 if and only if every bounded sequence in X has a weakly Cauchy subsequence.*

Remark 10.2.3. If X^* is separable, then X cannot contain a copy of ℓ_1. However, it is not easy to construct a separable Banach space for which X^* is non-separable but X fails to contain a copy of ℓ_1. This was done by James [87] who produced an example called the *James tree space, $\mathcal{JT}$.* We postpone the construction of this example to Chapter 13.

If X is separable there is a very fine distinction between the conditions that (a) X^* is separable and (b) X does not contain ℓ_1. Let us illustrate this. If X^* is separable then the weak* topology on $B_{X^{**}}$ is a metrizable topology and thus Goldstine's theorem guarantees that for every $x^{**} \in B_{X^{**}}$ there is a *sequence* $(x_n)_{n=1}^{\infty}$ in B_X converging to x^{**} weak* (this sequence is, of course, a weakly Cauchy sequence in X).

If X does not contain ℓ_1 but X^* is not separable then the weak* topology is no longer metrizable, yet remarkably the same conclusion holds (this is due to Odell and Rosenthal [160]):

Theorem 10.2.4. *Let X be a separable Banach space. Then ℓ_1 does not embed into X if and only if every $x^{**} \in X^{**}$ is the weak* limit of a sequence $(x_n)_{n=1}^{\infty}$ in X.*

10.3 Tsirelson space

The question we want to address in this section is whether every Banach space contains a copy of one of the spaces ℓ_p for $1 \leq p < \infty$, or c_0. The motivation behind this question is that these spaces (which are prime!) appear to be in a certain sense the fundamental blocks from which all Banach spaces are constructed. Indeed every space we have met so far contains one of these blocks. For example, every subspace of ℓ_p contains a copy of ℓ_p. We also have seen that every subspace of L_p for $p > 2$ contains a copy of one of the spaces ℓ_p or ℓ_2 (Theorem 6.4.8). The case of subspaces of L_p for $1 \leq p < 2$ is much more difficult and was not resolved until 1981 by Aldous. He showed [2] that every subspace of L_p for $1 \leq p < 2$ also contains a copy of some ℓ_q; Krivine and Maurey [121] subsequently gave an alternative argument based on the notion of stability. Nevertheless the result is still not so easy and is beyond the scope of this book.

It was quite a surprise when in 1974 Tsirelson gave the first example of a Banach space not containing some ℓ_p $(1 \leq p < \infty)$ or c_0. Nowadays the dual of the space constructed by Tsirelson has become known as *Tsirelson space*. Despite its apparently strange definition it has turned out to be a remarkable springboard for further research.

Before getting to Tsirelson space we will need a result of James from 1964 [83]. He showed that if ℓ_1 embeds in a Banach space, then it must embed very well (close to isometrically). This result, although quite simple, is also very significant as we will discuss later.

Theorem 10.3.1 (James's ℓ_1 distortion theorem). *Let* $(x_n)_{n=1}^{\infty}$ *be a normalized basic sequence in a Banach space* X *which is equivalent to the canonical* ℓ_1-*basis. Then given* $\epsilon > 0$ *there is a normalized block basic sequence* $(y_n)_{n=1}^{\infty}$ *of* $(x_n)_{n=1}^{\infty}$ *such that*

$$\left\| \sum_{k=1}^{N} a_k y_k \right\| \geq (1 - \epsilon) \sum_{k=1}^{N} |a_k|$$

for any sequence of scalars $(a_k)_{k=1}^{N}$.

Proof. For each n let M_n be the least constant so that if $(a_k)_{k=1}^{\infty} \in c_{00}$ with $a_k = 0$ for $k \leq n$ then

$$\sum_{k=1}^{\infty} |a_k| \leq M_n \left\| \sum_{k=1}^{\infty} a_k x_k \right\|.$$

Then $(M_n)_{n=1}^{\infty}$ is a decreasing sequence with $\lim_{n \to \infty} M_n = M \geq 1$. Thus, for n large enough $M_n < (1 - \epsilon)^{-\frac{1}{2}} M$.

Now we can pick a normalized block basic sequence $(y_n)_{n=1}^{\infty}$ of the form

$$y_n = \sum_{j=p_{n-1}+1}^{p_n} b_j x_j$$

such that

$$\sum_{j=p_{n-1}+1}^{p_n} |b_j| \geq (1 - \epsilon)^{\frac{1}{2}} M, \qquad n = 1, 2, \ldots$$

and so that $M_{p_0} < (1 - \epsilon)^{-\frac{1}{2}} M$. Then,

$$\sum_{j=1}^{N} |a_j| \leq (1 - \epsilon)^{-\frac{1}{2}} M^{-1} \sum_{j=1}^{N} |a_j| \sum_{i=p_{j-1}+1}^{p_j} |b_i|$$

$$\leq (1 - \epsilon)^{-\frac{1}{2}} M^{-1} M_{p_0} \left\| \sum_{j=1}^{N} a_j y_j \right\|$$

$$\leq (1 - \epsilon)^{-1} \Big\| \sum_{j=1}^{N} a_j y_j \Big\|,$$

and the result is proved.

□

Next we construct Tsirelson's space. This is, as mentioned above, not the original space constructed by Tsirelson in 1974 [217] but its dual as constructed by Figiel and Johnson [59].

Theorem 10.3.2. *There is a reflexive Banach space T which contains no copy of ℓ_p for $1 \leq p < \infty$, or c_0.*

Proof. Suppose $(I_1, \ldots, I_m)$ is a set of disjoint intervals of natural numbers. We say $(I_1, \ldots, I_m)$ is *admissible* if $m < I_k$ for $k = 1, 2 \ldots, m$, i.e., each I_k is contained in $[m + 1, \infty)$.

We will adopt the convention that if E is a subset of $\mathbb{N}$ (in particular, if E is an interval of integers) and $\xi \in c_{00}$ we will write $E\xi$ for the sequence $(\chi_E(n)\xi(n))_{n=1}^{\infty}$, i.e., the sequence whose coordinates are $E\xi(n) = \xi(n)$ if $n \in E$ and $E\xi(n) = 0$ otherwise.

We define a norm, $\| \cdot \|_T$, on c_{00} by the formula

$$\|\xi\|_T = \max \left\{ \|\xi\|_{c_0}, \sup \frac{1}{2} \sum_{j=1}^{m} \|I_j\xi\|_T \right\}, \tag{10.1}$$

the supremum being taken over all admissible families of intervals. This definition is implicit and we need to show that there is such a norm. But that follows by a relatively easy inductive argument. Let $\|\xi\|_0 = \|\xi\|_{c_0}$ and then define inductively for $n = 1, 2, \ldots$

$$\|\xi\|_n = \max \left\{ \|\xi\|_{c_0}, \sup \sum_{j=1}^{m} \|I_j\xi\|_{n-1} \right\},$$

where, again, the supremum is taken over all admissible families of intervals. The sequence $(\|\xi\|_n)_{n=1}^{\infty}$ is increasing and bounded above by $\|\xi\|_{\ell_1}$. Hence it converges to some $\|\xi\|_T$, and it follows readily that $\| \cdot \|_T$ has all the required properties of norm.

It is necessary also to show that the definition uniquely determines $\| \cdot \|_T$. Indeed, suppose $\| \cdot \|_{T'}$ is another norm on c_{00} satisfying (10.1). It is clear from the induction argument that $\|\xi\|_{T'} \geq \|\xi\|_T$ for all $\xi \in c_{00}$. For $\alpha > 1$ let $S = \{\xi \in c_{00} : \|\xi\|_{T'} > \alpha\|\xi\|_T\}$. If S is nonempty it has a member with minimal support. But an appeal to (10.1) gives a contradiction. Hence there is a unique norm on c_{00} that is the solution of (10.1).

Let T be the completion of $(c_{00}, \| \cdot \|_T)$. The canonical unit vectors $(e_n)_{n=1}^{\infty}$ form a 1-unconditional basis of T.

Suppose ℓ_p for some $1 < p < \infty$, or c_0 embeds in T. Then, by the Bessaga-Pełczyński selection principle (Proposition 1.3.10), there is a normalized block basic sequence $(\xi_n)_{n=1}^{\infty}$ with respect to the canonical basis of T equivalent to the canonical basis. Suppose we fix m and choose n so that ξ_n is supported in $[m+1, \infty)$. Then

$$\|\xi_n + \cdots + \xi_{n+m-1}\|_T \geq \frac{1}{2}m$$

by the definition of $\|\cdot\|_T$. This contradicts the equivalence with the ℓ_p-basis (or the c_0-basis).

Let us show that ℓ_1 cannot be embedded in T. Assume it embeds. Then we can find a normalized block basic sequence equivalent to the ℓ_1-basis. If $\epsilon < \frac{1}{4}$, by James's ℓ_1 distortion theorem (Theorem 10.3.1) we pass to a sequence of blocks and assume we have a normalized block basic sequence $(\xi_n)_{n=0}^{\infty}$ so that

$$\left\| \sum_{j=0}^{n} a_j \xi_j \right\|_T \geq (1-\epsilon) \sum_{j=0}^{n} |a_j|$$

for any scalars $(a_j)_{j=0}^n$.

Suppose ξ_0 is supported on $[1, r]$. For every n we have

$$\left\| \xi_0 + \frac{1}{n} \sum_{j=1}^{n} \xi_j \right\|_T \geq 2(1-\epsilon).$$

It is clear that

$$\left\| \xi_0 + \frac{1}{n} \sum_{j=1}^{n} \xi_j \right\|_T > \left\| \xi_0 + \frac{1}{n} \sum_{j=1}^{n} \xi_j \right\|_{c_0},$$

so we must be able to find an admissible collection of intervals $(I_1, \ldots, I_k)$ such that

$$\left\| \xi_0 + \frac{1}{n} \sum_{i=1}^{n} \xi_i \right\|_T = \frac{1}{2} \sum_{j=1}^{k} \left\| I_j \left(\xi_0 + \frac{1}{n} \sum_{i=1}^{n} \xi_i \right) \right\|_T.$$

If $I_j \xi_0 = 0$ for every j then

$$\frac{1}{2} \sum_{j=1}^{k} \left\| I_j \left(\xi_0 + \frac{1}{n} \sum_{i=1}^{n} \xi_i \right) \right\|_T = \frac{1}{2} \sum_{i=1}^{k} \left\| I_i \left(\frac{1}{n} \sum_{j=1}^{n} \xi_i \right) \right\|_T \leq 1,$$

so we can assume that $I_j \xi_0 \neq 0$ for some j. But this means, by admissibility, that $k \leq r$. Note that

$$\frac{1}{2} \sum_{j=1}^{k} \left\| I_j \left(\xi_0 + \frac{1}{n} \sum_{i=1}^{n} \xi_i \right) \right\|_T \leq \frac{1}{2} \sum_{j=1}^{k} \| I_j \xi_0 \| + \frac{1}{2n} \sum_{j=1}^{k} \left\| I_j \left(\sum_{i=1}^{n} \xi_i \right) \right\|_T.$$

The first term is estimated by $\|\xi_0\|_T = 1$. For the second term we have

$$\frac{1}{2n}\sum_{j=1}^{k}\left\|I_j\Big(\sum_{i=1}^{n}\xi_i\Big)\right\|_T \leq \frac{1}{2n}\sum_{i=1}^{n}\sum_{j=1}^{k}\|I_j\xi_i\|_T.$$

There are at most $k \leq r$ values of i such that the support of ξ_i meets at least two I_j. For such values of i we have

$$\frac{1}{2n}\sum_{j=1}^{k}\|I_j\xi_i\|_T \leq \frac{1}{n}\|\xi_i\|_T = \frac{1}{n}.$$

For all values of i we have

$$\frac{1}{2n}\sum_{j=1}^{k}\|I_j\xi_i\|_T \leq \frac{1}{2n}.$$

Hence,

$$\left\|\xi_0 + \frac{1}{n}\sum_{i=1}^{n}\xi_i\right\|_T \leq 1 + \frac{k}{n} + \frac{n-k}{2n} = 1 + \frac{n+r}{2n}.$$

The right-hand side converges to $3/2$ as $n \to \infty$ and, as $3/2 < 2(1-\epsilon)$, we have a contradiction.

By James's theorem (Theorem 3.3.3), since T contains no copy of c_0 or ℓ_1, it must be reflexive.

□

The construction of Tsirelson space was thus a disappointment to those who expected a nice structure theory for Banach spaces. It was, however, far from the end of the story. Tsirelson space (and its modifications) as an example has continued to play an important role in the area since 1974. See the book by Casazza and Shura from 1989 [29].

The major problem left open was the unconditional basic sequence problem, which was discussed at the end of Chapter 3. Tsirelson space played a significant role in the solution of this problem.

There is a curious and deep relationship between the unconditional basic sequence problem and James's ℓ_1 distortion theorem (Theorem 10.3.1). James's result implies that if we put an equivalent norm $||| \cdot |||$ on ℓ_1 then we will always be able to find an infinite-dimensional subspace on which this norm is a close multiple of the original norm. Thus, given $\epsilon > 0$ we can find an infinite-dimensional subspace Y of ℓ_1 and a constant $c > 0$ so that

$$c(1-\epsilon)\|\xi\|_1 \leq |||\xi||| \leq c(1+\epsilon)\|\xi\|_1, \qquad \xi \in Y.$$

Here $\|\cdot\|_1$ denotes the usual norm on ℓ_1. James also showed the same property for c_0, and a problem arose as to whether a similar result might hold for arbitrary Banach spaces. The construction of Tsirelson space showed this to be false, using an earlier result of Milman [151]. However, it was left unresolved at this time whether one could specify a constant M with the property that for every Banach space X and every equivalent norm $|||\cdot|||$ there is an infinite-dimensional subspace Y and a constant $c > 0$ so that

$$cM^{-1}\|x\| \leq |||x||| \leq cM\|x\|, \qquad x \in Y.$$

This was solved negatively by Schlumprecht in 1991. He constructed an example (known nowadays as *Schlumprecht space*) which is a variant of Tsirelson's construction. Using this space, Odell and Schlumprecht [161] in 1994 showed that this property even fails in Hilbert spaces (and most other spaces). The Schlumprecht space was also a key ingredient in the Gowers-Maurey solution of the unconditional basic sequence problem [71].

Problems

10.1. Show that if X is a topological space and $\mathcal{V}$ is a Borel subset of X, then there is a dense G_δ-set $\mathcal{G}$, and an open set $\mathcal{U}$ such that $\mathcal{V} \cap \mathcal{G} = \mathcal{U} \cap \mathcal{G}$ (see Problem 4.7).

10.2 (Johnson). Let $(x_n)_{n=1}^\infty$ be a sequence in a Banach space X with the property that every subsequence $(x_{n_k})_{k=1}^\infty$ contains a further subsequence $(x_{n_{k_j}})_{j=1}^\infty$ such that

$$\sup_{n\geq 1}\left\|\sum_{j=1}^n (-1)^j x_{n_{k_j}}\right\| < \infty.$$

Show that $(x_n)_{n=1}^\infty$ has a subsequence $(y_n)_{n=1}^\infty$ such that $\left(\sum_{j=1}^n y_j\right)_{n=1}^\infty$ is a WUC series. In particular, if $(x_n)_{n=1}^\infty$ is normalized deduce that $(x_n)_{n=1}^\infty$ has a subsequence equivalent to the canonical basis of c_0.

10.3. James distortion theorem for c_0.
Let $(x_n)_{n=1}^\infty$ be a normalized basic sequence in a Banach space X equivalent to the canonical c_0-basis. Show that given $\epsilon > 0$ there is a normalized block basic sequence $(y_n)_{n=1}^\infty$ of $(x_n)_{n=1}^\infty$ such that

$$\left\|\sum_{k=1}^N a_k y_k\right\| \geq (1-\epsilon)\max_k |a_k|$$

for any sequence of scalars $(a_k)_{k=1}^N$.

10.4. (a) Let X be a nonreflexive Banach space and suppose $x^{**} \in X^{**} \setminus X$. Show that if $\epsilon > 0$, V is a weak* neighborhood of x^{**}, and $x_1,\ldots,x_n \in X$ there exists $x \in V \cap X^{**}$ so that

$$\left|\|x_j + x^{**}\| - \|x_j + x\|\right| < \epsilon, \qquad j = 1, 2, \ldots, n.$$

(b) Show that if X is a nonreflexive Banach space such that for some $x^{**} \in X^{**}$ we have $\|x^{**} + x\| = \|x^{**} - x\|$ for every $x \in X$ then X contains a copy of ℓ_1.

[*Hint*: Use (a) and an inductive construction to find a basic sequence equivalent to the canonical ℓ_1-basis.]

Part (b) is due to Maurey [145], who also proved the more difficult converse: if X is *separable* and contains a copy of ℓ_1 then there exists $x^{**} \in X^{**}$ with $\|x^{**} + x\| = \|x^{**} - x\|$ for all $x \in X$.

10.5. Show that Tsirelson space contains no symmetric basic sequence.

10.6. Let $||| \cdot |||$ be the norm on c_{00} obtained by the implicit formula

$$|||\xi||| = \max\left(\|\xi\|_\infty, \sup \sum_{j=1}^{2n} |||I_j\xi|||\right),$$

where the supremum is over all n and all collections of intervals $(I_j)_{j=1}^{2n}$ with $n < I_1 < I_2 < \cdots < I_{2n}$ (i.e., using $2n$ instead of n in the definition of T).

At the same time define two associated norms by

$$\|\xi\|_{T,1} = \sup\left\{\sum_{j=1}^{3} \|I_j\xi\|_T\right\},$$

where $(I_j)_{j=1}^{3}$ ranges over all triples of intervals $I_1 < I_2 < I_3$, and

$$\|\xi\|_{T,2} = \sup\left\{\sum_{j=1}^{8k} \|I_j\xi\|_T\right\},$$

the supremum being taken over all k and all collections of intervals $(I_j)_{j=1}^{8k}$ such that $k < I_1 < I_2 < \cdots < I_{8k}$.
(a) Show that $\|\xi\|_{T,2} \le \|\xi\|_{T,1} \le 3\|\xi\|_T$.
(b) Show by induction on the size of the support that

$$|||\xi||| \le \|\xi\|_{T,1}$$

and deduce that

$$\|\xi\|_T \le |||\xi||| \le 3\|\xi\|_T.$$

(c) Show that T is isomorphic to T^2.

10.7 (Casazza, Johnson, and Tzafriri [25]). Let $J_1, \ldots, J_m$ be disjoint intervals and suppose $\xi, \eta \in c_{00}$ are supported on $\bigcup_{j=1}^{m} J_k$ and satisfy $\|J_j\xi\|_T = \|J_j\eta\|_T$ for $1 \le j \le m$. The goal of this exercise is to show the following inequality:

$$\frac{1}{6}\|\xi\|_T \le \|\eta\|_T \le 6\|\xi\|_T. \tag{10.2}$$

To this end, first we will show by induction on m that $\|\xi\|_T \le 2|||\eta|||$, where $||| \cdot |||$ is the norm we introduced in 10.6. Suppose then this is proved for all collections of $m - 1$ intervals, and ξ and η are given as above.

(a) Consider an admissible collection of intervals $n < I_1 < \cdots < I_n$. Let $\mathbb{A}$ be the set of all j such that J_j meets more than one I_k, together with the first l such that J_l meets at least one I_k.

Show that $|\mathbb{A}| \leq n$, and that for each $j \in \mathbb{A}$,

$$\sum_{k=1}^{n} \|(I_k \cap J_j)\xi\|_T \leq 2\|J_j\eta\|_T.$$

(b) Let $I_k' = I_k \setminus \cup_{j \in \mathbb{A}} J_j$ and

$$I_k'' = I_k' \cup \bigcup \{J_j : J_j \cap I_k' \neq \emptyset\}.$$

Show that $(I_k'')_{k=1}^n$ is admissible and using the inductive hypothesis show that

$$\|I_k''\xi\|_T \leq 2\||I_k''\eta\||, \qquad k = 1, 2, \ldots, n.$$

(c) Complete the inductive proof that $\|\xi\|_T \leq 2\||\eta\||$.

(d) Prove the inequality (10.2).

10.8 (Casazza, Johnson, and Tzafriri [25]). Show that every block basic sequence in T is complemented. [*Hint:* Use the previous problem.]

11

Finite Representability of ℓ_p-Spaces

We are now going to switch gear and study local properties of infinite-dimensional Banach spaces. In Banach space theory the word *local* is used to denote finite-dimensional. We can distinguish between properties of a Banach space that are determined by its finite-dimensional subspaces and properties which require understanding of the whole space. For example, one cannot decide that a space is reflexive just by looking at its finite-dimensional subspaces, but properties like type and cotype which depend on inequalities with only finitely many vectors are local in character.

The key idea of the chapter is that, while a Banach space need not contain any subspace isomorphic to a space ℓ_p $(1 \leq p < \infty)$ or c_0 (as was shown by the existence of Tsirelson space), it will always contain such a space *locally*. The precise meaning of this statement will be made clear shortly.

There are two remarkable results of this nature due to Dvoretzky (1961) [49] and Krivine (1976) [119] which are the highlights of the chapter. The methods we use in this chapter are curiously infinite-dimensional in nature, although the results are local. In the following chapter we will consider a local and more quantitative approach to Dvoretzky's theorem.

11.1 Finite representability

In this section we present the notions of finite representability and ultraproducts. Finite representability emerged as a concept in the Banach space scene in the late 1960s; it was originally introduced by James [85].

Definition 11.1.1. Let X and Y be infinite-dimensional Banach spaces. We say that X is *finitely representable* in Y if given any finite-dimensional subspace E of X and $\epsilon > 0$ there is a finite-dimensional subspace F of Y with $\dim F = \dim E$, and a linear isomorphism $T : E \to F$, satisfying $\|T\|\|T^{-1}\| < 1 + \epsilon$; that is, in terms of the Banach-Mazur distance, $d(E, F) < 1 + \epsilon$.

Example 11.1.2. Every Banach space X (not necessarily separable) is finitely representable in c_0. Indeed, given any finite-dimensional subspace E of X and $\epsilon > 0$, pick ν so that $\frac{1}{1-\nu} < 1 + \epsilon$ and $\{e_1^*, \ldots, e_N^*\}$ a ν-net in B_{E^*}. Consider the mapping $T : E \to \ell_\infty^N$ defined by $T(e) = (e_j^*(e))_{j=1}^N$. Then, if we let $F = T(E)$, it is straightforward to check that $d(E, F) < 1 + \epsilon$.

Remark 11.1.3. (a) In Definition 11.1.1 we can assume that $\|T\| = 1$ and $\|T^{-1}\| < 1 + \epsilon$ by replacing T by a suitable multiple.

(b) If X is finitely representable in Y, X need not be isomorphic to a subspace of Y. For instance, ℓ_∞ is finitely representable in c_0 from Example 11.1.2 but it does not embed in c_0. Another example is provided by L_p $(1 \le p < \infty)$, which, despite the fact that does not embed in ℓ_p, is finitely representable in ℓ_p as we will see in Proposition 11.1.7.

Proposition 11.1.4. *If X is finitely representable in Y and Y is finitely representable in Z then X is finitely representable in Z.*

Proof. Suppose E is a finitely dimensional subspace of X and $\epsilon > 0$. Then there exists a finite-dimensional subspace F of Y and an isomorphism $T : E \to F$ with $\|T\| = 1$ and $\|T^{-1}\| < (1 + \epsilon)^{1/2}$. Similarly we can find a finite-dimensional subspace G of Z and an isomorphism $S : F \to G$ with $\|S\| = 1$ and $\|S^{-1}\| < (1 + \epsilon)^{\frac{1}{2}}$. Then $\|ST\|\|(ST)^{-1}\| < 1 + \epsilon$. $\qquad\square$

Definition 11.1.5. An infinite-dimensional Banach space X is said to be *crudely finitely representable (with constant λ)* in an infinite-dimensional Banach space Y if there is a constant $\lambda > 1$ such that given any finite-dimensional subspace E of X there is a finite-dimensional subspace F of Y with $\dim F = \dim E$ and a linear isomorphism $T : E \to F$ satisfying $\|T\|\|T^{-1}\| < \lambda$.

Thus X is finitely representable in Y if and only if X is crudely finitely representable in Y with constant λ for every $\lambda > 1$.

Lemma 11.1.6. *Suppose X is a separable Banach space and that $(E_n)_{n=1}^\infty$ is an increasing sequence of subspaces of X such that $\cup_{n=1}^\infty E_n$ is dense in X.*

(i) *X is finitely representable in a Banach space Y if and only if given $n \in \mathbb{N}$ and $\epsilon > 0$ there is a finite-dimensional subspace F of Y with $\dim F = \dim E_n$ and a linear isomorphism $T_n : E_n \to F$ satisfying $\|T_n\|\|T_n^{-1}\| < 1 + \epsilon$.*

(ii) *Let $\lambda > 1$ and suppose that X has the property that given $n \in \mathbb{N}$ there is a finite-dimensional subspace F of Y with $\dim F = \dim E_n$ and a linear isomorphism $T_n : E_n \to F$ satisfying $\|T_n\|\|T_n^{-1}\| \le \lambda$. Then, given any $\epsilon > 0$, X is crudely finitely representable in Y with constant $\lambda + \epsilon$.*

Proof. It is enough to prove (ii). Suppose X satisfies the property in the hypothesis, that E is any finite-dimensional subspace of X and that $(e_j)_{j=1}^N$ is a basis of E. Since E is finite-dimensional there is a constant $C = C(E)$ so that for any scalars $(a_j)_{j=1}^N$,

$$\frac{1}{C} \max_{1 \le j \le N} |a_j| \le \left\| \sum_{j=1}^N a_j e_j \right\| \le C \max_{1 \le j \le N} |a_j|.$$

Let us pick $\nu > 0$ small enough to ensure that $\lambda(1 + CN\nu)^2 < \lambda + \epsilon$. Then we can find an n so that there exist $x_j \in E_n$ for $1 \le j \le N$ with $\|x_j - e_j\| < \nu$. Define $S : E \to Y$ to be the linear map given by $Se_j = T_n x_j$. Then

$$\left\| \sum_{j=1}^N a_j e_j - \sum_{j=1}^N a_j x_j \right\| \le N\nu \max_{1 \le j \le N} |a_j| \le CN\nu \left\| \sum_{j=1}^N a_j e_j \right\|.$$

Hence

$$(1 + CN\nu)^{-1} \|e\| \le \|Se\| \le \lambda(1 + CN\nu)\|e\|, \qquad e \in E.$$

If we let $F = S(E)$, it is clear that $\|S\| \|S^{-1}\| < \lambda + \epsilon$.

□

One of the reasons for the idea of finite representability to develop is that we can express the obvious connection between the function spaces L_p and the sequence spaces ℓ_p in this language:

Proposition 11.1.7. L_p *is finitely representable in* ℓ_p *for* $1 \le p < \infty$.

Proof. For each p, just take E_n to be the subspace generated in L_p by the characteristic functions $\chi_{((k-1)/2^n, k/2^n)}$ for $1 \le k \le 2^n$. E_n is then isometric to a subspace of ℓ_p.

□

In fact a converse statement is also true:

Theorem 11.1.8. *Let* X *be a separable Banach space. If* X *is finitely representable in* ℓ_p *($1 \le p < \infty$) then* X *is isometric to a subspace of* L_p.

Proof. Let $(x_n)_{n=1}^\infty$ be a dense sequence in B_X; by making a small perturbation where necessary we can assume this sequence to be linearly independent in X. Let q be the conjugate index of p.

By hypothesis, for each $n \in \mathbb{N}$ there is a linear operator $T_n : E_n \to \ell_p$, where $E_n = [x_1, \ldots, x_n]$, satisfying

$$\|x\| \le \|T_n x\| \le \left(1 + \frac{1}{n}\right)\|x\|, \qquad x \in E_n.$$

Let $S : \ell_q \to X$ [respectively, $S : c_0 \to X$ if $q = \infty$] be the operator defined by

$$S\xi = \sum_{k=1}^{\infty} 2^{-k/p} \xi(k) x_k,$$

and for each n let $V_n : \ell_q \to \ell_p$ [respectively, $V_n : c_0 \to \ell_p$ when $p = 1$] be given by

$$V_n\xi = \sum_{k=1}^{n} 2^{-k/p} \xi(k) T_n(x_k).$$

We would like to estimate the quantity

$$\sum_{i=1}^{l} \|V_n\xi_i\|^p - \sum_{i=1}^{m} \|V_n\eta_i\|^p$$

for any $\xi_1, \ldots, \xi_l, \eta_1, \ldots, \eta_m \in c_{00}$.

Let $K = B_{\ell_q^*}$ [respectively, $K = B_{c_0^*}$ when $q = \infty$] with the weak* topology, and F the continuous function on K defined by

$$F(\xi^*) = \sum_{i=1}^{l} |\xi^*(\xi_i)|^p - \sum_{i=1}^{m} |\xi^*(\eta_i)|^p. \tag{11.1}$$

Note that $F(0) = 0$, so $\max_{s\in K} F(s) \geq 0$. Then, if we let (e_n^*) denote the biorthogonal functionals associated to the canonical basis (e_n) of ℓ_p, we have

$$\sum_{i=1}^{l} \|V_n\xi_i\|^p - \sum_{i=1}^{m} \|V_n\eta_i\|^p = \sum_{j=1}^{\infty} \left(\sum_{i=1}^{l} |e_j^*(V_n\xi_i)|^p - \sum_{i=1}^{m} |e_j^*(V_n\eta_i)|^p \right)$$

$$= \sum_{j=1}^{\infty} \left(\sum_{i=1}^{l} |V_n^* e_j^*(\xi_i)|^p - \sum_{i=1}^{m} |V_n^* e_j^*(\eta_i)|^p \right)$$

$$\leq \left(\sum_{j=1}^{\infty} \|V_n^* e_j^*\|^p \right) \max_{s\in K} F(s).$$

Now

$$\sum_{j=1}^{\infty} \|V_n^* e_j^*\|^p = \sum_{j=1}^{\infty} \sum_{k=1}^{\infty} |V_n^* e_j^*(e_k)|^p$$

$$= \sum_{j=1}^{\infty} \sum_{k=1}^{\infty} |e_j^*(V_n e_k)|^p$$

$$= \sum_{k=1}^{n} \|V_n e_k\|^p$$

$$= \sum_{k=1}^{n} 2^{-k} \|T_n e_k\|^p$$

$$\leq \left(1 + \frac{1}{n}\right)^p \sum_{k=1}^{\infty} 2^{-k} = \left(1 + \frac{1}{n}\right)^p.$$

Hence

$$\sum_{i=1}^{l} \|V_n \xi_i\|^p - \sum_{i=1}^{m} \|V_n \eta_i\|^p \leq \left(1 + \frac{1}{n}\right)^p \max_{s \in K} F(s).$$

If we let $n \to \infty$, the left-hand side converges to $\sum_{i=1}^{l} \|S\xi_i\|^p - \sum_{i=1}^{m} \|S\eta_i\|^p$, and so

$$\sum_{i=1}^{l} \|S\xi_i\|^p - \sum_{i=1}^{m} \|S\eta_i\|^p \leq \max_{s \in K} F(s). \tag{11.2}$$

The set of all F of the form (11.1) forms a linear subspace $\mathcal{V}$ of $\mathcal{C}(K)$. It follows from (11.2) that we can unambiguously define a linear functional φ on $\mathcal{V}$ by

$$\varphi(F) = \sum_{i=1}^{l} \|S\xi_i\|^p - \sum_{i=1}^{m} \|S\eta_i\|^p,$$

and that $\varphi(F) \leq \max_{s \in K} F(s)$. By the Hahn-Banach theorem there is a probability measure μ on K such that

$$\varphi(F) = \int_K F \, d\mu, \qquad F \in \mathcal{V}.$$

Now suppose $x \in E = \cup_{n=1}^{\infty} E_n$. Then $S^{-1}x \in c_{00}$ is well-defined since the sequence $(x_n)_{n=1}^{\infty}$ was chosen linearly independent. Define $Ux \in \mathcal{C}(K)$ by

$$Ux(\xi^*) = \xi^*(S^{-1}x).$$

U is a linear map from E into $\mathcal{C}(K)$ but we also have

$$\|Ux\|_{L_p(K,\mu)} = \|x\|,$$

so U is an isometry of E into $L_p(K, \mu)$ which extends by density to an isometry of X into $L_p(K, \mu)$.

$\square$

Proposition 11.1.9 (L_q-subspaces of L_p).
(i) For $1 \leq p \leq 2$, L_q embeds in L_p if and only if $p \leq q \leq 2$.
(ii) For $2 < p < \infty$, L_q embeds in L_p if and only if $q = 2$ or $q = p$.
Moreover, if L_q embeds in L_p then it embeds isometrically.

Proof. Let $1 \leq p, q < \infty$ and suppose that L_q embeds in L_p. Then, since ℓ_q embeds in L_q, it follows that ℓ_q embeds in L_p. This implies, by Theorem 6.4.19, that either $q = p$, or $q = 2$, or $1 \leq p < q < 2$.

It remains to be shown that L_q embeds in L_p for $1 \leq p < q < 2$. We know that L_q is finitely representable in ℓ_q for each q (Proposition 11.1.7) and that

ℓ_q embeds in L_p for $1 \leq p < q < 2$ (Theorem 6.4.19). Hence L_q is finitely representable in L_p if $1 \leq p < q < 2$. Since, in turn, L_p is finitely representable in ℓ_p, it follows that L_q is finitely representable in ℓ_p for $1 \leq p < q < 2$. By Theorem 11.1.8, L_q is isomorphic to a subspace of L_p.

$\square$

Next we are going to introduce ultraproducts of Banach spaces. This idea was crystallized by Dacunha-Castelle and Krivine [33] and serves as an appropriate vehicle to study finite representability by infinite-dimensional methods. Let us recall, first, a few definitions.

If $\mathcal{I}$ is any infinite set, a *filter* on $\mathcal{I}$ is a subset $\mathcal{F}$ of $\mathcal{PI}$ satisfying the properties:

- $\emptyset \notin \mathcal{F}$.
- If $A \subset B$ and $A \in \mathcal{F}$ then $B \in \mathcal{F}$.
- If $A, B \in \mathcal{F}$ then $A \cap B \in \mathcal{F}$.

A function $f : \mathcal{I} \to \mathbb{R}$ is said to *converge to ξ through $\mathcal{F}$*, and we write $\lim_{\mathcal{F}} f(x) = \xi$, if $f^{-1}(U) \in \mathcal{F}$ for every open set U containing ξ.

We will be primarily interested in the case $\mathcal{I} = \mathbb{N}$ so that a function of $\mathbb{N}$ is simply a sequence.

Example 11.1.10. Let us single out two examples of filters on $\mathbb{N}$:

(a) For each $n \in \mathbb{N}$ we can define the filter $\mathcal{F}_n = \{A : n \in A\}$. Then a sequence $(\xi_k)_{k=1}^{\infty}$ converges to ξ through $\mathcal{F}_n$ if and only if $\xi_n = \xi$.

(b) Let us consider the filter $\mathcal{F}_{\infty} = \{A : \exists n \in \mathbb{N} : [n, \infty) \subset A\}$. Then $\lim_{\mathcal{F}_{\infty}} \xi_n = \xi$ if and only if $\lim_{n \to \infty} \xi_n = \xi$.

An *ultrafilter* $\mathcal{U}$ is a maximal filter with respect to inclusion, i.e., a filter which is not properly contained in any larger filter. By Zorn's lemma, every filter is contained in an ultrafilter. Ultrafilters are characterized by one additional property:

- If $A \in \mathcal{PI}$ then either $A \in \mathcal{U}$ or $\tilde{A} = \mathcal{I} \setminus A \in \mathcal{U}$.

If $\mathcal{U}$ is an ultrafilter then any bounded function on $\mathcal{I}$ converges through $\mathcal{U}$. Indeed, suppose $|f(x)| \leq M$ for all x and f does not converge through $\mathcal{U}$. Then for every $\xi \in [-M, M]$ we can find an open set U_{ξ} containing ξ so that $f^{-1}(U_{\xi}) \notin \mathcal{U}$. Using compactness we can find a finite set $\xi_1, \ldots, \xi_n \in [-M, M]$ so that $[-M, M] \subset \cup_{j=1}^{n} U_{\xi_j}$. Now $f^{-1}(\tilde{U}_{\xi_j}) \in \mathcal{U}$ for each j since it is an ultrafilter. But then the properties of filters imply that the intersection $\cap_{j=1}^{n} f^{-1}(\tilde{U}_{\xi_j}) \in \mathcal{U}$; however, this set is empty and we have a contradiction.

Let us restrict again to $\mathbb{N}$. The filters $\mathcal{F}_n$ are in fact ultrafilters; these are called the *principal ultrafilters*. Any other ultrafilter must contain $\mathcal{F}_{\infty}$; these are the *nonprincipal ultrafilters*.

Suppose X is a Banach space and $\mathcal{U}$ is a nonprincipal ultrafilter on $\mathbb{N}$. We consider the ℓ_{∞}-product $\ell_{\infty}(X)$ and define on it a seminorm by

$$\|(x_n)_{n=1}^\infty\|_{\mathcal U} = \lim_{\mathcal U}\|x_n\|.$$

Then $\|(x_n)_{n=1}^\infty\|_{\mathcal U} = 0$ if and only if $(x_n)_{n=1}^\infty$ belongs to the subspace $c_{0,\mathcal U}(X)$ of $\ell_\infty(X)$ of all $(x_n)_{n=1}^\infty$ such that $\lim_{\mathcal U}\|x_n\| = 0$. It is readily verified that $\|\cdot\|_{\mathcal U}$ induces the quotient norm on the quotient space $X_{\mathcal U} = \ell_\infty(X)/c_{0,\mathcal U}(X)$. This space is called an *ultraproduct* of X.

It is, of course, possible to define ultraproducts using ultrafilters on sets $\mathcal I$ other than $\mathbb N$ and this is useful for nonseparable Banach spaces. For our purposes the natural numbers will suffice.

We will frequently make use of the following lemma:

Lemma 11.1.11. *Let E be a finite-dimensional normed space and suppose $(x_j)_{j=1}^N$ is an ϵ-net in the surface of the unit ball $\{e : \|e\| = 1\}$, where $0 < \epsilon < 1$. Suppose $T : E \to X$ is a linear map such that $1 - \epsilon \le \|Tx_j\| \le 1 + \epsilon$ for $1 \le j \le N$. Then for every $e \in E$ we have*

$$\left(\frac{1-3\epsilon}{1-\epsilon}\right)\|e\| \le \|Te\| \le \left(\frac{1+\epsilon}{1-\epsilon}\right)\|e\|.$$

Proof. First suppose $\|e\| = 1$. Pick j so that $\|e - x_j\| \le \epsilon$. Then

$$\|Te\| \le \|Te - Tx_j\| + (1 + \epsilon),$$

and so

$$\|T\| \le \|T\|\epsilon + (1 + \epsilon);$$

i.e.,

$$\|T\| \le \frac{1+\epsilon}{1-\epsilon}.$$

On the other hand we also have

$$\|Te\| \ge 1 - \epsilon - \|T\|\epsilon \ge \frac{1-3\epsilon}{1-\epsilon}.$$

$\square$

Proposition 11.1.12. *Let X, Y be infinite-dimensional Banach spaces.*

(i) The ultraproduct $X_{\mathcal U}$ is finitely representable in X.

(ii) If Y is separable then Y is finitely representable in X if and only if Y is isometric to a subspace of $X_{\mathcal U}$.

(iii) If Y is separable then Y is crudely finitely representable in X if and only if Y is isomorphic to a subspace of $X_{\mathcal U}$.

Proof. (i) Let E be a finite-dimensional subspace of $X_{\mathcal U}$ and suppose $\epsilon > 0$. We can (by selecting representatives for a basis in E) suppose $E \subset \ell_\infty(X)$ and that $\|\cdot\|_{\mathcal U}$ is a norm on E. Choose $\nu > 0$ so small that $(1+\nu)(1-3\nu)^{-1} < 1 + \epsilon$. Then pick a finite ν-net $\mathcal N = \{\xi_1, \ldots, \xi_N\}$ in the unit ball of E. Thus $B_E \subset \mathcal N + \nu B_E$.

There exists $A \in \mathcal{U}$ such that

$$1 - \nu < \|\xi_j(k)\| < 1 + \nu, \qquad k \in A, \ 1 \leq j \leq N.$$

Pick any fixed $k \in A$ and define $T : E \to X$ by $T\xi = \xi(k)$. Let $T(E) = F$. Then by Lemma 11.1.11, $\|T\|\|T^{-1}\| < 1 + \epsilon$.

(ii) Let us suppose $(E_n)_{n=1}^{\infty}$ is an ascending sequence of finite-dimensional subspaces of Y with $E = \cup_{n=1}^{\infty} E_n$ dense in Y, and let $T_n : E_n \to X$ be operators satisfying

$$(1 - \frac{1}{n})\|e\| \leq \|T_n e\| \leq \|e\|, \qquad e \in E_n,$$

for all $n \in \mathbb{N}$.

We define a map $L : E \to \ell_\infty(X)$ by setting $L(y) = \xi$, where

$$\xi(k) = \begin{cases} 0 & y \notin E_k \\ T_k(y) & y \in E_k. \end{cases}$$

L is nonlinear, but is linear as a map into $X_{\mathcal{U}}$ since

$$L(x + y) - L(x) - L(y) \in c_{00}(X) \subset c_{0,\mathcal{U}}(X).$$

If $y \in \cup_{n=1}^{\infty} E_n$ then $\lim_{n\to\infty} \|\xi(n)\| = \|y\|$, whence it is clear that L induces an isometry of Y into $X_{\mathcal{U}}$.

(iii) This is similar to (ii).

$\square$

An immediate deduction is the following:

Proposition 11.1.13. *Y is crudely finitely representable in X if and only if there is an equivalent norm on Y so that Y is finitely representable in X.*

The next theorem is an application of the basic idea of an ultraproduct. Note that we prove it only for real scalars; the proof for complex scalars would require some extra work.

Theorem 11.1.14. *Let X be a Banach space. Then*

(i) X fails to have type $p > 1$ if and only if ℓ_1 is finitely representable in X.
(ii) X fails to have cotype $q < \infty$ if and only if ℓ_∞ is finitely representable in X.

Proof. We will use Lemma 7.2.5. For (i) it suffices to note that $\alpha_N(X) = \sqrt{N}$ for every N. Thus for fixed N and all n we can find $(x_{nk})_{k=1}^{N}$ so that

$$\left(\sum_{k=1}^{N} \|x_{nk}\|^2 \right)^{1/2} = \sqrt{N},$$

but

$$N - \frac{1}{n} < \left(\mathbb{E} \left\| \sum_{k=1}^{N} \varepsilon_k x_{nk} \right\|^2 \right)^{1/2} \leq \sum_{k=1}^{N} \|x_{nk}\| \leq N.$$

Consider the elements

$$\xi_k(n) = (x_{nk})_{n=1}^{\infty}$$

in the ultraproduct $X_{\mathcal{U}}$. Then

$$\left(\sum_{k=1}^{N} \|\xi_k\|_{\mathcal{U}}^2 \right)^{\frac{1}{2}} = \sqrt{N},$$

$$\left(\mathbb{E} \left\| \sum_{k=1}^{N} \varepsilon_k \xi_k \right\|_{\mathcal{U}}^2 \right)^{\frac{1}{2}} \geq N, \text{ and}$$

$$\sum_{k=1}^{N} \|\xi_k\|_{\mathcal{U}} \geq N.$$

Using the Cauchy-Schwarz inequality we see that the last inequalities are equalities and we must have $\|\xi_k\|_{\mathcal{U}} = 1$ for all k. Furthermore, it follows that

$$\left\| \sum_{k=1}^{N} \epsilon_k \xi_k \right\|_{\mathcal{U}} = N,$$

whenever $\epsilon_k = \pm 1$.

Now suppose $-1 \leq a_k \leq 1$ and let $\epsilon_k = -1$ if $a_k < 0$ and $\epsilon_k = 1$ if $a_k \geq 0$. Then

$$\left\| \sum_{k=1}^{N} a_k \xi_k \right\|_{\mathcal{U}} \geq \left\| \sum_{k=1}^{N} \epsilon_k \xi_k \right\|_{\mathcal{U}} - \left\| \sum_{k=1}^{N} (\epsilon_k - a_k) \xi_k \right\|_{\mathcal{U}}$$

$$\geq N - \sum_{k=1}^{N} (1 - |a_k|)$$

$$= \sum_{k=1}^{N} |a_k|.$$

Thus $(\xi_k)_{k=1}^{N}$ is isometrically equivalent to the canonical basis of ℓ_1^N, and it follows that ℓ_1 is finitely representable in X.

(ii) is similar using again Lemma 7.2.5, and we leave the details to the Problems.

$\square$

11.2 The Principle of Local Reflexivity

The main result in this section is the very important result of Lindenstrauss and Rosenthal from 1969 [133] called the Principle of Local Reflexivity; it asserts that in a local sense every Banach space is reflexive. More precisely, for any infinite-dimensional Banach space X, its second dual X^{**} is finitely representable in X. Our proof is based on one given by Stegall [209]; see also [36] for an interpretation of the Principle in terms of spaces of operators.

Let $T : X \to Y$ be a bounded operator. If the range $T(X)$ is closed, T is sometimes called *semi-Fredholm*. This is equivalent to the requirement that T factors to an isomorphic embedding on $X/\ker(T)$ (i.e., the canonical induced map $T_0 : X/\ker(T) \to Y$ is an isomorphic embedding), which in turn is equivalent to the statement that for some constant C we have

$$d(x, \ker(T)) \leq C\|Tx\|, \qquad x \in X.$$

Proposition 11.2.1. *Let $T : X \to Y$ be an operator with closed range. Suppose $y \in Y$ is such that the equation $T^{**}x^{**} = y$ has a solution $x^{**} \in X^{**}$ with $\|x^{**}\| < 1$. Then the equation $Tx = y$ has a solution $x \in X$ with $\|x\| < 1$.*

Proof. This is almost immediate. We must show that $y \in T(U_X)$, where U_X is the open unit ball of X.

First suppose $y \notin T(X)$. In this case there exists $y^* \in Y^*$ with $T^*y^* = 0$ but $y^*(y) = 1$. This is impossible since $T^{**}x^{**}(y^*) = y^*(y) = 1$.

Next suppose $y \in T(X) \backslash T(U_X)$. By the Open Mapping theorem $T(U_X)$ is open relative to $T(X)$ and so, using the Hahn-Banach separation theorem, we can find $y^* \in Y^*$ with $y^*(y) \geq 1$ but $y^*(Tx) < 1$ for $x \in U_X$. Thus $\|T^*y^*\| \leq 1$ and so $|x^{**}(T^*y^*)| < 1$, i.e., $|y^*(y)| < 1$, which is a contradiction. $\square$

Proposition 11.2.2. *Let $T : X \to Y$ be an operator with closed range and suppose $K : X \to Y$ is a finite-rank operator. Then $T + K$ also has closed range.*

Proof. Suppose $T + K$ does not have closed range. Then there is a bounded sequence $(x_n)_{n=1}^{\infty}$ with $\lim_{n\to\infty}(T + K)(x_n) = 0$ but $d(x_n, \ker(T + K)) \geq 1$ for all n. We can pass to a subsequence and assume that $(Kx_n)_{n=1}^{\infty}$ converges to some $y \in Y$ and hence $\lim_{n\to\infty} Tx_n = -y$. This implies that there exists $x \in X$ with $Tx = -y$ and thus $\lim_{n\to\infty} \|Tx_n - Tx\| = 0$. Hence $\lim_{n\to\infty} d(x_n - x, \ker(T)) = 0$. It follows that $y - Kx \in K(\ker T)$.

Let $y - Kx = Ku$, where $u \in \ker(T)$. Then

$$\lim_{n\to\infty} d(x_n - x - u, \ker(T)) = 0,$$

and

$$\lim_{n\to\infty} \|Kx_n - Kx - u\| = 0.$$

Since $K|_{\ker(T)}$ has closed range this means that

$$\lim_{n\to\infty} d(x_n - x - u, \ker(T) \cap \ker(K)) = 0.$$

But $T(x + u) = -y = -K(x + u)$, so $x + u \in \ker(T + K)$ and therefore

$$\lim_{n\to\infty} d(x_n, \ker(T + K)) = 0,$$

contrary to our assumption.

$\square$

Theorem 11.2.3. *Let X be a Banach space, $A = (a_{jk})_{j,k=1}^{m,n}$ be an $m \times n$ real matrix, and $B = (b_{jk})_{j,k=1}^{p,n}$ be a $p \times n$ real matrix. Let $y_1, \ldots, y_m \in X$, $y_1^*, \ldots, y_p^* \in X^*$, and $\xi_1, \ldots, \xi_p \in \mathbb{R}$. Suppose there exist vectors $x_1^{**}, \ldots, x_n^{**}$ in X^{**} with $\max_{1 \le k \le n} \|x_k^{**}\| < 1$ satisfying the following equations:*

$$\sum_{k=1}^n a_{jk} x_k^{**} = y_j, \qquad 1 \le j \le m$$

and

$$y_j^* \left(\sum_{k=1}^n b_{jk} x_k^{**} \right) = \xi_j, \qquad 1 \le j \le p.$$

Then there exist vectors $x_1, \ldots, x_n$ in X with $\max_{1 \le k \le n} \|x_k\| < 1$ satisfying the (same) equations:

$$\sum_{k=1}^n a_{jk} x_k = y_j, \qquad 1 \le j \le m$$

and

$$y_j^* \left(\sum_{k=1}^n b_{jk} x_k \right) = \xi_j, \qquad 1 \le j \le p.$$

Proof. Consider the operator $T_0 : \ell_\infty^n(X) \to \ell_\infty^m(X)$ defined by

$$T_0(x_1, \ldots, x_n) = \left(\sum_{k=1}^n a_{jk} x_k \right)_{j=1}^m.$$

We claim that T_0 has closed range. This is an immediate consequence of the fact the matrix A can be written in the form $A = PDQ$, where P and Q are nonsingular, and D is in the form

$$D = \begin{pmatrix} I_r & 0 \\ 0 & 0 \end{pmatrix},$$

where r is the rank of A. This allows a factorization of T_0 in the form $T_0 = USV$ where U, V are invertible and S is given by the matrix D, and therefore trivially it has closed range.

Now define $T : \ell_\infty^n(X) \to \ell_\infty^m(X) \oplus_\infty \ell_\infty^p$ by

$$T(x_1, \ldots, x_n) = \Big(T_0(x_1, \ldots, x_n), \big(x_j^*(\sum_{k=1}^n b_{jk}x_k)\big)_{j=1}^p\Big).$$

By Proposition 11.2.2 it is clear that T also has closed range. The theorem then follows directly from Proposition 11.2.1.

$\square$

Theorem 11.2.4 (The Principle of Local Reflexivity). *Let X be a Banach space. Suppose that F is a finite-dimensional subspace of X^{**} and G is a finite-dimensional subspace of X^*. Then given $\epsilon > 0$ there is a subspace E of X containing $F \cap X$ with $\dim E = \dim F$, and a linear isomorphism $T : F \to E$ with $\|T\|\|T^{-1}\| < 1 + \epsilon$ such that*

$$Tx = x, \qquad x \in F \cap X$$

and

$$x^*(Tx^{**}) = x^{**}(x^*), \qquad x^* \in G, \; x^{**} \in F.$$

In particular X^{**} is finitely representable in X.

Proof. Given $\epsilon > 0$ let us take $\nu > 0$ so that $(1+\nu)(1-3\nu)^{-1} < 1+\epsilon$ and pick a ν-net $(x_j^{**})_{j=1}^N$ in $\{x^{**} \in F : \|x^{**}\| = 1\}$. Let $S : \mathbb{R}^N \to F$ be the operator defined by

$$S(\xi_1, \ldots, \xi_N) = \sum_{j=1}^N \xi_j x_j^{**}.$$

Let $H = S^{-1}(F \cap X)$ and suppose $(a^{(j)})_{j=1}^m$ is a basis for H. Let $S(a^{(j)}) = y_j \in F \cap X$ and define the matrix $A = (a_{jk})_{j=1,k=1}^{m,N}$ by $a^{(j)} = (a_{j1}, \ldots, a_{jN})$.

Next pick $x_1^*, \ldots, x_N^* \in X^*$ so that $\|x_j^*\| = 1$ and $x_j^*(x_j^*) > 1 - \nu$, and finally pick a basis $\{g_1^*, \ldots, g_l^*\}$ of G.

We consider the system of equations in $(x_1, \ldots, x_N)$:

$$\sum_{k=1}^N a_{jk}x_k = y_j, \qquad j = 1, 2, \ldots, m$$

$$x_j^*(x_j) = x_j^{**}(x_j^*), \qquad j = 1, 2, \ldots, N$$

and

$$g_j^*(x_j) = x_j^{**}(g_j^*), \qquad j = 1, 2, \ldots, l.$$

This system has a solution in X^{**}, namely, $(x_1^{**}, \ldots, x_N{**})$, and $\max_j \|x_j^{**}\| = 1$. It follows from Theorem 11.2.3 that it has a solution $(x_1, \ldots, x_N)$ in X with $\max \|x_j\| < 1 + \nu$.

If we define $S_1 : \mathbb{R}^N \to X$ by

$$S_1(\xi_1, \ldots, \xi_N) = \sum_{j=1}^{N} \xi_j x_j,$$

then it is clear from the construction that $S(\xi) = 0$ implies that $S_1(\xi) = 0$, and so $S_1 = TS$ for some operator $T : F \to X$. Let $E = T(F)$. Note that for $1 \le j \le N$ we have

$$1 - \nu < \|x_j\| < 1 + \nu$$

since $\|x_j\| \ge x_j^*(x_j) > 1 - \nu$. Hence, by Lemma 11.1.11, $\|T\|\|T^{-1}\| < 1 + \epsilon$. The other properties are clear from the construction.

$\square$

11.3 Krivine's theorem

In this section we will use the term *sequence space* to denote the completion $\mathcal{X}$ of c_{00} under some norm $\|\cdot\|_{\mathcal{X}}$ such that the basis vectors $(e_n)_{n=1}^{\infty}$ have norm one.

Definition 11.3.1. A sequence $(x_n)_{n=1}^{\infty}$ in a Banach space X is *spreading* if it has the property that for any integers $0 < p_1 < p_2 < \cdots < p_n$ and any sequence of scalars $(a_i)_{i=1}^{n}$ we have

$$\left\| \sum_{j=1}^{n} a_j x_{p_j} \right\| = \left\| \sum_{j=1}^{n} a_j x_j \right\|.$$

Notice that if $(x_n)_{n=1}^{\infty}$ is an unconditional basic sequence in a Banach space X the previous definition means that $(x_n)_{n=1}^{\infty}$ is subsymmetric (Definition 9.2.4).

Definition 11.3.2. A sequence space $\mathcal{X}$ is *spreading* if the canonical basis $(e_n)_{n=1}^{\infty}$ of $\mathcal{X}$ is spreading.

Definition 11.3.3. Let $(x_n)_{n=1}^{\infty}$ be a bounded sequence in a Banach space X, and $(y_n)_{n=1}^{\infty}$ be a bounded sequence in a Banach space Y. We will say that $(y_n)_{n=1}^{\infty}$ is *block finitely representable* in $(x_n)_{n=1}^{\infty}$ if given $\epsilon > 0$ and $N \in \mathbb{N}$ there exist a sequence of *blocks* of $(x_n)_{n=1}^{\infty}$,

$$u_j = \sum_{p_{j-1}+1}^{p_j} a_j x_j, \qquad j = 1, 2, \ldots, N,$$

where (p_j) are integers with $0 = p_0 < p_1 < \cdots < p_N$, and (a_n) are scalars, and an operator $T : [y_j]_{j=1}^{N} \to [u_j]_{j=1}^{N}$ with $Ty_j = u_j$ for $1 \le j \le N$ such that $\|T\|\|T^{-1}\| < 1 + \epsilon$.

Note here that we do not assume that $(x_n)_{n=1}^{\infty}$ or $(y_n)_{n=1}^{\infty}$ is a basic sequence, although usually they are.

Definition 11.3.4. Let $(x_n)_{n=1}^{\infty}$ be a bounded sequence in a Banach space X. A sequence space $\mathcal{X}$ is said to be *block finitely representable* in $(x_n)_{n=1}^{\infty}$ if the canonical basis vectors $(e_n)_{n=1}^{\infty}$ in $\mathcal{X}$ are block finitely representable in $(x_n)_{n=1}^{\infty}$.

Obviously if $\mathcal{X}$ is block finitely representable in $(x_n)_{n=1}^{\infty}$ it is also true that $\mathcal{X}$ is finitely representable in X. We are thus asking for a strong form of finite representability.

Definition 11.3.5. A sequence space $\mathcal{X}$ is said to be *block finitely representable* in another sequence space $\mathcal{Y}$ if it is block finitely representable in the canonical basis of $\mathcal{Y}$.

Proposition 11.3.6. *Suppose $(x_n)_{n=1}^{\infty}$ is a nonconstant spreading sequence in a Banach space X.*

(i) *If $(x_n)_{n=1}^{\infty}$ fails to be weakly Cauchy then $(x_n)_{n=1}^{\infty}$ is a basic sequence equivalent to the canonical ℓ_1-basis.*
(ii) *If $(x_n)_{n=1}^{\infty}$ is weakly null then it is an unconditional basic sequence with suppression constant $K_s = 1$.*
(iii) *If $(x_n)_{n=1}^{\infty}$ is weakly Cauchy then $(x_{2n-1} - x_{2n})_{n=1}^{\infty}$ is weakly null and spreading.*

Proof. (i) If $(x_n)_{n=1}^{\infty}$ is not weakly Cauchy then no subsequence can be weakly Cauchy (by the spreading property) and so, by Rosenthal's theorem (Theorem 10.2.1), some subsequence is equivalent to the canonical ℓ_1-basis; but then again this means the entire sequence is equivalent to the ℓ_1-basis.

(ii) It is enough to show that if $a_1, \ldots, a_n \in \mathbb{R}$ and $1 \leq m \leq n$ then

$$\left\| \sum_{j<m} a_j x_j + \sum_{m<j\leq n} a_j x_j \right\| \leq \left\| \sum_{j=1}^{n} a_j e_j \right\|.$$

Suppose $\epsilon > 0$. By Mazur's theorem we can find $c_j \geq 0$ for $1 \leq j \leq l$, say, so that $\sum_{j=1}^{l} c_j = 1$ and

$$\left\| \sum_{j=1}^{l} c_j x_j \right\| < \epsilon.$$

Now consider

$$x = \sum_{j=1}^{m-1} a_j x_j + a_m \sum_{j=m}^{m+l-1} c_{j-m+1} x_j + \sum_{j=m+l}^{m+l-1} a_{j-l+1} x_j.$$

Then

$$x = \sum_{i=1}^{l} c_i \left(\sum_{j<m} a_j x_j + a_m x_{i+m} + \sum_{j=m+1}^{n} a_j x_{l+j-1} \right),$$

and so

$$\|x\| \le \left\| \sum_{j=1}^{n} a_j x_j \right\|.$$

But

$$\left\| \sum_{j<m} a_j x_j + \sum_{m<j\le n} a_j x_j \right\| \le \|x\| + |a_m|\epsilon,$$

and so

$$\left\| \sum_{j<m} a_j x_j + \sum_{m<j\le n} a_j x_j \right\| \le \left\| \sum_{j=1}^{n} a_j x_j \right\| + |a_m|\epsilon.$$

Since $\epsilon > 0$ is arbitrary, we are done.

(iii) This is immediate since $(x_{2n-1} - x_{2n})_{n=1}^{\infty}$ is weakly null and spreading (obviously, it cannot be constant). $\qquad\square$

Theorem 11.3.7. *Suppose $(x_n)_{n=1}^{\infty}$ is a normalized sequence in a Banach space X such that $\{x_n\}_{n=1}^{\infty}$ is not relatively compact. Then there is a spreading sequence space which is block finitely representable in $(x_n)_{n=1}^{\infty}$. More precisely, there is a subsequence $(x_{n_k})_{k=1}^{\infty}$ of $(x_n)_{n=1}^{\infty}$ and a spreading sequence space $\mathcal{X}$ so that if we let $M = \{n_k\}_{k=1}^{\infty}$ then*

$$\lim_{\substack{(p_1,\dots,p_r)\in\mathcal{F}_r(M)\\ p_1<\cdots<p_r}} \left\| \sum_{j=1}^{r} a_j x_{p_j} \right\| = \left\| \sum_{j=1}^{r} a_j e_j \right\|_{\mathcal{X}}.$$

Proof. This is a neat application of Ramsey's theorem due to Brunel and Sucheston [19]. We first observe that by taking a subsequence we can assume that $(x_n)_{n=1}^{\infty}$ has no convergent subsequence.

Let us fix some finite sequence of real numbers $(a_j)_{j=1}^{r}$. According to Theorem 10.1.1, given any infinite subset M of $\mathbb{N}$ we can find a further infinite subset M_1 so that

$$\lim_{\substack{(p_1,\dots,p_r)\in\mathcal{F}_r(M_1)\\ p_1<\cdots<p_r}} \left\| \sum_{j=1}^{r} a_j x_{p_j} \right\| \quad \text{exists}.$$

Let $(a_1^{(k)},\dots,a_{r_k}^{(k)})_{k=1}^{\infty}$ be an enumeration of all finitely nonzero sequences of rationals, and let us construct a decreasing sequence $(M_k)_{k=1}^{\infty}$ of infinite subsets of $\mathbb{N}$ so that

$$\lim_{\substack{(p_1,\dots,p_r)\in\mathcal{F}_r(M_k)\\ p_1<\cdots<p_r}} \left\| \sum_{j=1}^{r} a_j^{(k)} x_{p_j} \right\| \quad \text{exists}.$$

A diagonal procedure allows us to pick an infinite subset M_∞ which is contained in each M_k up to a finite set. It is not difficult to check that

$$\lim_{\substack{(p_1,\ldots,p_r)\in\mathcal{F}_r(M_\infty)\\p_1<\cdots<p_r}} \left\|\sum_{j=1}^r a_j x_{p_j}\right\| \text{ exists}$$

for every finite sequence of reals $(a_j)_{j=1}^r$.

Given $\xi = (\xi(j))_{j=1}^\infty \in c_{00}$ put

$$\|\xi\|_{\mathcal{X}} = \lim_{\substack{(p_1,\ldots,p_r)\in\mathcal{F}_r(M_\infty)\\p_1<\cdots<p_r}} \left\|\sum_{j=1}^r \xi(j) x_{p_j}\right\|.$$

$\|\cdot\|_{\mathcal{X}}$ satisfies the spreading property, but we need to check that it is a norm on c_{00} (it obviously is a seminorm). If $\|\xi\|_{\mathcal{X}} = 0$ and $\xi = \sum_{j=1}^r a_j e_j$ with $a_r \neq 0$ then we also have $\|\sum_{j=1}^{r-1} a_j e_j + a_r e_{r+1}\|_{\mathcal{X}} = 0$. Hence

$$\|e_1 - e_2\|_{\mathcal{X}} = \|e_r - e_{r+1}\|_{\mathcal{X}} = 0.$$

Returning to the definition we see that this implies

$$\lim_{(p_1,p_2)\in\mathcal{F}_2(M_\infty)} \|x_{p_1} - x_{p_2}\| = 0,$$

which can only mean that the subsequence $(x_j)_{j\in M_\infty}$ is convergent, contrary to our construction.

$\square$

Definition 11.3.8. The spreading sequence space $\mathcal{X}$ introduced in Theorem 11.3.7 is called a *spreading model* for the sequence $(x_n)_{n=1}^\infty$.

We now turn to Krivine's theorem. This result was obtained by Krivine in 1976, and, although the main ideas of the proof we include here are the same as Krivine's original proof, we have used ideas from two subsequent expositions of Krivine's theorem by Rosenthal [198] and Lemberg [123].

Krivine's theorem should be contrasted with Tsirelson space, which we constructed in Section 10.3. The existence of Tsirelson space implies that there is a Banach space with a basis so that no (infinite) block basic sequence can be equivalent to one of the spaces ℓ_p or c_0. However, if we are content with *finite* block basic sequences then we can always find a good copy of one of these spaces! This difference of behavior between *infinite* and *arbitrarily large but finite* is a recurrent theme in modern Banach space theory.

Theorem 11.3.9 (Krivine's Theorem). *Let $(x_n)_{n=1}^\infty$ be a normalized sequence in a Banach space X such that $\{x_n\}_{n=1}^\infty$ is not relatively compact. Then, either c_0 is block finitely representable in $(x_n)_{n=1}^\infty$, or there exists $1 \leq p < \infty$ so that ℓ_p is block finitely representable in $(x_n)_{n=1}^\infty$.*

In order to simplify the proof of Theorem 11.3.9 let us start by making some observations.

We first claim it suffices to prove the theorem when $(x_n)_{n=1}^{\infty}$ is replaced by the canonical basis $(e_n)_{n=1}^{\infty}$ of a spreading model $\mathcal{X}$; this is a direct consequence of Theorem 11.3.7. We next claim that we can suppose that the canonical basis $(e_n)_{n=1}^{\infty}$ of the spreading model $\mathcal{X}$ is unconditional with suppression constant $K_s = 1$ (and hence 2-unconditional). Indeed, if the canonical basis of the spreading model fails to be weakly Cauchy then it is equivalent to the canonical ℓ_1-basis, and the fact that ℓ_1 is block finitely representable in $\mathcal{X}$ is simply the content of James's distortion theorem (Theorem 10.3.1). If $(e_n)_{n=1}^{\infty}$ is weakly Cauchy but not weakly null, we use Proposition 11.3.6 and replace it by the spreading sequence

$$ f_k = \frac{e_{2k} - e_{2k+1}}{\|e_{2k} - e_{2k+1}\|}, \qquad k = 1, 2, \ldots . $$

In this way we reduce the proof to showing the result for the canonical basis $(e_n)_{n=1}^{\infty}$ of some spreading sequence space $\mathcal{X}$.

We also observe at this point that the James distortion theorem for c_0 (see Problem 10.3) implies that if the spreading model is isomorphic to c_0 then c_0 is finite-representable in it. This reduction will be used later.

Now we will introduce some notation. Suppose $\mathcal{X}$ is a spreading sequence space whose canonical basis is unconditional with suppression constant $K_s = 1$. The norm of each $\xi \in \mathcal{X}$ depends only on its nonzero entries and their order of appearance. We shall say that the sequences ξ and η in c_{00} are *equivalent* if their nonzero entries and their order of appearance are identical. We will say that ξ and η are *ϵ-equivalent* if there exist $u, v \in c_{00}$ so that $u + \xi$ and $v + \eta$ are equivalent and $\|u\|_{\mathcal{X}} + \|v\|_{\mathcal{X}} < \epsilon$.

If $\xi, \eta \in c_{00}$ we define $\xi \oplus \eta$ to be any vector where the nonzero entries of ξ (in correct order) precede the nonzero entries of η (in correct order). For example, $\xi \oplus \eta$ could be obtained by writing first the entries of ξ in order and then the nonzero entries of η in order. Thus, if n is the largest integer so that $\xi(n) \neq 0$ we could take

$$ \xi \oplus \eta = \sum_{j=1}^{n} \xi(j) e_j + \sum_{j=n+1}^{\infty} \eta(j-n) e_j. $$

We will say that ξ is *replaceable* by η if

$$ \|u \oplus \xi \oplus v\|_{\mathcal{X}} = \|u \oplus \eta \oplus v\|_{\mathcal{X}}, \qquad u, v \in c_{00}, $$

and that ξ is *ϵ-replaceable* by η if

$$ \left| \|u \oplus \xi \oplus v\|_{\mathcal{X}} - \|u \oplus \eta \oplus v\|_{\mathcal{X}} \right| < \epsilon, \qquad u, v \in c_{00}. $$

Let us notice that if ξ and η are equivalent then ξ is replaceable by η. Similarly, if ξ and η are ϵ-equivalent then ξ is ϵ-replaceable by η.

To complete the proof of Krivine's theorem we will need the following two lemmas.

Lemma 11.3.10. *Suppose $\mathcal{X}$ is a spreading sequence space. Then there is a spreading sequence space $\mathcal{Y}$ which is block finitely representable in $\mathcal{X}$ so that the canonical basis of $\mathcal{Y}$ is unconditional with unconditional basis constant $K_u = 1$.*

Proof. By the previous remarks we can assume that the canonical basis $(e_n)_{n=1}^\infty$ of $\mathcal{X}$ is 2-unconditional, and that $\mathcal{X}$ is not isomorphic to c_0. Thus, if we let $y_n = \sum_{j=1}^n (-1)^j e_j$ we have $\|y_n\| \to \infty$. For each k let $u_k = y_k/\|y_k\|$. u_k is ϵ_k-equivalent to $-u_k$ for $\epsilon_k = 2/\|y_k\|$.

If we take a block basic sequence $(z_n)_{n=1}^\infty$ with respect to $(e_n)_{n=1}^\infty$, where each z_n is equivalent to u_k, we obtain a spreading sequence where $-z_n$ is ϵ_k-replaceable by z_n. Define $\mathcal{Y}_k$ by

$$\|\xi\|_{\mathcal{Y}_k} = \left\| \sum_{j=1}^\infty \xi(j) z_j \right\|_{\mathcal{X}}.$$

We can pass to a subsequence $(k_m)_{m=1}^\infty$ in such a way that $\lim_{m\to\infty} \|\xi\|_{\mathcal{Y}_{k_m}}$ exists for all $\xi \in c_{00}$. This is done by a standard diagonal argument for those ξ with rational coefficients, and then extended to all ξ by a routine approximation argument. This formula defines a spreading sequence space, still block finitely representable in $\mathcal{X}$ but such that e_1 is replaceable by $-e_1$. This shows that the canonical basis of $\mathcal{Y}$ is 1-unconditional.

□

Lemma 11.3.11. *Suppose $\mathcal{X}$ is a spreading sequence space whose canonical basis $(e_n)_{n=1}^\infty$ is 1-unconditional.*

(i) If $2^{1/p} e_1$ is replaceable by $e_1 + e_2$ for some $1 \le p < \infty$ then the norm on $\mathcal{X}$ is equivalent to the canonical ℓ_p-norm.

(ii) If for some $1 \le p < \infty$, $2^{1/p} e_1$ is replaceable by $e_1 + e_2$, and $3^{1/p} e_1$ is replaceable by $e_1 + e_2 + e_3$ then the norm on $\mathcal{X}$ coincides with the ℓ_p-norm.

Proof. (i) Suppose $(k_j)_{j=1}^\infty$ is a sequence of non-negative integers. If for each n we let $N = \sum_{j=1}^n 2^{k_j}$ we have

$$\left\| \sum_{j=1}^n 2^{k_j/p} e_j \right\|_{\mathcal{X}} = \left\| \sum_{j=1}^N e_j \right\|_{\mathcal{X}}.$$

Notice also that

$$\left\| \sum_{j=1}^{2^r} e_j \right\|_{\mathcal{X}} = 2^{r/p},$$

and so

$$2^{-1/p} N^{1/p} \le \left\| \sum_{j=1}^N e_j \right\|_{\mathcal{X}} \le 2^{1/p} N^{1/p}.$$

Suppose now that (a_j) are scalars with $\sum_{j=1}^n |a_j|^p = 1$, and let α be the least nonzero value of $|a_j|$. For each j pick a nonnegative integer k_j with $2^{k_j/p} \le |a_j|\alpha^{-1} \le 2^{(k_j+1)/p}$. Then, if $N = \sum_{j=1}^n 2^{k_j}$ we have

$$\left\|\sum_{j=1}^N e_j\right\|_{\mathcal{X}} \le \alpha^{-1}\left\|\sum_{j=1}^n a_j e_j\right\|_{\mathcal{X}} \le \left\|\sum_{j=1}^{2N} e_j\right\|_{\mathcal{X}},$$

and so $N\alpha^p \le 1 \le 2N\alpha^p$. Thus

$$2^{-1/p}N^{1/p}\alpha \le \left\|\sum_{j=1}^n a_j e_j\right\|_{\mathcal{X}} \le 2^{2/p}N^{1/p}\alpha,$$

which implies

$$2^{-2/p} \le \left\|\sum_{j=1}^n a_j e_j\right\|_{\mathcal{X}} \le 2^{2/p}.$$

The proof of (ii) is similar to (i). Here we use that the set of real numbers of the form $2^l 3^m$ with $l, m \in \mathbb{Z}$ is *dense* in $(0, +\infty)$, which is a consequence of the fact that $\log 3 / \log 2$ is irrational.

If $l, m \ge 0$ and $N = 2^l 3^m$ we have

$$\left\|\sum_{j=1}^N e_j\right\|_{\mathcal{X}} = N^{1/p}.$$

For any N pick $r, s \in \mathbb{Z}$ so that $N - \epsilon \le 2^r 3^s \le N$. Then

$$\left\|\sum_{j=1}^{2^{|r|}3^{|s|}N} e_j\right\|_{\mathcal{X}} = 2^{|r|/p}3^{|s|/p}\left\|\sum_{j=1}^N e_j\right\|_{\mathcal{X}}$$

$$\ge \left\|\sum_{j=1}^{2^{r+|r|}3^{s+|s|}} e_j\right\|_{\mathcal{X}}$$

$$= 2^{(r+|r|)/p}3^{(s+|s|)/p},$$

so

$$\left\|\sum_{j=1}^N e_j\right\|_{\mathcal{X}} \ge 2^{r/p}3^{s/p} \ge (N - \epsilon)^{1/p}.$$

Hence

$$\left\|\sum_{j=1}^N e_j\right\|_{\mathcal{X}} \ge N^{1/p}.$$

Conversely, we can find r, s in $\mathbb{Z}$ so that $N < 2^r 3^s < N + \epsilon$, and a similar argument yields

$$\left\|\sum_{j=1}^{N} e_j\right\|_{\mathcal{X}} \le N^{1/p}.$$

Thus we obtain

$$\left\|\sum_{j=1}^{N} e_j\right\|_{\mathcal{X}} = N^{1/p}, \qquad N = 1, 2, \ldots.$$

Suppose $a_1, a_2, \ldots, a_n$ are scalars of the form $|a_j| = 2^{l_j/p} 3^{m_j/p}$ for some $l_j, m_j \in \mathbb{Z}$. Pick $L, M \in \mathbb{N}$ so that $L + l_j \ge 0, M + m_j \ge 0$ for all $1 \le j \le n$. Then,

$$2^{L/p} 3^{M/p} \left\|\sum_{j=1}^{n} a_j e_j\right\|_{\mathcal{X}} = \left\|\sum_{j=1}^{N} e_j\right\|_{\mathcal{X}},$$

where

$$N = 2^L 3^M \sum_{j=1}^{n} |a_j|^p.$$

This implies

$$\left\|\sum_{j=1}^{n} a_j e_j\right\|_{\mathcal{X}} = \left(\sum_{j=1}^{n} |a_j|^p\right)^{1/p}.$$

A density argument implies the conclusion of the lemma for all sequences of scalars $(a_i)_{i=1}^{n}$.

$\square$

We are almost ready to complete the proof of Theorem 11.3.9. Suppose $\mathcal{X}$ is a 1-unconditional spreading sequence space; we will define a variant of $\mathcal{X}$ modeled on $\mathbb{Q}_0 = \mathbb{Q} \cap [0, 1)$ rather than $\mathbb{N}$.

Consider the space $c_{00}(\mathbb{Q})$ of all finitely nonzero sequences on $\mathbb{Q}$. For $\xi \in c_{00}(\mathbb{Q}_0)$ of the form $\xi = \sum_{j=1}^{n} a_j e_{q_j}$, where $q_1 < q_2 < \cdots < q_n$, we define

$$\left\|\sum_{j=1}^{n} a_j e_{q_j}\right\|_{\mathcal{X}(\mathbb{Q}_0)} = \left\|\sum_{j=1}^{n} a_j e_j\right\|_{\mathcal{X}}.$$

On $\mathcal{X}(\mathbb{Q}_0)$ we consider two bounded operators given by

$$T_2 e_q = e_{q/2} + e_{(q+1)/2}, \qquad q \in \mathbb{Q}_0$$

and

$$T_3 e_q = e_{q/3} + e_{(q+1)/3} + e_{(q+2)/3}, \qquad q \in \mathbb{Q}_0.$$

It is clear that $1 \le \|T_2\| \le 2$ and $1 \le \|T_3\| \le 3$. We consider the spectral radius of T_2 and define $0 \le \theta \le 1$ by

$$2^\theta = \lim_{n \to \infty} \|T_2^n\|^{\frac{1}{n}}.$$

Lemma 11.3.12. *Suppose $\mathcal{X}$ is a 1-unconditional spreading sequence space. Then*

(i) There exists a sequence $(\xi_n)_{n=1}^\infty$ in $\mathcal{X}(\mathbb{Q}_0)$ with $\|\xi_n\|_{\mathcal{X}(\mathbb{Q}_0)} = 1$ and such that $\lim_{n\to\infty} \|T_2\xi_n - 2^\theta\xi_n\|_{\mathcal{X}(\mathbb{Q}_0)} = 0$.

(ii) If the norm on $\mathcal{X}$ is equivalent to the ℓ_p-norm for some $1 \le p < \infty$ then $\theta = 1/p$, and there is a sequence $(\xi_n)_{n=1}^\infty$ in $\mathcal{X}(\mathbb{Q}_0)$ so that $\|\xi_n\|_{\mathcal{X}(\mathbb{Q}_0)} = 1$, $\lim_{n\to\infty} \|T_2\xi_n - 2^{1/p}\xi_n\|_{\mathcal{X}(\mathbb{Q}_0)} = 0$, and $\lim_{n\to\infty} \|T_3\xi_n - 3^{1/p}\xi_n\|_{\mathcal{X}(\mathbb{Q}_0)} = 0$.

Proof. (*i*) Let us start by observing that

$$\lim_{n\to\infty} \|T_2^n\|^{\frac{1}{n}} = \inf_n \|T_2^n\|^{\frac{1}{n}}, \tag{11.3}$$

and so

$$\|T_2^n\| \ge 2^{n\theta}, \qquad n = 1, 2, \dots.$$

Thus

$$\lim_{n\to\infty} \|(n+1)2^{-n\theta}T_2^n\| = \infty,$$

and, by the Uniform Boundedness principle, we can find $\eta \in \mathcal{X}$ with $\|\eta\| = 1$ so that the sequence $((n+1)2^{-n\theta}T_2^n\eta)_{n=1}^\infty$ is unbounded. Let us note that we can assume that η has only nonnegative entries. If we define $|\eta|$ by $|\eta|(q) = |\eta(q)|$ then $T_2^n|\eta|(q) \ge |T_2^n\eta(q)|$ for every q. Therefore we assume $\eta \ge 0$, i.e., $\eta(q) \ge 0$ for all q.

If $r < 2^{-\theta}$ then $(1 - rT_2)$ is invertible and we can expand $(I - rT_2)^{-2}$ in its binomial series (which converges). Thus

$$(1 - rT_2)^{-2}(\eta) = \sum_{n=0}^\infty (n+1)r^n T_2^n(\eta).$$

Since $\eta \ge 0$, it is immediate that

$$\lim_{r\to 2^{-\theta}} \|(1 - rT_2)^{-2}\eta\|_{\mathcal{X}(\mathbb{Q}_0)} = \infty.$$

Hence we can find a sequence (r_n) with $r_n \to 2^{-\theta}$ so that either

$$\lim_{n\to\infty} \frac{\|(I - r_nT_2)^{-2}\eta\|_{\mathcal{X}(\mathbb{Q}_0)}}{\|(I - r_nT_2)^{-1}\eta\|_{\mathcal{X}(\mathbb{Q}_0)}} = \infty$$

or

$$\lim_{n\to\infty} \|(I - r_nT_2)^{-1}\eta\|_{\mathcal{X}(\mathbb{Q}_0)} = \infty.$$

In either case we can determine ξ_n with $\|\xi_n\|_{\mathcal{X}(\mathbb{Q}_0)} = 1$ and $\lim_{n\to\infty} \|(I - r_nT_2)\xi_n\|_{\mathcal{X}(\mathbb{Q}_0)} = 0$, which implies (*i*).

(*ii*) This is easier. We work in the equivalent ℓ_p-norm on $\mathcal{X}(\mathbb{Q}_0)$. Then $\|T_2^n\|_{\ell_p(\mathbb{Q}_0)\to\ell_p(\mathbb{Q}_0)} = 2^{n/p}$ and $\|T_3^n\|_{\ell_p(\mathbb{Q}_0)\to\ell_p(\mathbb{Q}_0)} = 3^{n/p}$. Let

$$\xi_n = n^{-2/p} \sum_{j=1}^{n} \sum_{k=1}^{n} 2^{-j/p} 3^{-k/p} T_2^j T_3^k e_0, \qquad n = 1, 2, \ldots.$$

Then $\|\xi_n\|_p = 1$ and (since T_2 and T_3 commute!),

$$\|2^{-1/p} T_2 \xi_n - \xi_n\|_p = 2^{\frac{1}{p}} n^{-\frac{1}{p}},$$
$$\|3^{-1/p} T_2 \xi_n - \xi_n\|_p = 3^{\frac{1}{p}} n^{-\frac{1}{p}}.$$

Renormalizing in the $\mathcal{X}$-norm gives the result. $\qquad\qquad\square$

Conclusion of the proof of Theorem 11.3.9. We have reduced the proof to the case when X is a spreading sequence space $\mathcal{X}$ with 1-unconditional canonical basis. Using (i) of Lemma 11.3.12 we can find a sequence (u_n) in c_{00} so that $\|u_n\|_{\mathcal{X}} = 1$ and $2^\theta u_n$ is ϵ_n-equivalent to $u_n \oplus u_n$ where $\epsilon_n \to 0$. Indeed, we may assume the ξ_n given by the lemma have finite support and then we simply take u_n to have the same nonzero entries in the same order as ξ_n. Then $u_n \oplus u_n$ is, similarly, equivalent to $T_2 \xi_n$.

For each n we can define a new spreading sequence space $\mathcal{Y}_n$ by

$$\left\| \sum_{j=1}^{N} a_j e_j \right\|_{\mathcal{Y}_n} = \|a_1 u_n \oplus a_2 u_n \oplus \cdots \oplus a_N u_n\|_{\mathcal{X}},$$

and then passing to a subsequence we can form a limit $\mathcal{Y}$ (as in Lemma 11.3.10). $\mathcal{Y}$ is then block finitely representable in $\mathcal{X}$ and $2^\theta e_1$ is replaceable by $e_1 + e_2$. If $\theta = 0$ then $\|e_1 + \cdots + e_n\|_{\mathcal{Y}} = 1$ so $\mathcal{Y}$ is isometric to c_0 and we are done.

If $\theta > 0$, let $1/p = \theta$ and observe that Lemma 11.3.11 implies that $\mathcal{Y}$ has a norm equivalent to the ℓ_p-norm. Now use Lemma 11.3.12 (ii) and repeat the procedure to produce spreading sequence space $\mathcal{Z}$ with 1-unconditional canonical basis, still block finitely representable in $\mathcal{X}$ but this time with both the properties that $2^{1/p} e_1$ is replaceable by $e_1 + e_2$ and $3^{1/p} e_1$ is replaceable by $e_1 + e_2 + e_3$. Lemma 11.3.11 ensures that $\mathcal{Z}$ is isometric to ℓ_p. $\qquad\square$

Theorem 11.3.13 (Dvoretzky's Theorem). *ℓ_2 is finitely representable in every infinite-dimensional Banach space.*

Proof. An immediate conclusion from Krivine's theorem is that some ℓ_p ($1 \le p < \infty$) or c_0 is finitely representable in any infinite-dimensional Banach space X. In the case of c_0 this implies that ℓ_∞ is finitely representable in X, and hence so is every separable Banach space. If ℓ_p is finitely representable then so is L_p (Proposition 11.1.7) and, since ℓ_2 is isometric to a subspace of L_p (Theorem 6.4.13), we obtain the theorem.

$\qquad\square$

Dvoretzky's theorem is one of the most celebrated results in Banach space theory, but the above proof is not the first or the usual proof. Dvoretzky proved the theorem in 1961 [49] well before the techniques of Krivine's theorem were known. The form we have proved implies a quantitative version. More precisely, given $\epsilon > 0$ and $n \in \mathbb{N}$ there exists $N = N(n, \epsilon)$ so that if X is a Banach space of dimension N then it has a subspace E of dimension n with $d(E, \ell_2^n) < 1 + \epsilon$ (see the Problems). However, the *infinite-dimensional* method of proof prevents us from using this approach to gain any information about the function $N(n, \epsilon)$. In the last chapter we will look at quantitative *finite-dimensional* arguments which give more precise information.

There is much more to say about Krivine's theorem. It is of interest, for instance, to determine which ℓ_p is obtained in the theorem. For example, if we can find spreading model $\mathcal{X}$ with 1-unconditional canonical basis $(e_n)_{n=1}^{\infty}$ satisfying a lower estimate

$$\|e_1 + \cdots + e_n\|_{\mathcal{X}} \geq cn^{\frac{1}{p}}, \qquad n = 1, 2, \ldots,$$

one can show that that ℓ_p is finitely representable in $\mathcal{X}$ (see the Problems). By more delicate considerations we can obtain the following theorem essentially due to Maurey and Pisier [147]:

Theorem 11.3.14. *Let X be an infinite-dimensional Banach space and suppose $p_X = \inf\{p : X \text{ has type } p\}$ and $q_X = \sup\{q : X \text{ has cotype } q\}$. Then both ℓ_{p_X} and ℓ_{q_X} are finitely representable in X.*

The reader who wishes to know more should consult either the books of Milman and Schechtman [154] or Benyamini and Lindenstrauss [11].

Problems

11.1. Prove Theorem 11.1.14 *(ii)*.

11.2. Suppose X is a Banach space of type p [respectively, cotype q]. Show that X^{**} has type p [respectively, cotype q] with the same constants.

11.3. We recall that a Banach space X is said to be *strictly convex* if for any $x, y \in X$ with $\|x\| = \|y\| = 1$ such that $\|x + y\| = 2$ we have $x = y$, and that X is said to be *uniformly convex* if given $\epsilon > 0$ there exists $\delta(\epsilon) > 0$ so that if $\|x\| = \|y\| = 1$ and $\|x + y\| > 2 - \delta$ then $\|x - y\| < \epsilon$.
Show that a Banach space X is uniformly convex if and only if every Banach space finitely representable in X is strictly convex.

11.4. (a) Show that the L_p-spaces for $1 < p < \infty$ are strictly convex.

(b) Show that for any $f \in L_p$ with $\|f\|_p = 1$ and $|f(s)| > 0$ a.e. there is an isometric isomorphism $T_f : L_p \to L_p$ with $T_f f = 1$ (the constantly one function).

(c) Show that if $(f_n)_{n=1}^{\infty}$, $(g_n)_{n=1}^{\infty}$ are two sequences in L_p for $1 < p < \infty$ with $f_n + g_n = c_n 1$ where $\lim_{n\to\infty} c_n = 2$ then $\lim_{n\to\infty} \|f_n - g_n\|_p = 0$. [*Hint:* Use reflexivity.]

(d) Combine (a), (b), and (c) to show that the L_p-spaces for $1 < p < \infty$ are uniformly convex. Also note that we can deduce this from Problem 11.3.

11.5. James criterion for reflexivity ([84]).
(a) If X is a nonreflexive Banach space and $0 < \theta < 1$ show that we can find a sequence $(x_n)_{n=1}^{\infty}$ in the unit ball of X so that

$$\|x\| \geq \theta, \qquad x \in \mathrm{co}\{x_j\}_{j=1}^{\infty} \qquad (11.4)$$

and

$$\|y - z\| \geq \theta, \qquad y \in \mathrm{co}\{x_j\}_{j=1}^{n}, \ z \in \mathrm{co}\{x_j\}_{j=n+1}^{\infty}, \ n = 1, 2, \ldots. \qquad (11.5)$$

(b) Show, conversely, that the existence of a sequence in the unit ball satisfying (11.5) implies that X is nonreflexive.

(c) Deduce that a uniformly convex space is reflexive.

11.6. Superreflexivity ([85], [86]).
A Banach space X is said to be *superreflexive* if every Banach space Y which is finitely representable in X is reflexive.

(a) Give an example of a reflexive space which is not superreflexive.

(b) Show that X is superreflexive if and only if given $\epsilon > 0$ there exists $N = N(\epsilon)$ so that if $x_j \in B_X$ for $1 \leq j \leq N$ then there exists $1 \leq n \leq N$ and $y \in \mathrm{co}\{x_1, \ldots, x_n\}$, $z \in \mathrm{co}\{x_{n+1}, \ldots, x_N\}$ with $\|y - z\| < \epsilon$.

(c) Show that a uniformly convex space is superreflexive.

It is a result of Enflo [53] and Pisier [182] that superreflexive spaces always have an equivalent uniformly convex norm. The subject of renorming is a topic in itself and we refer the reader to [38].

11.7. Show that a Banach space X has nontrivial type if and only if given
$\epsilon > 0$ there exists N so that if $x_j \in B_X$ for $1 \leq j \leq N$ with $\|x_j\| = 1$ there exists a subset A of $\{1, 2, \ldots, N\}$ and $y \in \mathrm{co}\{x_j\}_{j \in A}$, $z \in \mathrm{co}\{x_j\}_{j \notin A}$ with $\|y - z\| < \epsilon$.

Compare with Problem 11.6; this criterion is simply an unordered version of the criterion for superreflexivity. However, James showed the existence of a nonreflexive Banach space with type 2 [88]!

11.8. Let X be a separable Banach space such that X^* is separable and
has (BAP). Show that X has (BAP) and indeed (MAP) (see Problems 1.8 and 1.9). [*Hint:* The problem here is that there exist finite-rank operators $T_n : X^* \to X^*$ so that $T_n x^* \to x^*$ for $x^* \in X^*$ but the T_n need not be adjoints of operators on X. Use the Principle of Local Reflexivity.]

11.9. Let $\mathcal{U}$ be a nonprincipal ultrafilter on $\mathbb{N}$. Show that if X is superreflexive then the dual of $X_{\mathcal{U}}$ can be naturally identified with $(X^*)_{\mathcal{U}}$.

11.10. Prove the equality (11.3) in Lemma 11.3.12.

11.11. Dvoretzky's theorem (quantitative version).
Prove that given $\epsilon > 0$ and $n \in \mathbb{N}$ there exists $N = N(n, \epsilon)$ so that if X is a Banach space of dimension N then it has a subspace E of dimension n with $d(E, \ell_2^n) < 1 + \epsilon$. [*Hint:* Use an ultraproduct.]

11.12. Suppose X is a Banach space with an unconditional basis $(x_n)_{n=1}^{\infty}$ such that for some $1 \leq p \leq 2$ we have

$$c|A|^{1/p} \leq \left\| \sum_{j \in A}^{n} x_j \right\|$$

for every finite subset of $\mathbb{N}$. Show that ℓ_p is finitely representable in X.

An Introduction to Local Theory

The aim of this chapter is to provide an introduction to the ideas of the local theory and a quantitative proof of Dvoretzky's theorem. Dvoretzky's theorem asserts that every n-dimensional normed space contains a subspace F of dimension $k = k(n, \epsilon)$ with $d_F = d(F, \ell_2^k) < 1 + \epsilon$, where $k(n, \epsilon) \to \infty$ as $n \to \infty$. Dvoretzky's original paper [49] gave this without the optimal estimates for $k(n, \epsilon)$. We present a proof due to Milman [152] which gives the estimate

$$k(n, \epsilon) \geq c\epsilon^2 |\log \epsilon|^{-1} \log n.$$

This is optimal in dependence on n but not on ϵ; in 1985, Gordon [69] showed that the $|\log \epsilon|$ term can be removed so that $k(n, \epsilon) \geq c\epsilon^2 \log n$.

The study of finite-dimensional normed spaces is a very rich area and Dvoretzky's theorem is only the beginning of this subject, which flowered remarkably during the 1980s and early 1990s. Since then there has been an evolution of the area with more emphasis on the geometry of convex sets; nowadays it continues to be an important area.

As a prelude we introduce the John ellipsoid and prove the Kadets-Snobar theorem that every n-dimensional subspace of a Banach space is $\sqrt{n}$-complemented.

Finally we return to the complemented subspace problem and present a complete proof that a Banach space in which every subspace is complemented is a Hilbert space (Lindenstrauss-Tzafriri [135]).

We emphasize that throughout this chapter we treat only *real* scalars, although much of the theory does permit an easy extension to complex scalars.

12.1 The John ellipsoid

Definition 12.1.1. Suppose X is an n-dimensional normed space. An *ellipsoid* $\mathcal{E}$ in X is the unit ball of some Euclidean norm on X (i.e., a norm on X induced by an inner product). The *John ellipsoid* of X is defined to be the ellipsoid of maximal volume contained in B_X.

The John ellipsoid was introduced by John in 1948 [89]. Its existence follows by compactness of the unit ball of a finite-dimensional space. Let us indicate one way to reach this. Introduce some inner product structure on X (i.e., identify X with $\mathbb{R}^n$ with its canonical inner product, where $n = \dim X$). Each ellipsoid $\mathcal{E}$ contained in B_X corresponds to a linear map $S : \mathbb{R}^n \to X$ (where $n = \dim X$) so that $S(B_{\ell_2^n}) = \mathcal{E}$. The volume of $\mathcal{E}$ is measured by the determinant of $S : \mathbb{R}^n \to \mathbb{R}^n$. To be precise,

$$\frac{\text{vol } \mathcal{E}}{\text{vol } B_{\ell_2^n}} = |\det S|,$$

where $B_{\ell_2^n} = \{\xi = (\xi(i))_{i=1}^n \in \mathbb{R}^n : \sum_{i=1}^n |\xi(i)|^2 \leq 1\}$. We are thus maximizing $\det S$ over the set of S with $\|S\|_{\ell_2^n \to X} \leq 1$. It is also true but irrelevant to the remainder of the chapter that the John ellipsoid is *unique*; in fact we only need its existence.

Once we have agreed on the existence of the John ellipsoid in X, it is natural to insist that our inner product structure on X coincides with that induced by $\mathcal{E}$. We then denote by $\| \cdot \|_E$ the Euclidean norm induced on X by its John ellipsoid. Put

$$E = (X, \| \cdot \|_E).$$

Now, X has an associated inner product $\langle \, , \rangle$ and corresponding norm $\| \cdot \|_E$ so that $\|I\|_{E \to X} \leq 1$, and

$$|\det T| \leq 1 \ \text{ if } \|T\|_{E \to X} \leq 1.$$

Next we are going to show that the John ellipsoid has some remarkable and important properties.

Lemma 12.1.2. *If $T : E \to X$ then*

$$|tr\ T| \leq n\|T\|_{E \to X},$$

where $tr\ T$ is the trace of T.

Proof. First we note that if $T \in \mathcal{L}(X)$,

$$|\det T| \leq \|T\|_{E \to X}^n.$$

Thus

$$\det (I + tT) \leq \|(1 + tT)\|_{E \to X}^n, \qquad t \in \mathbb{R}.$$

Now

$$\lim_{t \to 0^+} \frac{\det (I + tT) - 1}{t} = tr\ T,$$

and so

$$tr\ T \leq n\|T\|_{E \to X}.$$

$\square$

Theorem 12.1.3. *We have*

$$\|I\|_{X \to E} \leq \pi_2(I_{X \to E}) \leq \sqrt{n}.$$

Proof. Let us identify the dual of X^* using the inner product. Thus for $x \in X$ we define

$$\|x\|_{X^*} = \sup\{|\langle x, y \rangle| : y \in X, \ \|y\|_X \leq 1\}.$$

It is then clear that

$$\|x\|_X \leq \|x\|_E \leq \|x\|_{X^*}.$$

Suppose $x_1, \ldots, x_k \in X$. Let T be the operator $T = \sum_{i=1}^{k} x_i \otimes x_i$, that is,

$$Tu = \sum_{i=1}^{k} \langle x_i, u \rangle x_i.$$

We note that

$$\operatorname{tr} T = \sum_{i=1}^{k} \langle x_i, x_i \rangle = \sum_{i=1}^{k} \|x_i\|_E^2.$$

We also have that if $\|u\|_{X^*}, \|v\|_{X^*} \leq 1$ then

$$|\langle Tu, v \rangle| = \left| \sum_{i=1}^{k} \langle x_i, u \rangle \langle x_i, v \rangle \right|$$

$$\leq \left(\sum_{i=1}^{k} |\langle x_i, u \rangle|^2 \right)^{1/2} \left(\sum_{i=1}^{k} |\langle x_i, v \rangle|^2 \right)^{1/2}.$$

Hence

$$\|T\|_{E \to X} \leq \|T\|_{X^* \to X} \leq \max_{\|u\|_{X^*} \leq 1} \sum_{i=1}^{k} |\langle x_i, u \rangle|^2.$$

By Lemma 12.1.2 we conclude that

$$\sum_{i=1}^{k} \|x_i\|_E^2 \leq n \max_{\|u\|_{X^*} \leq 1} \sum_{i=1}^{k} |\langle x_i, u \rangle|^2.$$

This is exactly the statement that

$$\pi_2(I_{X \to E}) \leq \sqrt{n}.$$

$\square$

This theorem has immediate applications. We denote by d_X the Euclidean distance of X, i.e., $d_X = d(X, \ell_2^n)$ where $n = \dim X$. If X is an infinite-dimensional Banach space, $d_X = d(X, H)$ where H is a Hilbert space of the same density character of X.

Theorem 12.1.4 (John). *If X is n-dimensional then $d_X \leq \sqrt{n}$.*

Proof. We have $\|I\|_{E \to X} = 1$ and $\|I\|_{X \to E} \leq \sqrt{n}$.

$\square$

The estimate given by this theorem is the best possible:

Proposition 12.1.5. *If* $X = \ell_\infty^n$ *(or* $X = \ell_1^n$*) then* $d_X = \sqrt{n}$.

Proof. Let $S : \ell_\infty^n \to \ell_2^n$ be an operator which realizes the optimal isomorphism, that is,

$$\|x\|_\infty \leq \|Sx\|_2 \leq d\|x\|_\infty,$$

where $d = d_{\ell_\infty^n}$.

For each choice of signs $(\epsilon_i)_{i=1}^n$, the operator $U_{\epsilon_1,\ldots,\epsilon_n}(x) = (\epsilon_1 x_1, \ldots, \epsilon_n x_n)$ is an isometry on ℓ_∞^n, so $SU_{\epsilon_1,\ldots,\epsilon_n}$ is another optimal embedding. Considering choices of signs as outcomes of a Rademacher sequence $\varepsilon_1, \ldots, \varepsilon_n$ on some probability space $(\Omega, \mathbb{P})$, we may define $T : \ell_\infty^n \to L_2(\Omega, \mathbb{P}; \ell_2^n)$ by

$$Tx(\omega) = U_{\varepsilon_1(\omega),\ldots,\varepsilon_n(\omega)}x.$$

Then

$$\|x\|_\infty \leq \|Tx\|_{L_2(\mathbb{P})} \leq d\|x\|_\infty.$$

But

$$\|Tx\|^2 = \mathbb{E}\Big\| \sum_{i=1}^n \varepsilon_i x_i Se_i \Big\|^2 = \sum_{i=1}^n |x_i|^2 \|Se_i\|^2,$$

and this makes it clear that our optimal choice must satisfy $\|Se_i\| = 1$ for $1 \leq i \leq n$, and so

$$\|Tx\| = \|x\|_2.$$

Hence $\|T\| = \sqrt{n} = d$.

$\square$

The following result is due to Kadets and Snobar [100].

Theorem 12.1.6 (The Kadets-Snobar Theorem). *Let* F *be a Banach space of dimension* n. *Then for any Banach space* X *containing* F *as a subspace there is a projection* P *of* X *onto* F *with* $\|P\| \leq \sqrt{n}$.

Proof. According to Theorem 12.1.3, there is an operator $S : F \to \ell_2^n$ where $n = \dim F$ so that $\|S^{-1}\| = 1$ and $\pi_2(S) \leq \sqrt{n}$. Using Theorem 8.2.13, S extends to a bounded operator $T : X \to \ell_2^n$ with $\pi_2(T) = \pi_2(S)$. Hence $\|T\| \leq \sqrt{n}$ and if $P = S^{-1}T$ we have our desired projection.

$\square$

This result is not optimal (but very nearly is). We refer to the Handbook article [114] for more details. We also mention that the example of Pisier [184] cited in Chapter 8 gives a Banach space X with the property that there is a constant $c > 0$ so that whenever F is a finite-dimensional subspace and $P : X \to F$ is a projection then $\|P\| \geq c\sqrt{n}$.

12.2 The concentration of measure phenomenon

We are now en route to Dvoretzky's theorem, which will be deduced from a principle which has become known as the *concentration of measure phenomenon*. Roughly speaking this says that a Lipschitz function on the Euclidean sphere in dimension n behaves more and more like a constant as the dimension grows. More precisely, the set where it deviates from its average by some fixed ϵ has measure converging to zero at a very rapid rate.

This type of result is usually derived from Lévy's isoperimetric inequality [124]. We follow an alternative approach due to Maurey and Pisier [187], and [154] Appendix V which has the advantage of using Gaussians.

We shall consider $\mathbb{R}^n$ with its canonical Euclidean norm, $\|\cdot\|$. We denote by σ_n the normalized invariant measure on the surface of the sphere $\mathcal{S}^{n-1} = \{\xi = (\xi(j))_{j=1}^n : \sum_{j=1}^n |\xi_j|^2 = 1\}$. Thus σ_n is simply a normalized surface measure and it is invariant under orthogonal transformations. It can be obtained by the formula

$$\int_{\mathcal{S}^{n-1}} f(\xi)d\sigma_n(\xi) = \int_{\mathcal{O}_n} f(U\xi_0)d\mu(U), \qquad f \in \mathcal{C}(\mathcal{S}^{n-1}),$$

where μ is normalized Haar measure on the orthogonal group $\mathcal{O}_n$ and ξ_0 is some fixed vector in $\mathcal{S}^{n-1}$.

Let $(g_1, \ldots, g_n)$ be a sequence of mutually independent Gaussians on some probability space, and let G be the vector-valued Gaussian $G = \sum_{j=1}^n g_j e_j$, where $(e_j)_{j=1}^n$ is the canonical basis of $\mathbb{R}^n$. The distribution of G on $\mathbb{R}^n$ is given by the density function

$$\frac{1}{(2\pi)^{n/2}} e^{-(|\xi_1|^2 + \cdots + |\xi_n|^2)/2} = \frac{1}{(2\pi)^{n/2}} e^{-\|x\|^2/2}.$$

It is clear that the distribution of $G/\|G\|$ is given by the unique orthogonally invariant probability measure on $\mathcal{S}^{n-1}$, that is, σ_n.

Theorem 12.2.1. *Let f be a Lipschitz function on $\mathbb{R}^n$ with Lipschitz constant 1. Then for each $t > 0$,*

$$\mathbb{P}(|f(G) - \mathbb{E}f(G)| > t) \leq 2e^{-2t^2/\pi^2}.$$

Proof. We can suppose, by approximation, that f is continuously differentiable and, by adjusting the constant, that $\mathbb{E}f(G) = 0$.

Let us introduce an independent copy G' of G. For every θ put

$$G_\theta = G \sin \theta + G' \cos \theta$$

and

$$G'_\theta = G \cos \theta - G' \sin \theta.$$

Using the orthogonal invariance of (G, G') in $\mathbb{R}^{2n}$ it is then clear that (G_θ, G'_θ) has the same distribution as (G, G') for every choice of θ.

Suppose $\lambda > 0$. We note that, since $\mathbb{E}f(G') = 0$, by Jensen's inequality

$$\mathbb{E}(e^{\lambda f(G)}) \leq \mathbb{E}(e^{\lambda f(G) - \lambda f(G')}).$$

Now

$$f(G) - f(G') = \int_0^{\pi/2} \frac{d}{d\theta} f(G_\theta) d\theta$$

$$= \int_0^{\pi/2} \langle \nabla f(G_\theta), G'_\theta \rangle \, d\theta,$$

where $\nabla f(G_\theta)$ is the gradient of f at G_θ. Using Jensen's inequality again,

$$\mathbb{E}e^{\lambda\big(f(G) - f(G')\big)} = \mathbb{E} \exp \left(\lambda \int_0^{\pi/2} \left\langle \frac{\pi}{2} f(G_\theta), G'_\theta \right\rangle \frac{2}{\pi} \, d\theta \right)$$

$$\leq \frac{2}{\pi} \int_0^{\pi/2} \mathbb{E} \exp \left(\lambda \left\langle \frac{\pi}{2} \nabla f(G_\theta), G'_\theta \right\rangle \right) d\theta$$

$$= \frac{2}{\pi} \int_0^{\pi/2} \mathbb{E} \exp \left(\lambda \left\langle \frac{\pi}{2} \nabla f(G), G' \right\rangle \right) d\theta$$

$$= \mathbb{E} \exp \left(\lambda \left\langle \frac{\pi}{2} \nabla f(G), G' \right\rangle \right).$$

Now,

$$\mathbb{E}_{G'} \exp \left(\lambda \left\langle \frac{\pi}{2} \nabla f(G), G' \right\rangle \right) = \exp \left(\frac{\lambda^2 \pi^2 \|\nabla f(G)\|^2}{8} g \right),$$

where g is a standard scalar Gaussian. But $\mathbb{E}e^{\alpha g} = e^{\alpha^2/2}$, and so

$$\mathbb{E}_{G'} \exp \left(\lambda \left\langle \frac{\pi}{2} \nabla f(G), G' \right\rangle \right) = \exp \left(\frac{\lambda^2 \pi^2 \|\nabla f(G)\|^2}{8} \right) \leq \exp \left(\frac{\lambda^2 \pi^2}{8} \right).$$

Thus

$$\mathbb{E}\big(\exp(\lambda f(G)) \big) \leq \exp \left(\frac{\lambda^2 \pi^2}{8} \right).$$

By symmetry,

$$\mathbb{E} \left(\exp(\lambda |f(G)|) \right) \leq 2 \exp \left(\frac{\lambda^2 \pi^2}{8} \right),$$

and hence (by Chebyshev's inequality),

$$\mathbb{P}(|f(G)| > t) \leq 2 \exp \left(\frac{\lambda^2 \pi^2 - 8\lambda t}{8} \right).$$

Choosing $\lambda = 4t/\pi^2$ we obtain

$$\mathbb{P}\big(|f(G)| > t\big) \leq 2 \exp \left(-\frac{2t^2}{\pi^2} \right).$$

□

The following theorem is due to Milman [152] and is generally referred to as the *Concentration of Measure Phenomenon*. The precise constants are irrelevant: the key point is that as $n \to \infty$ the estimate for $\sigma_n(|f - \overline{f}| > t)$ tends to zero very rapidly. In high dimensions, Lipschitz functions on $\mathcal{S}^{n-1}$ are almost constant!

Theorem 12.2.2 (The Concentration of Measure Phenomenon). *Let f be a Lipschitz function on $\mathcal{S}^{n-1}$ with Lipschitz constant 1. Then for $t > 0$,*

$$\sigma_n(|f - \overline{f}| > t) \leq 4e^{-nt^2/72\pi^2},$$

where

$$\overline{f} = \int_{\mathcal{S}^{n-1}} f \, d\sigma_n.$$

Proof. We shall assume that $\overline{f} = 0$, and so $|f(x)| \leq 1$ for all $x \in \mathcal{S}^{n-1}$. Let us first extend f to $\mathbb{R}^n$ by putting

$$f(x) = \|x\| f(x/\|x\|), \qquad x \in \mathbb{R}^n.$$

Then if $x, y \in \mathbb{R}^n$,

$$\left\| \frac{x}{\|x\|} - \frac{y}{\|y\|} \right\| \leq 2\frac{\|x - y\|}{\|x\|}.$$

If $\|x\| \geq \|y\|$ we therefore have

$$|f(x) - f(y)| \leq \|y\| \left| f\left(\frac{x}{\|x\|}\right) - f\left(\frac{y}{\|y\|}\right) \right| + (\|x\| - \|y\|) \left| f\left(\frac{x}{\|x\|}\right) \right|$$

$$\leq 3\|x - y\|.$$

Thus f extended to $\mathbb{R}^n$ has Lipschitz constant at most 3; note that $\mathbb{E}f(G) = 0$. We wish to estimate $\mathbb{P}(|f(G/\|G\|)| > t)$. First note that

$$\mathbb{E}\|G\| \geq \frac{1}{\sqrt{n}} \mathbb{E} \sum_{j=1}^n |g_j| = \sqrt{\frac{2n}{\pi}} > \frac{1}{2}\sqrt{n}.$$

By Theorem 12.2.1,

$$\mathbb{P}\left(\|G\| < \frac{1}{4}\sqrt{n}\right) \leq \mathbb{P}\left(\left|\|G\| - \mathbb{E}\|G\|\right| > \frac{1}{4}\sqrt{n}\right)$$

$$\leq 2e^{-n/8\pi^2}.$$

On the other hand,

$$\mathbb{P}\left(|f(G)| > t\sqrt{n}/4\right) \leq 2e^{-nt^2/72\pi^2}.$$

For $t \leq 1$ this is larger than $2e^{-n/8\pi^2}$. Thus

$$\mathbb{P}\left(|f(G/\|G\|)| > t\right) \leq 4e^{-nt^2/72\pi^2}.$$

□

12.3 Dvoretzky's theorem

Consider $\mathbb{R}^n$ with its canonical Euclidean norm, $\|\cdot\|$, and suppose that we are given a second norm $\|\cdot\|_X$ on $\mathbb{R}^n$ such that

$$\|x\|_X \le \|x\|, \qquad x \in \mathbb{R}^n.$$

Obviously, we can hope to use the results of the previous section for the function $f(x) = \|x\|_X$ which is 1-Lipschitz.

We will need the following lemma:

Lemma 12.3.1. *Let F be an m-dimensional normed space. Suppose $\epsilon > 0$. Then there is an ϵ-net $\{x_j\}_{j=1}^N$ for $\{x : \|x\|_F = 1\}$ with $N \le (1 + \frac{2}{\epsilon})^m$.*

Proof. Pick a maximal subset $\{x_j\}_{j=1}^N$ of $\{x : \|x\| = 1\}$ with the property that $\|x_i - x_j\| \ge \epsilon$ whenever $i \ne j$. It is clear that this is an ϵ-net. The open balls $\{x : \|x - x_j\| < \frac{1}{2}\epsilon\}$ are disjoint and contained in $(1 + \frac{1}{2}\epsilon)B_F$. Thus, by comparing volumes,

$$N\left(\frac{\epsilon}{2}\right)^m \le \left(1 + \frac{\epsilon}{2}\right)^m.$$

This gives the estimate on N.

$\square$

Theorem 12.3.2. *Suppose $\|\cdot\|_X$ is a norm on $\mathbb{R}^n$ with $\|x\|_X \le \|x\|$. Let*

$$\theta = \theta_X = \int_{\mathcal{S}^{n-1}} \|\xi\|_X d\sigma_n(\xi).$$

Suppose $0 < \epsilon < \frac{1}{3}$. Then there is a k-dimensional subspace F of $\mathbb{R}^n$ with

$$(1 - \epsilon)\theta\|x\| \le \|x\|_X \le (1 + \epsilon)\theta\|x\|, \qquad x \in F \qquad (12.1)$$

provided

$$k \le c\theta^2 n \frac{\epsilon^2}{|\log \epsilon|},$$

where $c > 0$ is a suitable absolute constant. Hence, we can find a subspace F of $\mathbb{R}^n$ with $\dim F \ge k$ such that $d_F \le 1 + \epsilon$, provided

$$k \le c_1 \theta^2 n^2 \frac{\epsilon^2}{|\log \epsilon|},$$

where c_1 is an absolute constant.

Proof. Let us fix some k-dimensional subspace of $\mathbb{R}^n$, say $G = [e_1, \ldots, e_k]$, and pick an $\epsilon/3$-net $\{x_j\}_{j=1}^N$ for $\{x \in G : \|x\| = 1\}$ with $N \le (1 + 6/\epsilon)^k$ (using Lemma 12.3.1).

Let $\mathcal{O}_n$ denote, as usual, the orthogonal group and μ its normalized Haar measure. We wish to estimate $\mu(A)$ where A is the set of $U \in \mathcal{O}_n$ so that

$$(1 - \epsilon/3)\theta \le \|Ux_j\|_X \le (1 + \epsilon/3)\theta, \qquad j = 1, 2 \ldots, N.$$

Let $\tilde{A}$ be the complementary set. Then

$$\mu(\tilde{A}) \le \sum_{j=1}^{N} \mu\left(U: \left|\|Ux_j\|_X - \theta\right| > \tfrac{1}{3}\epsilon\theta\right).$$

But,

$$\mu\left(U: \left|\|Ux_j\|_X - \theta\right| > \tfrac{1}{3}\epsilon\theta\right) = \sigma_n\left(x: \left|\|Ux_j\|_X - \theta\right| > \tfrac{1}{3}\epsilon\theta\right),$$

hence

$$\mu(\tilde{A}) \le 4Ne^{-n\epsilon^2\theta^2/648\pi^2}.$$

Now,

$$4N \le (7/\epsilon)^{(k+1)} \le e^{(k+1)(2 - \log \epsilon)},$$

and so $\mu(\tilde{A}) < 1$ provided

$$k + 1 < \frac{n\epsilon^2\theta^2}{648\pi^2(2 - \log \epsilon)}.$$

We are now in position to use Lemma 11.1.11, which yields that if $U \in A$,

$$(1 - \epsilon)\theta\|x\| \le \|Ux\|_X \le (1 + \epsilon)\theta\|x\| \qquad x \in G.$$

Taking $F = U(G)$ we obtain (12.1). This implies the theorem for a suitable $c > 0$.

The last statement of the theorem follows with a slightly different constant. □

Notice that, in this theorem, $0 < \theta_X \le 1$. In order to apply it in a non-trivial way one needs θ_X large compared with $n^{-1/2}$. We first use this to consider finite-dimensional ℓ_p-spaces. This result is due to Figiel, Lindenstrauss, and Milman [60].

Theorem 12.3.3. *Suppose $1 \le p < \infty$ and $n \in \mathbb{N}$. Then for $\epsilon > 0$, ℓ_p^n contains a subspace F with $\dim F = k$ and $d(F, \ell_2^n) \le 1 + \epsilon$, provided:*

(i) $k \le cn^{2/p}\epsilon^2|\log \epsilon|^{-1}$ if $p \ge 2$;
(ii) $k \le cn\epsilon^2|\log \epsilon|^{-1}$ if $1 \le p \le 2$,

where $c > 0$ is an absolute constant.

Proof. We consider $\mathbb{R}^n$ equipped with the norms $\|\cdot\|_p$, $1 \le p < \infty$.

If $p > 2$, by Hölder's inequality we have

$$\|x\|_p \le \|x\|_2 \le n^{\frac{1}{2} - \frac{1}{p}}\|x\|_p.$$

Let $\|\cdot\|_X = \|\cdot\|_p$. Then

$$\int_{\mathcal{S}^{n-1}} \|\xi\|_p d\sigma_n(\xi) \geq n^{\frac{1}{p}-\frac{1}{2}},$$

and so

$$\theta_X \geq n^{\frac{1}{p}-\frac{1}{2}}.$$

Now Theorem 12.3.2 gives the conclusion.

We do the cases $1 \leq p \leq 2$ simultaneously. Note that

$$\frac{1}{\sqrt{n}}\|x\|_1 \leq \frac{1}{n^{1/p-1/2}}\|x\|_p \leq \|x\|_2. \tag{12.2}$$

We will use the norm $\|\cdot\|_X = n^{-1/2}\|\cdot\|_1$. If $g_1, \ldots, g_n$ are independent (normalized) Gaussians and $G = \sum_{j=1}^n g_j e_j$ as before, note that $G/\|G\|_2$ and $\|G\|_2$ are independent. Thus

$$\theta_X = \frac{1}{\sqrt{n}}\mathbb{E}\frac{\|G\|_1}{\|G\|_2} = \frac{1}{\sqrt{n}}\frac{\mathbb{E}\|G\|_1}{\mathbb{E}\|G\|_2}.$$

Now

$$\mathbb{E}\|G\|_1 = n\sqrt{\frac{2}{\pi}},$$

and

$$\mathbb{E}\|G\|_2 \leq \left(\mathbb{E}\|G\|_2^2\right)^{\frac{1}{2}} = n^{\frac{1}{2}}.$$

We thus deduce that

$$\theta_X \geq \sqrt{\frac{2}{\pi}}, \tag{12.3}$$

independent of n. Using Theorem 12.3.2, we get the conclusion for $p = 1$. But for $1 < p < 2$, (12.2) allows us to show equally that (12.3) holds for the norms $\|\cdot\|_X = n^{1/2-1/p}\|\cdot\|_p$.

□

In order to prove Dvoretzky's theorem we need to take an arbitrary n-dimensional normed space and introduce coordinates or an inner product structure so that Theorem 12.3.2 can be applied. The problem is to find the right inner product structure. The John ellipsoid is a natural place to start. However, the best estimate for θ_X that we can obtain follows from Theorem 12.1.4, which says that

$$n^{-1/2}\|x\|_E \leq \|x\|_X \leq \|x\|_E,$$

and hence that $\theta_X \geq n^{-\frac{1}{2}}$. As already remarked, this is insufficient to get any real information from Theorem 12.3.2.

The trick is to use the John ellipsoid and then pass to a smaller subspace. In fact, this technique was originally devised by Dvoretzky and Rogers in their proof of the Dvoretzky-Rogers theorem in 1950 [50]. We remark that the following proposition is a slightly weaker form of the original lemma which is sufficient for our purposes (we found this version in [154] where it is attributed to Bill Johnson).

Proposition 12.3.4 (The Dvoretzky-Rogers Lemma). *Let X be an n-dimensional normed space and suppose that $\|\cdot\|_E$ is the norm induced on X by the John ellipsoid. Then there is an orthonormal basis $(e_j)_{j=1}^n$ of $(X, \|\cdot\|_E)$ with the property that*

$$\|e_j\|_X \geq 2^{-\frac{n}{n-j+1}}, \qquad j = 1, 2, \ldots, n.$$

In particular,

$$\|e_j\|_X \geq 1/4, \qquad j \leq \frac{n}{2} + 1.$$

Proof. We must recall the definition of the John ellipsoid of X as the ellipsoid of maximal volume contained in B_X. We pick $(e_j)_{j=1}^n$ inductively so that $\|e_1\|_X = 1$ and, subsequently, e_j such that $\|e_j\|_X$ is maximal subject to the requirement that $\langle e_j, e_i \rangle = 0$ for $i < j$ and $\|e_j\|_E = 1$.

Thus $\|x\| \leq \|e_j\|_X = t_j$, say, if $x \in [e_j, \ldots, e_n]$.

Fix $1 \leq j \leq n$. For $a, b > 0$ let us consider the ellipsoid $\mathcal{E}_{a,b}$ of all x such that

$$a^{-2} \sum_{i=1}^{j-1} |\langle x, e_i \rangle|^2 + b^{-2} \sum_{i=j}^{n} |\langle x, e_i \rangle|^2 \leq 1.$$

$\mathcal{E}_{a,b}$ is contained in B_X provided

$$a + bt_j \leq 1,$$

and it has volume $a^{j-1}b^{n-j+1}$ relative to the volume of $\mathcal{E}$. It follows that if $0 \leq b \leq t_j^{-1}$,

$$(1 - bt_j)^{j-1}b^{n-j+1} \leq 1.$$

Choosing $b = (2t_j)^{-1}$, we obtain

$$2^n t_j^{-(n-j+1)} \leq 1.$$

This gives the conclusion.

$\square$

We will need a lemma on the behavior of the maximum of m Gaussians.

Lemma 12.3.5. *There is an absolute constant $c > 0$ such that if $g_1, \ldots, g_m$ are (normalized) Gaussians then*

$$\mathbb{E} \max_{1 \leq j \leq m} |g_j| \geq c(\log m)^{1/2}.$$

Proof. If $t > 0$,

$$\mathbb{P}(|g_j| > t) = \sqrt{\frac{2}{\pi}} \int_t^\infty e^{-\frac{1}{2}s^2} ds \geq \sqrt{\frac{2}{\pi}} t e^{-2t^2}.$$

Thus if $m \geq 2$,

$$\mathbb{P}\Big(\max_{1\le j\le m}|g_j|\le t(\log m)^{1/2}\Big)\le\Big(1-\sqrt{\frac{2}{\pi}}t(\log m)^{\frac{1}{2}}m^{-2t^2}\Big)^m.$$

In particular, if $t<1/\sqrt{2}$,

$$\lim_{m\to\infty}\mathbb{P}\Big(\max_{1\le j\le m}|g_j|\le t(\log m)^{1/2}\Big)=0.$$

Since

$$\mathbb{E}\max_{1\le j\le m}|g_j|\ge t(\log m)^{1/2}\mathbb{P}\Big(\max_{1\le j\le m}|g_j|>t(\log m)^{1/2}\Big),$$

we have the lemma for some choice of c.

$\square$

We are finally ready to complete the proof of Dvoretzky's theorem, giving quantitative estimates as promised:

Theorem 12.3.6 (Dvoretzky's Theorem). *There is an absolute constant $c>0$ with the following property: If X is an n-dimensional normed space and $0<\epsilon<1/3$, then X has a subspace F with $\dim F=k$ and $d_F<1+\epsilon$ whenever*

$$k\le c\log n\frac{\epsilon^2}{|\log\epsilon|}.$$

Proof. Let $\|\cdot\|_E$ be the norm induced on X by the John ellipsoid. By the Dvoretzky-Rogers lemma, we can pass to a subspace X_0 of X with $m=\dim X_0\ge n/2$, and with the property that $(X_0,\|\cdot\|_E)$ has an orthonormal basis $(e_1,\ldots,e_m)$ such that $\|e_j\|_X\ge 1/4$ for $j=1,\ldots,m$.

Let $(g_j)_{j=1}^m$ be a sequence of independent Gaussians and $G=\sum_{j=1}^m g_j e_j$. For $m\ge 2$ we have

$$\mathbb{E}\|G\|_X=\mathbb{E}\Big\|\sum_{j=1}^m g_j e_j\Big\|_X$$

$$=\mathbb{E}\Big\|\sum_{j=1}^m \varepsilon_j g_j e_j\Big\|_X$$

$$\ge\mathbb{E}\max_{1\le j\le m}\|g_j e_j\|_X$$

$$\ge\frac{1}{4}\mathbb{E}\max_{1\le j\le m}|g_j|$$

$$\ge\frac{c}{4}(\log m)^{\frac{1}{2}}.$$

In this argument we used a sequence of Rademachers $(\varepsilon_j)_{j=1}^m$ independent of the $(g_j)_{j=1}^m$, and that

$$\mathbb{E}\Big\|\sum_{j=1}^m \varepsilon_j x_j\Big\|\ge\max_{1\le j\le m}\|x_j\|.$$

This, combined with the obvious fact that $\mathbb{E}\|G\|_E^2 = \mathbb{E}\sum_{j=1}^m g_j^2 = m$, yields

$$
\begin{aligned}
\theta_{X_0} &= \int_{\mathcal{S}^{m-1}} \|\xi\|_X d\sigma_m(\xi) \\
&= \sqrt{\frac{2}{\pi}} \frac{\mathbb{E}\|G\|_X}{\mathbb{E}\|G\|_E} \\
&\geq \sqrt{\frac{2}{\pi}} \frac{\mathbb{E}\|G\|_X}{(\mathbb{E}\|G\|_E^2)^{1/2}} \\
&\geq c_1 \frac{(\log m)^{1/2}}{m^{1/2}},
\end{aligned}
$$

for some absolute constant $c_1 > 0$. If we apply Theorem 12.3.2, we obtain Dvoretzky's theorem.

$\square$

Dvoretzky's theorem is, of course, just the beginning for a very rich theory which is still evolving. One of the interesting questions is to decide the precise dimension of the almost Hilbertian subspace of an n-dimensional space. The estimate of $\log n$ is, in fact, optimal for arbitrary spaces (see the Problems), but we have seen in Theorem 12.3.3 that for special spaces one can expect to do better and perhaps even obtain subspaces of *proportional dimension* cn as in the case ℓ_p^n where $1 \leq p < 2$. It turns out that this is related to the concept of cotype. Remarkably, the first part of Theorem 12.3.3 holds for any space of cotype two; this is due to Figiel, Lindenstrauss, and Milman [60]. Another remarkable result is Milman's theorem, which, roughly speaking, says that if one can take quotients as well as subspaces then one can find an almost Hilbertian space of proportional dimension [153]. Let us give the precise statement:

Theorem 12.3.7 (Milman's Quotient-Subspace Theorem). *There is an absolute constant c such that if $0 < \theta < 1$ and X is a finite-dimensional normed space then there is a quotient Y of a subspace of X with $\dim Y > \theta \dim X$ and $d_Y \leq c(1 - \theta)^{-2} \log(1 - \theta)$.*

The reader interested in this subject should consult the books of Milman and Schechtman [154], Pisier [188], and Tomczak-Jaegermann [216] as a starting point to learn about a rapidly evolving field.

12.4 The complemented subspace problem

Armed with Dvoretzky's theorem (which we have proved twice!) we can return to complete the complemented subspace problem, which we solved only partially in Chapter 9. Our proof follows a treatment given by Kadets and

Mitjagin [97] (using an observation of Figiel) and not the original proof of Lindenstrauss and Tzafriri [135].

To get the most precise result we will prove a strengthening of Dvoretzky's theorem which is of interest in its own right. Figiel's observation was based on a somewhat easier argument of Milman [151]. However, the proof we present is in the spirit of this chapter, and demonstrates a use of the concentration of measure phenomenon.

Theorem 12.4.1. *Let X be an infinite-dimensional Banach space. Suppose E is a finite-dimensional subspace of X. Then for any $m \in \mathbb{N}$ there is a norm $\| \cdot \|_Y$ on $Y = E \oplus \ell_2^m$ so that Y is isometric to a subspace of an ultraproduct of X and:*

$$\|(x,0)\|_Y = \|x\|, \qquad x \in E$$
$$\|(0,\xi)\|_Y = \|\xi\|, \qquad \xi \in \ell_2^m$$
$$\|(x,\xi)\|_Y = \|(x,-\xi)\|_Y, \qquad x \in E, \; \xi \in \ell_2^m.$$

Proof. Let us suppose $\nu > 0$ and $(x_j)_{j=1}^N$ be a ν-net for B_E. We also choose a ν-net $(\xi_j)_{j=1}^M$ for $\mathcal{S}^{m-1}$.

Let $n \in \mathbb{N}$, $n > m$; we regard ℓ_2^m as a subspace of ℓ_2^n. By Dvoretzky's theorem, there is a linear map $S : \ell_2^n \to X$ satisfying

$$(1-\nu)\|\xi\| \le \|S\xi\| \le \|\xi\|, \qquad \xi \in \ell_2^n.$$

For $1 \le j \le N$ and $1 \le k \le [\nu^{-1}]$, we consider the functions $f_{j,k} : \mathcal{S}^{n-1} \to \mathbb{R}$ defined by

$$f_{j,k}(\xi) = \|k\nu S\xi + x_j\|.$$

Note that each $f_{j,k}$ has Lipschitz constant at most one. Let

$$a_{j,k} = \overline{f}_{j,k} = \int_{\mathcal{S}^{n-1}} f_{j,k} \, d\sigma_n.$$

Using Theorem 12.2.2, we have

$$\sigma_n\big(|f_{j,k} - a_{j,k}| > \nu\big) \le 4e^{-n\nu^2/72\pi^2}.$$

Thus

$$\sigma_n\Big(\max_{1 \le j \le N} \max_{1 \le k \le [\nu^{-1}]} |f_{j,k} - a_{j,k}| > \nu \Big) \le 4N\nu^{-1}e^{-n\nu^2/72\pi^2}.$$

Put

$$A = \{U \in \mathcal{O}_n : \max_{1 \le i \le M} \max_{1 \le j \le N} \max_{1 \le k \le [\nu^{-1}]} |f_{j,k}(U\xi_i) - a_{j,k}| > \nu\},$$

where $\mathcal{O}_n$ is the orthogonal group and μ its Haar measure. Arguing as in Theorem 12.3.2 we obtain the following estimate for $\mu(A)$:

$$\mu(A) \le 4MN\nu^{-1}e^{-n\nu^2/72\pi^2}.$$

Hence, if n is chosen large enough, $\mu(A) < 1$ and there exists $U \notin A$. Let $T = SU : \ell_2^m \to X$. Then,

$$\left| \|x_j + k\nu T\xi_i\| - a_{j,k} \right| \le \nu, \qquad 1 \le i \le M,\ 1 \le j \le N,\ 1 \le k \le [\nu^{-1}].$$

It follows that

$$\left| \|x_j + k\nu T\xi\| - a_{j,k} \right| \le 2\nu, \qquad 1 \le j \le N,\ 1 \le k \le [\nu^{-1}], \xi \in \mathcal{S}^{m-1},$$

and so

$$\left| \|x_j + k\nu T\xi\| - \|x_j - k\nu T\xi\| \right| \le 4\nu, \qquad 1 \le j \le N,\ 1 \le k \le [\nu^{-1}], \xi \in \mathcal{S}^{m-1}.$$

Hence, approximating $\xi/\|\xi\|$ by some $k\nu$, we have

$$\left| \|x_j - T\xi\| - \|x_j + T\xi\| \right| \le 6\nu, \qquad 1 \le j \le N,\ \|\xi\| \le 1.$$

This, in turn, implies that

$$\left| \|x - T\xi\| - \|x + T\xi\| \right| \le 8\nu, \qquad \|x\| \le 1,\ \|\xi\| \le 1.$$

From the properties of T we deduce that if $F = T(\ell_2^m)$ we have

$$\left| \|x - f\| - \|x + f\| \right| \le 10\nu \max(\|x\|, \|f\|), \qquad x \in E,\ f \in F.$$

Since by the triangle law,

$$\|x - f\| + \|x - f\| \ge 2\max(\|x\|, \|f\|),$$

this yields

$$\|x - f\| \le \frac{1 + 5\nu}{1 - 5\nu}\|x + f\|, \qquad x \in E, f \in F.$$

Since $d_F < 1 + \nu$, and $\nu > 0$ is arbitrary, we are done.

$\square$

Theorem 12.4.2. *Let X be an infinite-dimensional Banach space with the property that there exists $\lambda \ge 1$ so that for every finite-dimensional subspace E of X there is a projection $P : X \to E$ with $\|P\| \le \lambda$. Then X is isomorphic to a Hilbert space, and $d_X \le 4\lambda^2$.*

Proof. Let E be a finite-dimensional subspace of X. Suppose $n = \dim E$ and $d = d_E$. Using Theorem 12.4.1 we may find a space $Y = E \oplus \ell_2^n$ isometric to a subspace of a space finitely representable in X, so that the norm on $E \oplus \ell_2^n$ satisfies

$$\|(x,\xi)\|_Y = \|(x,-\xi)\|_Y, \qquad x \in X, \ \xi \in \ell_2^n.$$

In particular this will imply that

$$\max(\|x\|,\|\xi\|) \le \|(x,\xi)\|_Y, \qquad x \in X, \ \xi \in \ell_2^n.$$

The space Y must also have the property that every subspace of it is λ-complemented.

Let $\theta^2 = d_E$, and choose an invertible operator $S : E \to \ell_2^n$ so that

$$\theta^{-1}\|x\| \le \|Sx\| \le \theta\|x\|, \qquad x \in E.$$

We define a subspace of Y by taking $Z = \{(x, Sx) : \ x \in E\}$. Let $R : Y \to Z$ be a projection with $\|R\| \le \lambda$.

We now define a second operator $T : E \to \ell_2^n$ by

$$R(x,0) = (S^{-1}Tx, Tx), \qquad x \in X.$$

It is clear that $\|T\| \le \lambda$.

Then we introduce an operator $V : E \to \ell_2^{2n} = \ell_2^n \oplus \ell_2^n$ given by $Vx = (\lambda Sx, \theta Tx)$. Let us estimate $\|V\|$. Clearly,

$$\|Vx\|^2 \le \lambda^2\|Sx\|^2 + \theta^2\|Tx\|^2 \le 2\lambda^2\theta^2\|x\|^2,$$

that is,

$$\|V\| \le \sqrt{2}\lambda\theta.$$

If $x \in E$ we have

$$R(0, Sx) = (x - S^{-1}Tx, Sx - Tx),$$

and so

$$\|x - S^{-1}Tx\| \le \lambda\|Sx\|, \qquad x \in E.$$

Hence

$$\begin{aligned}
\|x\| &\le \|x - S^{-1}Tx\| + \|S^{-1}Tx\| \\
&\le \lambda\|Sx\| + \theta\|Tx\| \\
&\le \sqrt{2}(\lambda^2\|Sx\|^2 + \theta^2\|Tx\|^2)^{1/2} \\
&= \sqrt{2}\|Vx\|.
\end{aligned}$$

This yields that V is an isomorphism onto its range, and that $\|V^{-1}\| \le \sqrt{2}$. Thus $\|V\|\|V^{-1}\| \le 2\lambda\theta$. But, by hypothesis, this means that $\theta^2 \le 2\lambda\theta$, i.e., $\theta \le 2\lambda$, or, equivalently, $d_E \le 4\lambda^2$.

Thus X is $4\lambda^2$-crudely finitely representable in a Hilbert space, which implies that $d_X \le 4\lambda^2$ (this is proved in Proposition 11.1.12).

$\square$

Lemma 12.4.3. *Let X be an infinite-dimensional Banach space with the property that every closed subspace is complemented. Then there exists $\lambda \geq 1$ so that every finite-dimensional subspace E of X is λ-complemented in X.*

Proof. For E a finite-dimensional subspace of X denote by $\lambda(E)$ the norm of the optimal projection (one may show that such a projection exists by compactness). Suppose $\sup\{\lambda(E) : \dim E < \infty\} = \infty$. We first argue that, then, for every subspace X_0 of finite codimension we have

$$\sup\{\lambda(E) : \dim E < \infty, \ E \subset X_0\} = \infty. \tag{12.4}$$

Indeed, suppose

$$\sup\{\lambda(E) : \dim E < \infty, \ E \subset X_0\} = M < \infty.$$

Let k be the codimension of X_0. Then suppose E is any finite-dimensional subspace of X. Let $E_0 = E \cap X_0$ and let P_0 be a projection of X onto E_0 with $\|P_0\| \leq M$. Let $F = \{x \in E : P_0 x = 0\}$. Then $\dim F \leq k$, so there is a projection P_1 of X onto F with $\|P_1\| \leq \sqrt{k}$ (Theorem 12.1.6). Let $P = P_0 + P_1 - P_1 P_0$; then P is a projection onto E with $\|P\| \leq (M+1)(\sqrt{k}+1)$. This establishes (12.4).

Next we note that if E is a finite-dimensional subspace of X and $\epsilon > 0$ then there is a finite codimensional subspace X_0 such that

$$\|e + x\| \geq (1 - \epsilon)\|e\|, \qquad e \in E, \ x \in X_0.$$

This is essentially the content of Lemma 1.5.1 in Chapter 1.

We now proceed by induction to construct a sequence of finite-dimensional subspaces $(E_n)_{n=1}^\infty$ and finite codimensional subspaces $(X_n)_{n=1}^\infty$ so that

- $\lambda(E_n) > n, \quad n \in \mathbb{N}$.
- $\|e + x\| \geq \frac{1}{2}\|e\|, \quad e \in E_n, \ x \in X_n$.
- $E_{n+1} \subset X_n, \quad n \in \mathbb{N}$.
- $X_{n+1} \subset X_n, \quad n \in \mathbb{N}$.

Let $Y = [\cup_{n=1}^\infty E_n]$, the closed linear span of $\cup_{n=1}^\infty E_n$. If $e_j \in E_j$ for $j = 1, 2, \ldots, N$, and $1 \leq m \leq N$, we have

$$\|e_1 + \cdots + e_m\| \leq 2\|e_1 + \cdots + e_N\|.$$

Hence,

$$\|e_m\| \leq 4\|e_1 + \cdots + e_N\|,$$

from which it follows that each E_m is 4-complemented in Y. Since, by assumption, Y is complemented, this implies $\sup_n \lambda(E_n) < \infty$, and we reached a contradiction.

$\square$

Combining these results we have proved:

Theorem 12.4.4 (Lindenstrauss-Tzafriri, 1971). *Let X be an infinite-dimensional Banach space in which every closed subspace is complemented. Then X is isomorphic to a Hilbert space.*

Problems

12.1. Auerbach's Lemma.
Let X be an n-dimensional normed space. Show that X has a basis $(e_j)_{j=1}^n$ with biorthogonal functions $(e_j^*)_{j=1}^n$ such that $\|e_j\| = \|e_j^*\| = 1$ for $1 \le j \le n$. [*Hint*: Maximize the volume of the parallelepiped generated by n vectors $x_1, \ldots, x_n$ in the unit ball.]

This basis is called an *Auerbach basis* and the result is due to Auerbach [6].

12.2. Let E be a subspace of ℓ_1^n of dimension k and suppose E is complemented by a projection of norm λ. Show that $k \le K_G \lambda^2 d_E^2$ where K_G is Grothendieck's constant.

12.3. Suppose $1 \le p < 2$. Let E be a subspace of ℓ_p^n of dimension k and suppose E is complemented by a projection of norm λ. By considering E as a subspace of ℓ_1^n, show that

$$k \le K_G \lambda^2 n^{2-2/p} d_E^2,$$

where K_G is Grothendieck's constant.

12.4. Suppose $d > 1$ and $2 < p < \infty$. Show that there is a constant $C = C(d, p)$ so that if E is a subspace of ℓ_2^n with $d_E \le d$ then $k \le Cn^{2/p}$. [*Hint*: Use the fact that ℓ_p^n has type 2, and duality.] This shows that Theorem 12.3.3 is (in a certain sense) best possible.

12.5. Let X be an n-dimensional normed space. Suppose $(x_j)_{j=1}^N$ is a set of points in X such that $\partial B_X \subset \cup_{j=1}^N (x_j + \nu B_X)$. Show that B_X is covered by the sets $A_{jk} = k\nu x_j + 2\nu B_X$ for $1 \le j \le N$ and $1 \le k \le [\nu^{-1}]$. Deduce that

$$N \ge 2^{-n} \nu^{1-n}.$$

12.6. Let H be a Hilbert space and suppose $x \in H$ with $\|x\| = 1$ is written as a convex combination

$$x = \sum_{j=1}^n c_j y_j,$$

where $c_1, \ldots, c_n \ge 0$, $c_1 + \cdots + c_n = 1$, and $\|y_j\| \le \alpha$ for $1 \le j \le n$. Show that there exists j so that

$$\|x - y_j\|^2 \le \alpha^2 - 1.$$

12.7. Let H be a k-dimensional Hilbert space and suppose $T : H \to \ell_\infty^N$ is a linear operator satisfying

$$\|x\| \le \|Tx\| \le (1 + \epsilon)\|x\|, \qquad x \in H.$$

(a) By considering the adjoint, show that

$$2N \geq 2^{-k}((1+\epsilon)^2 - 1)^{-(k-1)/2}.$$

(b) Deduce that if ℓ_∞^N contains a k-dimensional subspace E with $d_E < 11/10$ then $k \leq C \log N$ where C is some absolute constant.

12.8. Prove the Dvoretzky-Rogers theorem directly from Proposition 12.3.4.

12.9. Lozanovskii factorization.
Let $\| \cdot \|_X$ be a norm on $\mathbb{R}^n$ for which the canonical basis $(e_j)_{j=1}^n$ is 1-unconditional. Show that for any $u = (u(j))_{j=1}^n$ with $u(j) \geq 0$ and $\sum_{j=1}^n u(j) = 1$ we can find $\xi, \eta \in \mathbb{R}^n$ so that $\xi(j), \eta(j) \geq 0$ for $1 \leq j \leq n$, $\|\xi\|_X = \|\eta\|_{X^*} = 1$ and

$$\xi(j)\eta(j) = u(j), \qquad 1 \leq j \leq n.$$

[*Hint:* Maximize $\sum_{j=1}^n u(j) \log |\xi(j)|$ for $\|\xi\|_X \leq 1$.]

This result and infinite-dimensional generalizations are due to Lozanovskii [142]; see also [67].

12.10 (Figiel, Lindenstrauss, and Milman [60]). Let X be an infinite-dimensional Banach space of cotype $q < \infty$. Show that if $\epsilon > 0$ then every n-dimensional subspace F of X contains a subspace E with $\dim E \geq cn^{2/q}$ and $d_E < 1 + \epsilon$, where $c = c(\epsilon, X)$.

Important Examples of Banach Spaces

In the last, optional chapter, we construct some examples of Banach spaces that played an important role in the development of Banach space theory. These constructions are not elementary so we have preferred to remove them from the main text.

We first discuss a generalization of James space constructed by James [82] and improved by Lindenstrauss [130]. They show that for every separable Banach space X one can construct a separable Banach space $\mathcal{Z}$ so that $\mathcal{Z}^{**}/\mathcal{Z} \approx X$. Furthermore $\mathcal{Z}^*$ has a shrinking basis.

We then turn to tree-like constructions and use a tree method to construct Pełczyński's universal basis space [174] which was a fundamental example in basis theory. It shows that there is a Banach space U with a basis $(e_n)_{n=1}^{\infty}$ such that every basic sequence in U is equivalent to a *complemented* subsequence of $(e_n)_{n=1}^{\infty}$.

Finally we turn to the James tree space $\mathcal{JT}$ which was constructed in connection with Rosenthal's theorem (Chapter 10, Theorem 10.2.1). It is clear that if X is a Banach space with separable dual, X cannot contain ℓ_1. The *James tree space*, $\mathcal{JT}$, gives an example to show that the converse statement is not true. The key is that $\mathcal{JT}^{**}/\mathcal{JT}$ is shown to be a nonseparable Hilbert space and this is sufficient to show that ℓ_1 cannot embed into $\mathcal{JT}$.

13.1 A generalization of the James space

In this section we will give an exposition of the construction of a generalization of the James space whose idea originated in James's 1960 paper [82] but was given in final form by Lindenstrauss in 1971 [130].

We recall our convention that if E is a subset of $\mathbb{N}$ (in particular, any interval of integers) and $\xi = (\xi(n))_{n=1}^{\infty} \in c_{00}$ we write $E\xi$ for the sequence $(\chi_E(n)\xi(n))_{n=1}^{\infty}$, i.e., the sequence whose coordinates are $E\xi(n) = \xi(n)$ if $n \in E$ and $E\xi(n) = 0$ otherwise. We also remind the reader that if E, F are subsets of $\mathbb{N}$ we write $E < F$ to mean $m < n$ whenever $m \in E$ and $n \in F$.

Let X be any separable Banach space and suppose $(x_n)_{n=1}^{\infty}$ is any sequence so that $\{\pm x_n\}_{n=1}^{\infty}$ is dense in the surface of the unit ball of X, $\{x \in X : \|x\| = 1\}$. We define a norm on c_{00} by

$$\|\xi\|_{\mathcal{X}} = \sup \left(\sum_{j=1}^{n} \left\| \sum_{i \in I_j} \xi(i) x_i \right\|^2 \right)^{1/2},$$

where the supremum is taken over all $n \in \mathbb{N}$ and all intervals $I_1 < I_2 < \cdots < I_n$.

In the case when $X = \mathbb{R}$ we may take $x_n = 1$ for all n and then we recover the original James space $\mathcal{J}$ but with a different basis from the original one, as in Problem 3.10.

Let $\mathcal{X}$ be the completion of $(c_{00}, \|\cdot\|_{\mathcal{X}})$. The following proposition is quite trivial to see and we leave its proof as an exercise to the reader.

Proposition 13.1.1.

(i) The canonical unit vectors $(e_n)_{n=1}^{\infty}$ form a monotone basis for that $\mathcal{X}$. Hence $\mathcal{X}$ can be identified as the space of all sequences ξ such that

$$\|\xi\|_{\mathcal{X}} = \sup \left(\sum_{j=1}^{n} \left\| \sum_{i \in I_j} \xi(i) x_i \right\|^2 \right)^{1/2} < \infty.$$

(ii) $(e_n)_{n=1}^{\infty}$ is boundedly complete. Hence $(e_n^*)_{n=1}^{\infty}$ is a monotone basis for a subspace $\mathcal{Y}$ of $\mathcal{X}^*$ and so $\mathcal{X}$ can be identified (isometrically in this case) with $\mathcal{Y}^*$.

Proposition 13.1.2. There is a norm-one operator $T : \mathcal{X} \to X$ defined by $Te_n = x_n$ for $n \in \mathbb{N}$. T is a quotient map.

Proof. It is easy to see that $\xi \in \mathcal{X}$ implies that $\sum_{j=1}^{\infty} \xi(j) x_j$ must converge and that

$$\left\| \sum_{j=1}^{\infty} \xi(j) x_j \right\| \leq \|\xi\|_{\mathcal{X}}.$$

Thus T is well-defined and has norm one. Since $T(B_{\mathcal{X}})$ contains $(x_n)_{n=1}^{\infty}$ it follows that $\overline{T(B_{\mathcal{X}})}$ contains B_X and hence T is a quotient map. □

Therefore the adjoint of T, $T^* : X^* \to \mathcal{X}^*$ given by

$$\langle \xi, T^* x^* \rangle = \sum_{i=1}^{\infty} \xi(i) x^*(x_i),$$

is a isometric embedding.

Lemma 13.1.3. $T^*(X^*) \cap \mathcal{Y} = \{0\}$, and $T^* X^* + \mathcal{Y}$ is norm closed.

Proof. It is enough to note that if $x^* \in X^*$ and $\xi^* \in \mathcal{Y}$,

$$\|T^*x^*\|_\mathcal{X} = \|x^*\| \le \|T^*x^* + \xi^*\|_{\mathcal{X}^*}.$$

Once we have this, it follows that $T^*X^* + \mathcal{Y}$ splits as a direct sum. In fact,

$$\|x^*\| = \limsup_{n\to\infty} |x^*(x_n)|.$$

But

$$\lim_{n\to\infty} \xi^*(e_n) = 0,$$

and so

$$\limsup_{n\to\infty} |(T^*x^* + \xi^*)e_n| = \|x^*\|.$$

$\square$

Lemma 13.1.4. *Suppose $m < n$ and that $\xi^* \in B_{\mathcal{X}^*}$. Then we can decompose $\xi^* = \eta^* + \zeta^* + \psi^*$ where:*

$$\eta^*(e_j) = 0, \qquad 1 \le j \le m, \tag{13.1}$$

$$\zeta^*(e_j) = 0, \qquad n \le j < \infty, \tag{13.2}$$

$$(\|\eta^*\|_{\mathcal{X}^*}^2 + \|\zeta^*\|_{\mathcal{X}^*}^2)^{\frac{1}{2}} + \|\psi^*\|_{\mathcal{X}^*} \le 1, \tag{13.3}$$

and for some $x^* \in B_{X^*}$ we have

$$T^*x^*(e_j) = \psi^*(e_j), \qquad m \le j \le n. \tag{13.4}$$

Proof. The set of $\xi^* \in B_{\mathcal{X}^*}$ which satisfy (13.1)-(13.4) is clearly convex. It is also weak* closed. To see this, suppose that $\xi_k^* \to \xi^*$ weak*, where each ξ_k^* has a decomposition as prescribed $\xi_k^* = \eta_k^* + \zeta_k^* + \psi_k^*$ and $\psi_k^*(e_j) = x_k^*(e_j)$ for $m \le j \le n$ with $x_k^* \in B_{X^*}$. Then we can always pass to a subsequence so that $(\eta_k^*)_{k=1}^\infty, (\zeta_k^*)_{k=1}^\infty, (\psi_k^*)_{k=1}^\infty$ and $(x_k^*)_{k=1}^\infty$ are weak* convergent.

Now consider the set $\mathcal{S}$ of all ξ^* of the form

$$\xi^* = \sum_{k=1}^N I_k^*(T^*x_k^*),$$

where

$$\sum_{k=1}^n \|x_k^*\|^2 \le 1$$

and given intervals $I_1 < I_2 < \cdots < I_n$, I_k^* is the adjoint of I_k regarded as an operator. Then $\mathcal{S} \subset B_{\mathcal{X}^*}$. But if $\xi \in \mathcal{X}$ with $\|\xi\|_\mathcal{X} = 1$, and if $\epsilon > 0$, we can find $I_1 < I_2 < \cdots < I_N$ so that

$$\left(\sum_{k=1}^N \left\| \sum_{i \in I_k} \xi(i)x_i \right\|^2 \right)^{1/2} > 1 - \epsilon.$$

Hence we can find $x_1^*, x_2^*, \ldots, x_n^*$ with $\sum_{j=1}^n \|x_j^*\|^2 \le 1$ and

$$\sum_{k=1}^N x^* \left(\sum_{i \in I_k} \xi(i) x_k^*(x_i) \right) > 1 - \epsilon$$

or, equivalently,

$$\left\langle \xi, \sum_{k=1}^N I_k^* T^* x_k^* \right\rangle > 1 - \epsilon.$$

Thus the set $\mathcal{S}$ norms $\mathcal{X}$ and hence its weak* closed convex hull $\overline{\text{co}}^{w^*}(\mathcal{S})$ coincides with $B_{\mathcal{X}^*}$ by a simple Hahn-Banach argument.

It remains only to show that if $\xi^* \in \mathcal{S}$ then (13.1)-(13.4) hold. Suppose

$$\xi^* = \sum_{k=1}^N I_k^* x_k^*$$

with $\sum_{k=1}^N \|x_k^*\|^2 \le 1$. If one of the intervals I_k includes $[m, n]$ we just put $\eta^* = \zeta^* = 0$ and $\psi^* = \xi^*$. If not, we let

$$\eta^* = \sum_{m < I_k} I_k^* x_k^*$$

and $\zeta^* = \xi^* - \eta^*$, $\psi^* = 0$, and we are done.

$\square$

Lemma 13.1.5. $T^*(X^*) \oplus \mathcal{Y} = \mathcal{X}^*$.

Proof. Let us suppose that $\|\xi^*\|_{\mathcal{X}^*} = 1$ and let $d = d(\xi^*, \mathcal{Y}^* + T^*(X^*))$. For every pair $m \le n$ we can write $\xi^* = \eta_{m,n}^* + \zeta_{m,n}^* + \psi_{m,n}^*$ so that (13.1)-(13.4) hold for $\eta^* = \eta_{m,n}^*, \zeta^* = \zeta_{m,n}^*$, and $\psi^* = \psi_{m,n}^*$.

We observe that $\zeta_{m,n}^* \in \mathcal{Y}$, and so

$$\|\eta_{m,n}^*\|_{\mathcal{X}} + \|\psi_{m,n}^*\|_{\mathcal{X}^*} \ge d.$$

Now

$$(\|\eta_{m,n}^*\|_{\mathcal{X}}^2 + \|\zeta_{m,n}^*\|_{\mathcal{X}}^2)^{\frac{1}{2}} - \|\eta_{m,n}^*\|_{\mathcal{X}}^2 \le 1 - d$$

which yields

$$1 - (1 - \|\zeta_{m,n}^*\|_{\mathcal{X}^*}^2)^{1/2} \le 1 - d$$

or, equivalently,

$$\|\zeta_{m,n}^*\|_{\mathcal{X}^*} \le (1 - d^2)^{1/2}.$$

By compactness we can pick a subsequence $M = (n_k)_{k=1}^\infty$ so that, keeping m fixed,

$$\lim_{k \to \infty} \eta_{m,n_k}^* = \eta_m^*, \ \lim_{k \to \infty} \zeta_{m,n_k}^* = \zeta_m^*, \ \lim_{k \to \infty} \psi_{m,n_k}^* = \psi_m^*$$

all exist in the weak* topology.

It follows that $\|\zeta_m^*\|_{\mathcal{X}^*} \leq (1 - d^2)^{1/2}$. It is also elementary to see by Alaoglu's theorem that there exists $x^* \in B_{X^*}$ so that $\psi^*(e_j) = T^* x^*(e_j)$ for $m \leq j < \infty$. Hence $\psi^* - T^* x^* \in \mathcal{Y}$, i.e., $\psi^* \in T^* X^* + \mathcal{Y}$. Therefore,

$$d \leq \|\eta_m^*\|_{\mathcal{X}^*} + \|\zeta_m^*\|_{\mathcal{X}^*} \leq \|\eta_m^*\|_{\mathcal{X}^*} + (1 - d^2)^{1/2},$$

and so

$$\|\eta_m^*\|_{\mathcal{X}^*} \geq d - (1 - d^2)^{\frac{1}{2}}.$$

This yields

$$\|\psi_m^*\|_{\mathcal{X}^*} \leq 1 - d + (1 - d^2)^{1/2}.$$

The next step is to let $m \to \infty$; by passing again to a subsequence we can ensure that

$$\lim_{k \to \infty} \eta_{m_k}^* = \eta^*, \quad \lim_{k \to \infty} \zeta_{m_k}^* = \zeta^*, \quad \lim_{k \to \infty} \psi_{m_k}^* = \psi^*$$

all exist in the weak* topology. But it is clear from the construction that $\eta^* = 0$, so $\xi^* = \zeta^* + \psi^*$ and therefore

$$1 = \|\xi^*\|_{\mathcal{X}^*} \leq (1 - d) + 2(1 - d^2)^{\frac{1}{2}}.$$

Hence $5d^2 \leq 4$ or, equivalently, $d \leq 2/\sqrt{5} < 1$.

This is enough to show $T^*(X^*) + \mathcal{Y} = \mathcal{X}^*$ since, if not, there exists $\xi^* \in B_{X^*}$ with $d(\xi^*, T^*(X^*) + \mathcal{Y}) > 2/\sqrt{5}$.

$\square$

Theorem 13.1.6. *For every separable Banach space X there is a separable Banach space $\mathcal{Z}$ such that $\mathcal{Z}^{**}/\mathcal{Z}$ is isomorphic to X. Furthermore $\mathcal{Z}^*$ has a shrinking basis.*

Remark 13.1.7. The fact that $\mathcal{Z}^*$ has a basis implies that $\mathcal{Z}$ has a basis: this is deep result of Johnson, Rosenthal, and Zippin [94] which is beyond the scope of this book.

Proof. We take $\mathcal{Z} = \ker T$ in the above construction. We show that $\mathcal{X}$ can then be identified canonically with $\mathcal{Z}^{**}$. More precisely, we show that under the pairing between $\mathcal{X}$ and $\mathcal{Y}$ we can identify $\mathcal{Y}$ with $\mathcal{Z}^*$. The identification is not isometric, however.

Clearly, if $\eta^* \in \mathcal{Y}$ then $\eta^*|_{\mathcal{Z}} \in \mathcal{Z}^*$. Conversely, suppose $\zeta^* \in \mathcal{Z}^*$. By the Hahn-Banach theorem there exists $\xi^* \in \mathcal{X}^*$ such that $\xi^*|_{\mathcal{Z}} = \zeta^*$. By Lemma 13.1.5 there is a unique $x^* \in X^*$ such that $\eta^* = \xi^* - T^* x^* \in \mathcal{Y}$. Then $\eta^*|_{\mathcal{Z}} = \zeta^*$. Note that

$$\|\zeta^*\|_{\mathcal{Z}^*} \leq \|\eta^*\|_{\mathcal{Y}} \leq \|\xi^*\|_{\mathcal{X}^*} + \|x^*\| \leq 2\|\zeta^*\|_{\mathcal{Z}^*}.$$

This completes the proof as $\mathcal{Z}^{**}/\mathcal{Z}$ is isomorphic to $\mathcal{X}/\ker T$, i.e., to X.

Corollary 13.1.8.

(a) If X is a separable dual space then there is a Banach space Z with a shrinking basis such that $Z^{**} \approx Z \oplus X$.

(b) If X is a separable reflexive space then there is a Banach space Z with a boundedly-complete basis such that $Z^{**} \approx Z \oplus X$.

Proof. (a) If $X = Y^*$ construct $\mathcal{Z}$ as above so that $\mathcal{Z}^{**}/\mathcal{Z} \approx Y$ and then $\mathcal{Z}^{***}/\mathcal{Z}^* \approx X$. Let $Z = \mathcal{Z}^*$.

(b) In this case take $Z = \mathcal{Z}^{**}$.

$\square$

13.2 Constructing Banach spaces via trees

Let $\mathcal{F}\mathbb{N}$ denote the family of all finite subsets of $\mathbb{N}$. We introduce an ordering on $\mathcal{F}\mathbb{N}$: given $A = \{m_1, m_2, \ldots, m_j\}$ and $E = \{n_1, n_2, \ldots, n_k\}$ in $\mathcal{F}\mathbb{N}$, we write $A \preceq E$ if and only if we have $j \leq k$ and $m_i = n_i$ for $1 \leq i \leq j$. This means that A is the initial part of E. We will write $A \prec E$ if $A \preceq E$ and $A \neq E$.

The partially ordered set $(\mathcal{F}\mathbb{N}, \preceq)$ is an example of a *tree*. This means that for each $A \in \mathcal{F}\mathbb{N}$ the set $\{E : E \preceq A\}$ is both finite and totally ordered, and is empty for exactly one choice of A, namely, $A = \emptyset$; the empty set is then the *root* of the tree.

We will actually find it more convenient to consider the partially ordered set $\mathcal{F}^*\mathbb{N}$ of all *nonempty* sets in $\mathcal{F}\mathbb{N}$. This is not a tree as it has infinitely many roots (i.e., the singletons); it is perhaps a forest.

A *segment* in $\mathcal{F}^*\mathbb{N}$ is a subset of $\mathcal{F}^*\mathbb{N}$ of the form $S = S(A_0, A_1) = \{E : A_0 \subset E \subset A_1\}$. A subset $\mathcal{A}$ of $\mathcal{F}^*\mathbb{N}$ is called *convex* (for the partial order $\preceq$) if given $A_0, A_1 \in \mathcal{A}$ we also have $S(A_0, A_1) \subset \mathcal{A}$.

A *branch* B is a maximal totally ordered subset: this is easily seen to be a sequence $(A_n)_{n=1}^{\infty}$ of the form

$$A_n = \{m_1, \ldots, m_n\}, \qquad n = 1, 2, \ldots,$$

where $(m_n)_{n=1}^{\infty}$ is a subsequence of $\mathbb{N}$.

It will be convenient to introduce a coding, or labeling, of $\mathcal{F}^*\mathbb{N}$ by the natural numbers as follows. For $A = \{m_1, \ldots, m_n\}$ we define

$$\psi(A) = 2^{m_1 - 1} + 2^{m_2 - 1} + \cdots + 2^{m_n - 1}.$$

$\psi : \mathcal{F}^*\mathbb{N} \to \mathbb{N}$ is thus a bijection such that $A \preceq E \implies \psi(A) \leq \psi(E)$.

We can thus transport $\preceq$ to $\mathbb{N}$ and define

$$m \preceq n \Leftrightarrow \psi(m) \preceq \psi(n).$$

We then consider $(\mathbb{N}, \preceq)$ and we can similarly define segments, convex sets, and branches in this partially ordered set. Note that intervals $I = [m, n]$ for the usual order on $\mathbb{N}$ are convex for the ordering $\preceq$.

The key idea of our construction is that we want to make a norm on $c_{00} = c_{00}(\mathbb{N})$ which agrees with certain prescribed norms on $c_{00}(B)$ for every branch B. For this we require certain compatibility assumptions.

Let us suppose that for every branch B in $(\mathbb{N}, \preceq)$ we are given a norm $\|\cdot\|_B$ on $c_{00}(B)$ and the family of norms $\|\cdot\|_B$ satisfy the following conditions:

$$\|S\xi\|_B \leq \|\xi\|_B, \qquad S \subset B, \ S \text{ an initial segment,} \qquad (13.5)$$

and

$$\|\xi\|_B = \|\xi\|_{B'}, \qquad x \in c_{00}(B) \cap c_{00}(B'). \qquad (13.6)$$

Condition (13.5) simply asserts that $(e_n)_{n \in B}$ is a monotone basis of the completion X_B of $c_{00}(B)$. The second condition asserts that the family of norms is consistent on the intersections. We are next going to construct norms on c_{00}, such that $(e_n)_{n=1}^{\infty}$ is a monotone basis, and whose restrictions to each complete branch B reduce isometrically to the norms $\|\cdot\|_B$.

Our first, simplest definition will not solve our problem but leads in itself to an interesting example. We define

$$\|\xi\|_{\mathcal{X}} = \sup_{B \in \mathcal{B}} \|B\xi\|, \qquad \xi \in c_{00}, \qquad (13.7)$$

where $\mathcal{B}$ is the collection of all branches. Let $\mathcal{X}$ denote the completion of c_{00} under this norm.

The following proposition is quite trivial and we omit the proof.

Proposition 13.2.1. *In the space $\mathcal{X}$ we have:*

(i) $(e_n)_{n=1}^{\infty}$ *is a monotone basis.*
(ii) $\|B\xi\| \leq \|\xi\|$ *for each $B \in \mathcal{B}$, and so X_B is complemented in $\mathcal{X}$.*

Now let us try to use this. Let us suppose that X is a Banach space with a normalized monotone basis $(x_n)_{n=1}^{\infty}$. Consider the branch generated by the increasing sequence $(m_j)_{j=1}^{\infty}$, i.e., consisting of the sets $A_j = \{m_1, \ldots, m_j\}$ for $j = 1, 2, \ldots$. We define

$$\left\| \sum_{j=1}^{N} \xi(j) e_{\psi(A_j)} \right\|_B = \left\| \sum_{j=1}^{N} \xi(j) x_{m_j} \right\|_X.$$

Obviously the restriction that $(x_n)_{n=1}^{\infty}$ is monotone can be circumvented by simply renorming X. It is clear that we have:

Proposition 13.2.2. *If X is a Banach space with a basis $(x_n)_{n=1}^{\infty}$ there is a Banach space $\mathcal{X}$ with a basis $(e_n)_{n=1}^{\infty}$ so that for every increasing sequence $(m_j)_{j=1}^{\infty}$ the subsequence $(x_{m_j})_{j=1}^{\infty}$ of $(x_n)_{n=1}^{\infty}$ is equivalent to a complemented subsequence $(e_{n_j})_{j=1}^{\infty}$ of $(e_n)_{n=1}^{\infty}$.*

13.3 Pełczyński's universal basis space

We are in position to prove the following surprising result due to Pełczyński [174] (1969); our proof uses ideas of Schechtman [204]. We have seen by the Banach-Mazur theorem (Theorem 1.4.3) that every separable Banach space embeds in $C[0,1]$; however, very few spaces embed as a complemented subspace (for example, $C[0,1]$ has no complemented reflexive subspaces as we saw in Proposition 5.6.4). It is therefore rather interesting that we can construct a separable Banach space U with a basis so that every separable Banach space with a basis is isomorphic to a complemented subspace of U; moreover there is exactly one such space. At the time of Pełczyński's paper, the basis problem was unsolved and so it was not clear whether it might be that every separable Banach space was isomorphic to a complemented subspace of U; indeed there was hope that this space might lead to some resolution of the basis problem. Later, Johnson and Szankowski [95] showed, using the negative solution of the approximation property, that there is no separable Banach space which contains a complemented copy of all separable Banach spaces.

Theorem 13.3.1 (Pełczyński's universal basis space). *There is a unique separable Banach space U with a basis and with the property that every Banach space with a basis is isomorphic to a complemented subspace of U.*

Proof. To prove the existence of U it suffices to construct a Banach space X with a basis $(x_n)_{n=1}^{\infty}$ so that every normalized basic sequence (in any Banach space) is equivalent to a complemented subsequence of $(x_n)_{n=1}^{\infty}$. Then the existence of U follows from Proposition 13.2.2.

To construct X we first find a sequence $(f_n)_{n=1}^{\infty}$ which is dense in the surface of the unit ball of $C[0,1]$. We define a norm on c_{00} by

$$\|\xi\|_X = \sup_k \left\| \sum_{j=1}^{k} \xi(k) f_k \right\|_{C[0,1]}, \qquad \xi \in c_{00}.$$

X is the completion of $(c_{00}, \|\cdot\|_X)$.

One readily checks that the canonical basis $(e_n)_{n=1}^{\infty}$ is a monotone basis of X.

$C[0,1]$ is universal for separable spaces, and if $(g_j)_{j=1}^{\infty}$ is a basic sequence in $C[0,1]$ and $\epsilon > 0$, we can find an increasing sequence $(m_j)_{j=1}^{\infty}$ so that

$$\sum_{j=1}^{\infty} \|g_j - f_{m_j}\| < \epsilon.$$

Taking ϵ small enough we can ensure that $(f_{m_j})_{j=1}^{\infty}$ is a basic sequence equivalent to $(g_j)_{j=1}^{\infty}$. But then $(e_{m_j})_{j=1}^{\infty}$ is equivalent to $(f_{m_j})_{j=1}^{\infty}$. This yields the existence of U.

Uniqueness is an exercise in the Pełczyński decomposition technique. It is clear that $\ell_2(U)$ also has a basis, and so $\ell_2(U)$ is isomorphic to a complemented subspace of U. Hence for some Y we have

$$U \approx Y \oplus \ell_2(U) \approx Y \oplus \ell_2(U) \oplus \ell_2(U) \approx U \oplus \ell_2(U) \approx \ell_2(U).$$

If V is any other space with the same properties then V is isomorphic to a complemented subspace of U and U is isomorphic to a complemented subspace of V. Hence, by Theorem 2.2.3, $U \approx V$.

$\square$

Notice that the basis of U which we implicitly constructed above has the property that every normalized basic sequence in any Banach space is equivalent to a complemented subsequence.

There is an unconditional basis form of the universal basis space, also constructed by Pełczyński.

Theorem 13.3.2. *There is a unique Banach space U_1 with an unconditional basis $(u_n)_{n=1}^{\infty}$ and with the property that every Banach space with an unconditional basis is isomorphic to a complemented subspace of U_1.*

Proof. Suppose X is the space constructed in the preceding proof. Then we can define a norm on c_{00} by

$$\|\xi\|_{U_1} = \sup_{\epsilon_j = \pm 1} \left\| \sum_{j=1}^{\infty} \epsilon_j \xi(j) e_j \right\|_X.$$

We leave to the reader the remaining details. See [174] and [204].

$\square$

13.4 The James tree space

It is clear that if X is a separable Banach space with separable dual, then X cannot contain a copy of ℓ_1. The aim of this section is to give the example promised in Chapter 10 (Remark 10.2.3) of a separable Banach space which does not contain a copy of ℓ_1, but has nonseparable dual.

Let us start by introducing a definition that will be useful in the remainder of the section.

Definition 13.4.1. A basis $(x_n)_{n=1}^{\infty}$, with biorthogonal functionals $(x_n^*)_{n=1}^{\infty}$, in a Banach space X is said to satisfy a *lower 2-estimate on blocks* if there is a constant C so that whenever $I_1, \ldots, I_n$ are disjoint intervals of integers,

$$\sum_{j=1}^{n} \left\| \sum_{k \in I_j} x_k^*(x) x_k \right\|^2 \leq C \|x\|^2.$$

We say that $(x_n)_{n=1}^{\infty}$ satisfies an *exact lower 2-estimate on blocks* if we may take $C = 1$.

Proposition 13.4.2. *Suppose a basis $(x_n)_{n=1}^{\infty}$ of a Banach space X satisfies a lower 2-estimate on blocks. Then,*

(i) The formula

$$|||x||| = \max \left\{ \|x\|, \sup \left(\sum_{j=1}^{n} \| \sum_{k \in I_j} x_k^*(x) x_k \|^2 \right)^{1/2} \right\}, \qquad x \in X$$

defines an equivalent norm on X so that we have an exact lower 2-estimate on blocks.

(ii) $(x_n)_{n=1}^{\infty}$ is boundedly complete.

Thus, $X = [x_n]_{n=1}^{\infty}$ is isomorphic to the dual of the space $Y = [x_n^*]_{n=1}^{\infty}$.

Proof. We leave the verification of (i) to the reader. To show (ii), suppose

$$\sup_n \left\| \sum_{k=1}^{n} a_k x_k \right\| < \infty$$

but the series $\sum_{k=1}^{\infty} a_k x_k$ does not converge. Then we may find disjoint intervals $I_1 < I_2 < \ldots$ so that

$$\left\| \sum_{k \in I_j} a_k x_k \right\| \geq \delta > 0, \qquad j = 1, 2, \ldots.$$

But then, if $I_1, \ldots, I_n \subset \{1, 2, \ldots, N\}$,

$$n^{\frac{1}{2}} \delta \leq C \left\| \sum_{k=1}^{N} a_k x_k \right\|,$$

and we get a contradiction.

□

Remark 13.4.3. In the particular case that $(x_n)_{n=1}^{\infty}$ satisfies an exact lower 2-estimate on blocks in Proposition 13.4.2, then the basis $(x_n)_{n=1}^{\infty}$ is monotone, and hence X is isometrically identified with Y^*.

In order to provide the aforementioned example we need to modify our construction of $\mathcal{X}$. Returning to our conditions on the branch norms $\| \cdot \|_B$ in Section 13.2, we shall impose one further condition in addition to (13.5) and (13.6). We shall assume that for any disjoint segments $S_1, \ldots, S_n$,

$$\sum_{j=1}^{n} \|S_j \xi\|_B^2 \leq \|\xi\|_B^2, \qquad \xi \in c_{00}(B). \tag{13.8}$$

Thus we are assuming that for every branch B, the basis $(e_n)_{n \in B}$ of X_B satisfies an exact lower 2-estimate on blocks (for the obvious ordering). This, in turn, means by Proposition 13.4.2 that each such basis is boundedly-complete and that X_B can be identified isometrically with the dual of the space $Y_B = [e_n^*]_{n \in B}$.

Notice that for any segment S, by (13.6) all the branch norms $\| \cdot \|_B$ for which $S \subset B$ agree on if $c_{00}(S)$. Thus if $\xi \in c_{00}$, the value of $\|S\xi\|$ is well-defined for any segment S. We put

$$\|\xi\|_{\mathcal{X}} = \sup \left\{ \left(\sum_{j=1}^{n} \|S_j \xi\|^2 \right)^{1/2} : S_1, \ldots, S_n \text{ disjoint segments} \right\},$$

and let $\mathcal{X}$ be the completion of c_{00} with this norm.

We shall say that two subsets $E, F \subset \mathbb{N}$ are *mutually incomparable* (for the order $\preceq$) if $m \in E$ and $n \in F$ imply that neither $m \preceq n$ nor $n \preceq m$ can hold. It is easy to see that the union of a family of mutually incomparable convex sets is again convex.

Proposition 13.4.4. *The norm $\| \cdot \|_{\mathcal{X}}$ has the following properties:*

(i) For any $B \in \mathcal{B}$,
$$\|\xi\|_B = \|\xi\|_{\mathcal{X}}, \qquad \xi \in c_{00}(B).$$

(ii) If $E_1, \ldots, E_n$ are disjoint and convex,

$$\sum_{j=1}^{n} \|E_j \xi\|_{\mathcal{X}}^2 \leq \|\xi\|_{\mathcal{X}}^2, \qquad \xi \in c_{00}.$$

(iii) The basis $(e_n)_{n=1}^{\infty}$ of $\mathcal{X}$ satisfies an exact lower 2-estimate on blocks.
(iv) If $E_1, \ldots, E_n$ are convex and mutually incomparable then

$$\sum_{j=1}^{n} \|E_j \xi\|_{\mathcal{X}}^2 = \left\| \sum_{j=1}^{n} E_j \xi \right\|_{\mathcal{X}}^2 \leq \|\xi\|_{\mathcal{X}}^2, \qquad \xi \in c_{00}.$$

Proof. (i) follows directly from (13.8).

(ii) Given $\epsilon > 0$, pick disjoint segments $(S_{jk})_{k=1}^{m_n}$ for $j = 1, 2, \ldots, n$ so that

$$\sum_{j=1}^{n} \sum_{k=1}^{m_n} \|S_{jk} E_j \xi\|^2 \geq \sum_{j=1}^{n} \|E_j \xi\|_{\mathcal{X}}^2 - \epsilon.$$

Let $S_{jk}' = E_j \cap S_{jk}$. Then the family of segments $(S_{jk}')_{j=1,k=1}^{n,m_n}$ is disjoint, so

$$\sum_{j=1}^{n} \sum_{k=1}^{m_n} \|S_{jk}' \xi\|^2 \leq \|\xi\|_{\mathcal{X}}^2.$$

Hence

$$\sum_{j=1}^{n} \|E_j \xi\|_{\mathcal{X}}^2 - \epsilon \le \|\xi\|_{\mathcal{X}}^2.$$

As $\epsilon > 0$ is arbitrary, we are done.

(iii) Intervals are convex.

(iv) In this case, for $\epsilon > 0$ pick disjoint segments $S_1, \ldots, S_m$ so that

$$\sum_{k=1}^{m} \left\| S_k \sum_{j=1}^{n} E_j \xi \right\|^2 \ge \left\| \sum_{j=1}^{n} E_j \xi \right\|_{\mathcal{X}}^2 - \epsilon.$$

Let $S'_{jk} = E_j \cap S_k$. The assumption that the E_j's are mutually incomparable implies that, for each k, S'_{jk} is nonempty for at most one j. Thus,

$$\sum_{k=1}^{m} \left\| S_k \sum_{j=1}^{n} E_j \xi \right\|^2 = \sum_{k=1}^{m} \sum_{j=1}^{n} \|S'_{jk} \xi\|^2 \le \sum_{j=1}^{n} \|E_j \xi\|_{\mathcal{X}}^2.$$

Hence,

$$\left\| \sum_{j=1}^{n} E_j \xi \right\|_{\mathcal{X}}^2 - \epsilon \le \sum_{j=1}^{n} \|E_j \xi\|_{\mathcal{X}}^2.$$

Since $\epsilon > 0$, this establishes an inequality

$$\left\| \sum_{j=1}^{n} E_j \xi \right\|_{\mathcal{X}}^2 \le \sum_{j=1}^{n} \|E_j \xi\|_{\mathcal{X}}^2.$$

The reverse inequality follows from (ii).

Finally, since $E_1, \ldots, E_n$ are incomparable, the union $\cup_{j=1}^{m} E_j$ is also convex and, by (ii),

$$\left\| \sum_{j=1}^{n} E_j \xi \right\|_{\mathcal{X}} \le \|\xi\|_{\mathcal{X}}.$$

$\square$

Remark 13.4.5. By (iii) of Proposition 13.4.4, we see that the basis $(e_n)_{n=1}^{\infty}$ of $\mathcal{X}$ is boundedly-complete and that $\mathcal{X}$ can be isometrically identified with the dual of $\mathcal{Y} = [e_n^*]_{n=1}^{\infty} \subset \mathcal{X}^*$.

For $n \in \mathbb{N}$ let $T_n = \{m : n \preceq m\}$ and $T_n^+ = \{m : n \prec m\}$.

Lemma 13.4.6. *Suppose $\xi \in c_{00}$ is supported on $[1, N]$ and $\eta \in c_{00}$ is supported on $[N+1, \infty)$. Then*

$$\|\xi + \eta\|_{\mathcal{X}} \le (\|\xi\|_{\mathcal{X}}^2 + \|\eta\|_{\mathcal{X}}^2)^{\frac{1}{2}} + N^{\frac{1}{2}} \sup_{m \ge N+1} \|T_m \eta\|_{\mathcal{X}}.$$

Proof. Let $\delta = \sup_{m \geq N+1} \|T_m \eta\|_{\mathcal{X}}$. Suppose $\epsilon > 0$ and pick disjoint segments $(S_j)_{j=1}^m$ so that

$$\|\xi + \eta\|_{\mathcal{X}}^2 < \sum_{j=1}^m \|S_j(\xi + \eta)\|^2 + \epsilon.$$

We may assume the segments $(S_j)_{j=1}^m$ are such that $S_j \subset [1, N]$ for $1 \leq j < k$, $S_j \subset [N+1, \infty)$ for $l < j \leq m$, and that S_j meets both $[1, N]$ and $[N+1, \infty)$ for $k \leq j \leq l$ where $0 \leq k \leq l+1 \leq m+1$ (taking account of the possibilities that each collection might be empty!).

Then

$$\sum_{l < j \leq m} \|S_j(\xi + \eta)\|^2 \leq \|\eta\|_{\mathcal{X}}^2.$$

But, if $k \leq j \leq l$,

$$\|S_j(\xi + \eta)\| \leq \|S_j \xi\| + \|S_j \eta\| \leq \|S_j \xi\| + \delta.$$

Thus,

$$\left(\sum_{1 \leq j \leq l} \|S_j(\xi + \eta)\|^2 \right)^{1/2} \leq \left(\sum_{1 \leq j \leq l} \|S_j(\xi)\|^2 \right)^{1/2} + (l - k + 1)^{\frac{1}{2}} \delta$$

$$\leq \|\xi\|_{\mathcal{X}} + N^{\frac{1}{2}} \delta,$$

since $l - k + 1 \leq N$ as the sets S_j are disjoint. Hence,

$$\left(\sum_{j=1}^m \|S_j(\xi + \eta)\|^2 \right)^{1/2} \leq (\|\xi\|_{\mathcal{X}} + \|\eta\|_{\mathcal{X}}^2)^{\frac{1}{2}} + N^{\frac{1}{2}} \delta,$$

and this completes the proof.

$\square$

We now come to the main point of the construction. Let us recall that whenever $(X_i)_{i \in \mathcal{I}}$ is an uncountable family of Banach spaces, $\ell_\infty(X_i)_{i \in \mathcal{I}}$ is the Banach space of all $(x_i)_{i \in \mathcal{I}} \in \prod_{i \in \mathcal{I}} X_i$ such that $(\|x_i\|)_{i \in \mathcal{I}}$ is bounded, with the norm

$$\|(x_i)_{i \in \mathcal{I}}\|_\infty = \sup_{i \in I} \|x_i\|_{X_i}.$$

Similarly $\ell_2(X_i)_{i \in \mathcal{I}}$ is the Banach space of all $(x_i)_{i \in \mathcal{I}} \in \prod_{i \in \mathcal{I}} X_i$ such that $(\|x_i\|_{X_i})_{i \in \mathcal{I}} \in \ell_2(\mathcal{I})$ with the norm

$$\|(x_i)_{i \in \mathcal{I}}\|_2 = \left(\sum_{i \in \mathcal{I}} \|x_i\|_{X_i}^2 \right)^{1/2}.$$

Proposition 13.4.7. *Then* $\mathcal{Y}^{**}/\mathcal{Y}$ *is isometrically isomorphic to the space* $\ell_2(Y_B^{**}/Y_B)_{B \in \mathcal{B}}$.

Proof. Let us write $J_n = [n, \infty)$. Then if $\xi^* \in \mathcal{Y}^{**} = \mathcal{X}^*$ we have $\xi^* \in \mathcal{Y}$ if and only if $\lim_{n\to\infty} \|J_n^*\xi^*\| = 0$. Here we interpret J_n as an operator on $\mathcal{X}$.

We will repeatedly use the following fact: If $(A_n)_{n=1}^\infty$ is a sequence of mutually incomparable convex sets then $A = \cup_{n=1}^\infty$ is also convex and $A\mathcal{X} = \ell_2(A_n\mathcal{X})$; this follows directly from Proposition 13.4.4 (*iii*). Hence if $\xi^* \in \mathcal{X}^*$ we have

$$\|A^*\xi^*\| = \Big(\sum_{n=1}^\infty \|A_n\xi^*\|^2 \Big)^{1/2}.$$

Define a linear operator $V : \mathcal{X}^* \to \ell_\infty(X_B^*)_{B\in\mathcal{B}}$ naturally by setting $V\xi^* = (\xi^*|_{X_B})_{B\in\mathcal{B}}$. V is clearly a norm-one operator and $V(\mathcal{Y}) \subset \ell_\infty(Y_B)_{B\in\mathcal{B}}$.

The first step is to show that $V^{-1}(\ell_\infty(Y_B)_{B\in\mathcal{B}}) = \mathcal{Y}$. Suppose $\xi^* \in \mathcal{X}^*$ and $\xi^*|_{X_B} \in Y_B$ for every $B \in \mathcal{B}$. This means that

$$\lim_{n\to\infty} \|(J_n \cap B)^*\xi^*\| = 0$$

for every branch B.

Fix a branch B. For each $n \in B$ let $T_n' = T_n^+ \setminus T_{n'}$ where n' is the successor of n in the branch. Then the sequence $(T_n')_{n\in B}$ consists of mutually incomparable tree-convex sets. Hence

$$\|(\cup_{n\preceq m}T_m')^*\xi^*\| = \Big(\sum_{n\preceq m} \|(T_m')^*\xi^*\|^2 \Big)^{\frac{1}{2}}, \qquad n \in B,$$

and so

$$\lim_{\substack{n\to\infty \\ n\in B}} \|(\cup_{n\preceq m}T_m')^*\xi^*\| = 0.$$

Since $\cup_{n\preceq m}T_m' \cup (J_n \cap B) = T_n$, by the triangle law we have

$$\lim_{\substack{n\to\infty \\ n\in B}} \|T_n^*\xi^*\| = 0$$

for every branch B.

We next want to conclude that

$$\lim_{n\to\infty} \|T_n^*\xi^*\| = 0. \qquad (13.9)$$

Indeed, if there exists $\epsilon > 0$ and infinitely many n so that $\|T_n^*\xi^*\| \geq \epsilon$, then by the preceding reasoning we cannot find infinitely many belonging to one branch. Hence we can pass to an infinite subset A so that if $m, n \in A$ with $m < n$ then it is not true that $m \preceq n$. Then the sets $\{T_n\}_{n\in A}$ are mutually incomparable. Hence

$$\sum_{n\in A} \|T_n^*\xi^*\|^2 < \infty$$

and this gives a contradiction. Thus (13.9) holds.

Assuming (13.9), let $\delta_n = \sup_{m\geq n} \|T_m^*\xi^*\|$. Let us fix m and $\epsilon > 0$. Then we may find $\xi \in c_{00}$ with $\|\xi\|_\mathcal{X} = 1$ and $\langle \xi, J_m^*\xi^* \rangle > (1 - \epsilon)\|J_m^*\xi^*\|$. Choose r

such that $\xi(j) = 0$ for $j \geq r$. If $n \geq r$, let A be the set of $k \geq n$ such that the predecessor of k is less than or equal to n. There are at most n of such k. Then the sets $(T_k)_{k \in A}$ are mutually incomparable and convex and $\cup_{k \in A} T_k = J_n$. For $0 < \epsilon < \frac{1}{2}$ identifying $J_n \mathcal{X}$ with the ℓ_2-sum of the space $T_k \mathcal{X}$ for $k \in A$ we can find $\eta \in J_n \mathcal{X} \cap c_{00}$ with $\|\eta\|_{\mathcal{X}} = 1$ and

$$\langle \eta, J_n^* \xi^* \rangle > (1 - \epsilon) \| J_n^* \xi^* \|$$

in such a way that

$$\| T_k \eta \| \leq 2 \| T_n^* \xi^* \| \| J_n^* \xi^* \|^{-1}, \qquad k \in A.$$

Hence,

$$\sup_{k \in A} \| T_k \eta \| \leq 2 \delta_n \| J_n^* \xi^* \|^{-1}.$$

Therefore,

$$
\begin{aligned}
(1 - \epsilon) \left(\| J_m^* \xi^* \| + \| J_n^* \xi^* \| \right) &\leq \langle \xi + \eta, J_m^* \xi^* \rangle \\
&\leq \| J_m^* \xi^* \| \| \xi + \eta \|_{\mathcal{X}} \\
&\leq \| J_m^* \xi^* \| (2^{\frac{1}{2}} + r^{\frac{1}{2}} \sup_{l \geq r} \| T_l \eta \|) \\
&\leq \| J_m^* \xi^* \| (2^{\frac{1}{2}} + r^{\frac{1}{2}} \sup_{l \geq n} \| T_l \eta \|) \\
&\leq \| J_m^* \xi^* \| (2^{\frac{1}{2}} + 2 r^{\frac{1}{2}} \delta_n \| J_n^* \xi^* \|^{-1}).
\end{aligned}
$$

Assume $\lim_{n \to \infty} \| J_n^* \xi^* \| > 0$. Then, letting $n \to \infty$, and then $\epsilon \to 0$,

$$\| J_m^* \xi^* \| + \lim_{n \to \infty} \| J_n^* \xi^* \| \leq \sqrt{2} \| J_m^* \xi^* \|,$$

and so

$$\lim_{n \to \infty} \| J_n^* \xi^* \| \leq (\sqrt{2} - 1) \| J_m^* \xi^* \|, \qquad m \in \mathbb{N}.$$

Letting $m \to \infty$ shows that $\lim_{n \to \infty} \| J_n^* \xi^* \| = 0$ giving a contradiction. This concludes the proof of the first step, i.e., $V^{-1}(\ell_\infty(Y_B)_{B \in \mathcal{B}}) = \mathcal{Y}$.

This yields a naturally induced one-to-one map,

$$\tilde{V} : \mathcal{Y}^{**} / \mathcal{Y} \to \ell_\infty(Y_B^{**}/Y)_{B \in \mathcal{B}}.$$

Let us show $\tilde{V}$ maps into $\ell_2(Y_B^{**}/Y)_{B \in \mathcal{B}}$. Let Q be the quotient map of $\mathcal{Y}^{**}$ onto $\mathcal{Y}^{**}/\mathcal{Y}$ and Q_B be the corresponding quotient map of Y_B^{**} onto Y_B^{**}/Y. If $B_1, \ldots, B_n$ are distinct complete branches and $\xi^* \in \mathcal{X}^*$ then we may pick m large enough so that the branches $B_j \cap J_m$ are disjoint. Since they are tree-convex we have

$$\left\| \sum_{j=1}^{n} (B_j \cap J_m)^* \xi^* \right\|^2 = \sum_{j=1}^{m} \| (B_j \cap J_m)^* \xi^* \|^2 \leq \| \xi^* \|^2,$$

which yields

$$\sum_{j=1}^{n} \|Q_{B_j}\xi^*|_{X_{B_j}}\|^2 \leq \|\xi^*\|^2.$$

It follows that $\|\tilde{V}\| \leq 1$ as an operator from $\mathcal{Y}^{**}/\mathcal{Y}$ into $\ell_2(Y_B^{**}/Y)_{B\in\mathcal{B}}$.

Finally we check that $\tilde{V}$ is an onto isometry. Suppose we have a finitely supported element $u = (u_B)_{B\in\mathcal{B}}$ in $\ell_2(Y_B^{**}/Y)_{B\in\mathcal{B}}$. For $\epsilon > 0$ pick $\xi_B^* \in Y_B^* = B^*(\mathcal{X}^*)$ with $\|\xi_B^*\| \leq (1+\epsilon)\|\xi_B^*\|$ and $Q_B\xi_B^* = u_B$. Pick m large enough so that the branches $\{B\cap J_m : u_B \neq 0\}$ are disjoint. Then let $\xi^* = \sum_{u_B \neq 0} J_m^*\xi_B^*$; we have

$$\|\xi^*\| = (\sum_{u_B \neq 0} \|J_m^*\xi_B^*\|^2)^{\frac{1}{2}} \leq (1+\epsilon)(\sum_{u_B \neq 0} \|u_B\|^2)^{\frac{1}{2}} = (1+\epsilon)\|u\|.$$

Since $\tilde{V}Q\xi^* = u$, $\tilde{V}$ is an onto isometry.

□

In the following theorem we re-create an example due to James [87]. The space $\mathcal{X} = \mathcal{Y}^*$ is usually called the *James tree space* and it is denoted $\mathcal{JT}$. James showed that ℓ_1 does not embed into $\mathcal{JT}$ but that $\mathcal{JT}^*$ is not separable. Other examples were independently constructed by Lindenstrauss and Stegall [134]. The next theorem is, in fact, due to Lindenstrauss and Stegall [134]. A full account of James-type constructions can be found in [58].

Theorem 13.4.8. *There is a Banach space $\mathcal{Y}$ such that $\mathcal{Y}^*$ is separable and $\mathcal{Y}^{**}/\mathcal{Y}$ is isometric to $\ell_2(\mathcal{I})$ where $\mathcal{I}$ has the cardinality of the continuum.*

Proof. We use the space $\mathcal{J}$ but with the basis of Problem 3.10 which is a special case of the construction of Theorem 13.1.6. It is trivial to see that the basis $(e_n)_{n=1}^{\infty}$ of the space $\mathcal{X}$ constructed in Section 13.1 has an exact lower 2-estimate on blocks. To avoid confusion let us denote this norm now by $|||\cdot|||$.

Again we identify $(\mathbb{N}, \preceq)$ with $\mathcal{FN}$. Let B be the branch generated by the increasing sequence $(m_j)_{j=1}^{\infty}$, i.e., consisting of the sets $A_j = \{m_1, \ldots, m_j\}$. We define the branch norms on c_{00} by

$$\left\|\sum_{j=1}^{n} a_j e_{\psi(A_j)}\right\|_B = \left|\left|\left|\sum_{j=1}^{n} a_j e_j^*\right|\right|\right|.$$

Letting our construction run its course we see that each Y_B^{**}/Y is isometric to $\mathbb{R}$. The result is then immediate.

□

Theorem 13.4.9. *The space $\mathcal{Y}^* = \mathcal{JT}$ has nonseparable dual but ℓ_1 does not embed into $\mathcal{JT}$.*

Proof. Obviously, $\mathcal{JT}^*$ is nonseparable. Since $\mathcal{JT}$ is a dual space, it is complemented in its bidual, and so $\mathcal{JT}^{**} = \mathcal{JT} \oplus W$ where W can be identified as the dual of the space $\mathcal{JT}^*/\mathcal{JT}_*$, and $\mathcal{JT}_*$ is the predual $\mathcal{Y}$ given by the construction. Hence, using Theorem 13.4.8, we conclude that $W = \ell_2(\mathcal{I})$ for an uncountable set $(\mathcal{I})$.

If ℓ_1 embeds in $\mathcal{JT}$, then $\ell_1^{**} = \ell_\infty^*$ embeds in $\mathcal{JT}^{**}$. But $\ell_\infty = \mathcal{C}(K)$ for some uncountable compact Hausdorff space K, and hence using point masses, the space $\ell_1(\Gamma)$ embeds into $\mathcal{JT}^{**}$ for some uncountable set Γ. Let $T : \ell_1(\Gamma) \rightarrow \mathcal{JT} \oplus W$ be an embedding and assume it has the form $T = T_1 \oplus T_2$ where $T_1 : \ell_1(\Gamma) \rightarrow \mathcal{JT}$ and $T_2 : \ell_1(\Gamma) \rightarrow W$. Using the separability of $\mathcal{JT}$ we may find a sequence of basis vectors $(e_{\gamma_n})_{n=1}^\infty$ so that $(T_1 e_{\gamma_n})_{n=1}^\infty$ converges. Hence $\lim_{n\to\infty} \|T_1(e_{\gamma_{2n}} - e_{\gamma_{2n+1}})\| = 0$, so replacing the original sequence by a subsequence we can assume that $(T_2(e_{\gamma_{2n}} - e_{\gamma_{2n+1}}))_{n=1}^\infty$ is a basic sequence equivalent to the canonical basis of ℓ_1; this is absurd since W is a Hilbert space.

□

In his 1974 paper [87], James showed that every infinite-dimensional subspace of $\mathcal{JT}$ contains a subspace isomorphic to a Hilbert space and thus deduced Theorem 13.4.9.

Going back to Theorem 13.4.8 and using Theorem 13.1.6 it is clear we can also prove:

Theorem 13.4.10. *Let X be any separable dual space. Then there is a Banach space Z such that Z^{**}/Z is isomorphic to $\ell_2(X)_{i\in\mathcal{I}}$ where $\mathcal{I}$ has the cardinality of the continuum.*

Proof. Let $X = Y^*$ and construct $\mathcal{Z}$ as in Section 13.1 so that $\mathcal{Z}^{**}/\mathcal{Z} \approx Y$. Using the canonical basis of $\mathcal{Z}$ as in Theorem 13.4.8 will give us a space Z so that Z^{**}/Z is isomorphic to $\ell_2(Y^*)_{i\in\mathcal{I}}$.

□

A

Fundamental Notions

A *normed space* $(X, \| \cdot \|)$ is a linear space X endowed with a nonnegative function $\| \cdot \| : X \to \mathbb{R}$ called *norm* satisfying

(i) $\|x\| = 0$ if and only if $x = 0$;

(ii) $\|\alpha x\| = |\alpha| \|x\|$ $(\alpha \in \mathbb{R}, x \in X)$;

(iii) $\|x_1 + x_2\| \leq \|x_1\| + \|x_2\|$ $(x_1, x_2 \in X)$.

A *Banach space* is a normed linear space $(X, \|\cdot\|)$ that is complete in the metric defined by $\rho(x, y) = \|x - y\|$. B_X will denote the *closed unit ball of X*, that is, $\{x \in X : \|x\| \leq 1\}$. Similarly, the *open unit ball of X* is $\{x \in X : \|x\| < 1\}$ and $S_X = \{x \in X : \|x\| = 1\}$ is the *unit sphere of X*.

A.1. Completeness Criterion. *A normed space $(X, \| \cdot \|)$ is complete if and only if the (formal) series $\sum_{n=1}^{\infty} x_n$ in X converges in norm whenever $\sum_{n=1}^{\infty} \|x_n\|$ converges.*

A linear subspace Y of a Banach space $(X, \| \cdot \|)$ is closed in X if and only if $(Y, \| \cdot \|_Y)$ is a Banach space, where $\| \cdot \|_Y$ denotes the restriction of $\| \cdot \|$ to Y. If Y is a subspace of X, so is its closure $\overline{Y}$.

Two norms $\| \cdot \|$ and $\|x\|_0$ on a linear space X are *equivalent* if there exist positive numbers c, C such that for all $x \in X$ we have

$$c\|x\|_0 \leq \|x\| \leq C\|x\|_0. \tag{A.1}$$

An *operator* between two Banach spaces X, Y is a norm-to-norm continuous linear map. The following conditions are equivalent ways to characterize the continuity of a mapping $T : X \to Y$ with respect to the norm topologies of X and Y:

(i) T is bounded, meaning $T(B)$ is a bounded subset of Y whenever B is a bounded subset of X.

(ii) T is continuous at 0.

(iii) There is a constant $C > 0$ such that $\|Tx\| \leq C\|x\|$ for every $x \in X$.

(iv) T is uniformly continuous on X.

(v) The quantity $\|T\| = \sup\{\|Tx\| : \|x\| \le 1\}$ is finite.

The linear space of all continuous operators from a normed space X into a Banach space Y with the usual *operator norm*:

$$\|T\| = \sup\{\|Tx\| : \|x\| \le 1\}$$

is a Banach space that will be denoted by $\mathcal{L}(X, Y)$. When $X = Y$ we will put $\mathcal{L}(X) = \mathcal{L}(X, X)$.

The set of all *functionals* on a normed space X (that is, the continuous linear maps from X into the scalars) is a Banach space, denoted by X^* and called the *dual space of X*. The norm of a functional $x^* \in X^*$ is given by

$$\|x^*\| = \sup\{|x^*(x)| : x \in B_X\}.$$

Let $T : X \to Y$ be an operator. T is called *invertible* if there exists an operator $S : Y \to X$ so that TS is the identity operator on Y and ST is the identity operator on X. When this happens S is said to be the *inverse* of T and is denoted by T^{-1}.

A.2. Existence of inverse operator. *Let X be a Banach space. Suppose that $T \in \mathcal{L}(X)$ is such that $\|I_X - T\| < 1$ (I_X denotes the identity operator on X). Then T is invertible and its inverse is given by the Neumann series*

$$T^{-1}(x) = \lim_{n\to\infty} \left(I_X + (I_X - T) + (I_X - T)^2 + \cdots + (I_X - T)^n \right)(x), \quad x \in X.$$

An operator T between two normed spaces X, Y is an *isomorphism* if T is a continuous bijection whose inverse T^{-1} is also continuous. That is, an isomorphism between normed spaces is a linear homeomorphism. Equivalently, $T : X \to Y$ is an isomorphism if and only if T is onto and there exist positive constants c, C so that

$$c\|x\|_X \le \|Tx\|_Y \le C\|x\|_X$$

for all $x \in X$. In such a case the spaces X and Y are said to be *isomorphic* and we write $X \approx Y$. T is an *isometric isomorphism* when $\|Tx\|_Y = \|x\|_X$ for all $x \in X$.

An operator T is an *embedding of X into Y* if T is an isomorphism onto its image $T(X)$. In this case we say that X *embeds in Y* or that Y contains an isomorphic copy of X. If $T : X \to Y$ is an embedding such that $\|Tx\|_Y = \|x\|_X$ for all $x \in X$, T is said to be an *isometric embedding*.

A.3. Extension of operators by density. *Suppose that M is a dense linear subspace of a normed linear space X, that Y is a Banach space, and that $T : M \to Y$ is a bounded operator. Then there exists a unique continuous operator $\tilde{T} : X \to Y$ such that $\tilde{T}|_M = T$ and $\|\tilde{T}\| = \|T\|$. Moreover, if T is an isomorphism or isometric isomorphism then so is $\tilde{T}$.*

Given $T : X \to Y$, the operator $T^* : Y^* \to X^*$ defined as $T^*(y^*)(x) = y^*(T(x))$ for every $y^* \in Y^*$ and $x \in X$ is called the *adjoint of* T and has the property that $\|T^*\| = \|T\|$.

An operator $T : X \to Y$ between the Banach spaces X and Y is said to be *compact* if $T(B_X)$ is relatively compact, that is, $\overline{T(B_X)}$ is a compact set in Y. If $T : X \to Y$ is compact then it is continuous.

An operator $T : X \to Y$ has *finite rank* if the dimension of its range $T(X)$ is finite.

A.4. Schauder's Theorem. *A bounded operator T from a Banach space X into a Banach space Y is compact if and only if $T^* : Y^* \to X^*$ is compact.*

A bounded linear operator $P : X \to X$ is a *projection* if $P^2 = P$, i.e., $P(P(x)) = P(x)$ for all $x \in X$; hence $P(y) = y$ for all $y \in P(X)$. A subspace Y of X is *complemented* if there is a projection P on X with $P(X) = Y$. Thus complemented subspaces of Banach spaces are always closed.

A.5. Property. *Suppose Y is a closed subspace of a Banach space X. If Y is complemented in X then Y^* is isomorphic to a complemented subspace of X^*.*

Let us finish this section by recalling that the *codimension* of a closed subspace Y of a Banach space X is the dimension of the quotient space X/Y.

A.6. Subspaces of codimension one. *Any two closed subspaces of codimension 1 in a Banach space X are isomorphic.*

B

Elementary Hilbert Space Theory

An *inner product space* is a linear space X over the scalar field $\mathbb{K} = \mathbb{R}$ or $\mathbb{C}$ of X equipped with a function $\langle \cdot, \cdot \rangle : X \times X \to \mathbb{K}$ called an *inner product* or *scalar product* satisfying the following conditions:

(i) $\langle x, x \rangle \geq 0$ for all $x \in X$,
(ii) $\langle x, x \rangle = 0$ if and only if $x = 0$,
(iii) $\langle \alpha_1 x_1 + \alpha_2 x_2, y \rangle = \alpha_1 \langle x_1, y \rangle + \alpha_2 \langle x_2, y \rangle$ if $\alpha_1, \alpha_2 \in \mathbb{R}$ and $x_1, x_2, y \in X$,
(iv) $\langle x, y \rangle = \overline{\langle y, x \rangle}$ for all $x, y \in X$. (The bar denotes complex conjugation.)

An inner product on X gives rise to a norm on X defined by $\|x\| = \sqrt{\langle x, x \rangle}$. The axioms of a scalar product yield the **Schwarz Inequality**:

$$|\langle x, y \rangle| \leq \|x\| \|y\| \quad \text{for all } x \text{ and } y \in X,$$

as well as the **Parallelogram Law**:

$$\|x + y\|^2 + \|x - y\|^2 = 2\|x\|^2 + 2\|y\|^2, \qquad x, y \in X. \tag{B.1}$$

A *Hilbert space* is an infinite-dimensional inner product space which is complete in the metric induced by the scalar product. Hilbert spaces enjoy very nice properties to the extent of being the infinite-dimensional analogue of Euclidean spaces. It turns out that given a Banach space $(X, \| \cdot \|)$, there is an inner product $\langle \cdot, \cdot \rangle$ so that $(X, \langle \cdot, \cdot \rangle)$ is a Hilbert space with norm $\| \cdot \|$ if and only if $\| \cdot \|$ satisfies (B.1).

Two vectors x, y in a Hilbert space X are said to be *orthogonal*, and we write $x \perp y$, provided $\langle x, y \rangle = 0$. If M is a subspace of X, we say that x is orthogonal to M if and only if $\langle x, y \rangle = 0$ for all $y \in M$. The closed subspace $M^\perp = \{ x \in X : \langle x, y \rangle = 0 \text{ for all } y \in M \}$ is called the *orthogonal complement* of M.

A set S in X is said to be an *orthogonal system* when any two different elements x, y of S are *orthogonal*. The vectors in an orthogonal system are linearly independent. S is called *orthonormal* if it is orthogonal and $\|x\| = 1$ for each $x \in S$.

Assume that X is separable and let $\mathcal{C} = \{u_1, u_2, \ldots\}$ be a dense subset of X. Using the *Gram-Schmidt procedure*, from $\mathcal{C}$ we can construct an orthonormal sequence $(v_n)_{n=1}^{\infty} \subset X$ which has the added feature of being *complete* (or *total*): whenever $\langle x, v_k \rangle = 0$ for all k implies $x = 0$. A *basis* of a Hilbert space is a complete orthogonal sequence.

Let $(v_k)_{k=1}^{\infty}$ be an orthonormal (not necessarily complete) sequence in a Hilbert space X. The inner products $(\langle x, v_k \rangle)_{k=1}^{\infty}$ are the *Fourier coefficients* of x with respect to (v_k).

Suppose that $x \in X$ can be expanded as a series $x = \sum_{k=1}^{\infty} a_k v_k$ for some scalars (a_k). Then $a_k = \langle x, v_k \rangle$ for each $k \in \mathbb{N}$. In fact, for every $x \in X$, without any assumptions or knowledge about the convergence of the *Fourier series* $\sum_{k=1}^{\infty} \langle x, v_k \rangle v_k$, **Bessel's Inequality** always holds:

$$\sum_{k=1}^{\infty} |\langle x, v_k \rangle|^2 \leq \|x\|^2.$$

B.1. Parseval's Identity. *Let $(v_k)_{k=1}^{\infty}$ be an orthonormal sequence in an inner product space X. Then (v_k) is complete if and only if*

$$\sum_{k=1}^{\infty} |\langle x, v_k \rangle|^2 = \|x\|^2 \quad \text{for every } x \in X. \tag{B.2}$$

In turn, equation (B.2) is equivalent to saying that

$$x = \sum_{k=1}^{\infty} \langle x, v_k \rangle v_k$$

for each $x \in X$.

Bessel's inequality establishes that a necessary condition for a sequence of numbers $(a_k)_{k=1}^{\infty}$ to be the Fourier coefficients of an element $x \in X$ (relative to a fixed orthonormal system (v_k)) is that $\sum_{k=1}^{\infty} |a_k|^2 < \infty$. The Riesz-Fischer theorem tells us that, if (v_k) is complete, this condition is also sufficient.

B.2. The Riesz-Fischer Theorem. *Let X be a Hilbert space with complete orthonormal sequence $(v_k)_{k=1}^{\infty}$. Assume that $(a_k)_{k=1}^{\infty}$ is a sequence of real numbers such that $\sum_{k=1}^{\infty} |a_k|^2 < \infty$. Then there exists an element $x \in X$ whose Fourier coefficients relative to (v_k) are (a_k).*

Thus from the isomorphic classification point of view, ℓ_2 with the regular inner product of any two vectors $a = (a_n)_{n=1}^{\infty}$ and $b = (b_n)_{n=1}^{\infty}$:

$$\langle a, b \rangle = \sum_{n=1}^{\infty} a_n \overline{b_n}$$

is essentially the only separable Hilbert space. Indeed, combining B.1 with B.2, we obtain that the map from X onto ℓ_2 given by

$$x \mapsto (\langle x, v_k \rangle)_{k=1}^{\infty}$$

is a Hilbert space isomorphism (hence an isometry).

B.3. Representation of functionals on Hilbert spaces. *To every functional x^* on a Hilbert space X there corresponds a unique $x \in X$ such that $x^*(y) = \langle y, x \rangle$ for all $y \in X$. Moreover, $\|x^*\| = \|x\|$.*

Hilbert spaces are exceptional Banach spaces for many reasons. For instance, the Gram-Schmidt procedure and the fact that subsets of separable metric spaces are also separable yield that every subspace of a separable Hilbert space has an orthonormal basis. Another important property is that closed subspaces are always complemented, which relies upon the existence of unique minimizing vectors:

B.4. The Projection Theorem. *Let F be a nonempty, closed, convex subset of a Hilbert space X. For every $x \in X$ there exists a unique $\overline{y} \in F$ such that*

$$d(x, F) = \inf_{y \in F} \|x - y\| = \|x - \overline{y}\|.$$

In particular, every nonempty, closed, convex set in a Hilbert space contains a unique element of smallest norm.

If F is a nonempty, closed, convex subset of a Hilbert space X, for every $x \in X$ the point $\overline{y}$ given by B.4, called the *projection of x onto F*, is characterized by

$$\overline{y} \in F \quad \text{and} \quad \Re \langle x - \overline{y}, y - \overline{y} \rangle \leq 0 \quad \text{for all } y \in F.$$

The map $P_F : X \to F$ defined by $P_F(x) = \overline{y}$ is a contraction; that is:

$$\|P_F(x_1) - P_F(x_2)\| \leq \|x_1 - x_2\| \quad \text{for all } x_1, x_2 \in X,$$

therefore it is continuous.

If M is a closed subspace of X, then P_M is a linear operator from X onto M and $P_M(x)$ is the unique $y \in X$ such that $y \in M$ and $x - y \in M^{\perp}$. P_F is called the *orthogonal projection* from X onto M. Thus, if M is a closed subspace of a Hilbert space X then $X = M \oplus M^{\perp}$.

C

Main Features of Finite-Dimensional Spaces

Suppose that $\mathcal{S} = \{x_1, \ldots, x_n\}$ is a set of independent vectors in a normed space X of any dimension. Using a straightforward compactness argument it can be shown that there exists a constant $C > 0$ (depending only on $\mathcal{S}$) such that for every choice of scalars $\alpha_1, \ldots, \alpha_n$ we have

$$C\|\alpha_1 x_1 + \cdots + \alpha_n x_n\| \geq |\alpha_1| + \cdots + |\alpha_n|.$$

This is the basic ingredient to obtain both C.1 and C.2.

C.1. Operators on finite-dimensional normed spaces. *Suppose that $T : X \to Y$ is a linear operator between the normed spaces X and Y. If X has finite dimension then T is bounded. In particular any linear operator between normed spaces of the same finite dimension is an isomorphism.*

C.2. Isomorphic classification. *Any two finite-dimensional normed spaces (over the same scalar field) of the same dimension are isomorphic.*

From C.2 one easily deduces the following facts:

- **Equivalence of norms.** *If $\|\cdot\|$ and $\|\cdot\|_0$ are two norms on a finite-dimensional vector space X then they are equivalent. Consequently, if τ and τ_0 are the respective topologies induced on X by $\|\cdot\|$ and $\|\cdot\|_0$ then $\tau = \tau_0$.*

- **Completeness.** *Any finite-dimensional normed space is complete.*

- **Closedness of subspaces.** *The finite-dimensional linear subspaces of a normed space are closed.*

The **Heine-Borel Theorem** asserts that a subset of $\mathbb{R}^n$ is compact if and only if it is closed and bounded; combining this with C.2 we further deduce:

- **Compactness.** *Let X be a finite-dimensional normed space and A be a subset of X. Then A is compact if and only if A is closed and bounded.*

We know that the compact subsets of a Hausdorff topological space X are closed and bounded. A general topological space X is said to have the *Heine-Borel property* when the converse holds. The following lemma is not restricted to finite-dimensional spaces and it is a source of interesting results in functional analysis, as for instance the characterization of the normed spaces that enjoy the Heine-Borel property which we write as a corollary.

C.3. Riesz's Lemma. *Let X be a normed space and Y be a closed proper subspace of X. Then for each real number $\theta \in (0,1)$ there exists an $x_\theta \in S_X$ such that $\|y - x_\theta\| \geq \theta$ for all $y \in Y$.*

C.4. Corollary. *Let X be a normed space. X is finite-dimensional if and only if each closed bounded subset of X is compact.*

Taking into account that in metric spaces compactness and sequential compactness are equivalent we obtain:

C.5. Corollary. *Let X be a normed space. X is finite-dimensional if and only if every bounded sequence in X has a convergent subsequence.*

D

Cornerstone Theorems of Functional Analysis

D.1 The Hahn-Banach Theorem

D.1. The Hahn-Banach Theorem (Real Case). *Let X be a real linear space, $Y \subset X$ a linear subspace, and $p : X \to \mathbb{R}$ a sublinear functional, i.e.,*

(i) $p(x + y) \leq p(x) + p(y)$ for all $x, y \in X$ (p is subadditive), and
(ii) $p(\lambda x) \leq \lambda p(x)$ for all $x \in X$ and $\lambda \geq 0$ (p is nonnegatively sub-homogeneous).

Assume that we have a linear map $f : Y \to \mathbb{R}$ such that $f(y) \leq p(y)$ for all $y \in Y$. Then there exists a linear map $F : X \to \mathbb{R}$ such that $F|_Y = f$ and $F(x) \leq p(x)$ for all $x \in X$.

D.2. Normed-space version of the Hahn-Banach Theorem. *Let y^* be a bounded linear functional on a subspace Y of a normed space X. Then there is $x^* \in X^*$ such that $\|x^*\| = \|y^*\|$ and $x^*|_Y = y^*$.*

Let us note that this theorem says nothing about the *uniqueness* of the extension unless Y is a dense subspace of X. Note also that Y need not be closed.

D.3. Separation of points from closed subspaces. *Let Y be a closed subspace of a normed space X. Suppose that $x \in X \setminus Y$. Then there exists $x^* \in X^*$ such that $\|x^*\| = 1$, $x^*(x) = d(x, Y) = \inf\{\|x - y\| : y \in Y\}$, and $x^*(y) = 0$ for all $y \in Y$.*

D.4. Corollary. *Let X be a normed linear space and $x \in X$, $x \neq 0$. Then there exists $x^* \in X^*$ such that $\|x^*\| = 1$ and $x^*(x) = \|x\|$.*

D.5. Separation of points. *Let X be a normed linear space and $x, y \in X$, $x \neq y$. Then there exists $x^* \in X^*$ such that $x^*(x) \neq x^*(y)$.*

D.6. Corollary. *Let X be a normed linear space. For every $x \in X$ we have*

$$\|x\| = \sup \Big\{ |x^*(x)| : x^* \in X^*, \|x^*\| \leq 1 \Big\}.$$

D.7. Corollary. *Let X be a normed linear space. If X^* is separable then so is X.*

D.2 Baire's Theorem and its consequences

A subset E of a metric space X is *nowhere dense* in X (or *rare*) if its closure $\overline{E}$ has empty interior. Equivalently, X is nowhere dense in X if and only if $X \setminus \overline{E}$ is (everywhere) dense in X. The sets of the *first category in X* (or, also, *meager in X*) are those that are the union of countably many sets each of which is nowhere dense in X. Any subset of X that is not of the first category is said to be of the *second category in X* (or *nonmeager in X*). This density-based approach to give a topological meaning to the size of a set is due to Baire. Nowhere dense sets would be the "very small" sets in the sense of Baire whereas the sets of the second category would play the role of the "large" sets in the sense of Baire in a metric (or more generally in any topological) space.

D.8. Baire's Category Theorem. *Let X be a* complete *metric space. Then the intersection of every countable collection of dense open subsets of X is dense in X.*

Let $\{E_i\}$ be a countable collection of nowhere dense subsets of a complete metric space X. For each i the set $U_i = X \setminus \overline{E_i}$ is dense in X, hence by Baire's theorem it follows that $\cap U_i \neq \emptyset$. Taking complements we deduce that $X \neq \cup E_i$. That is, *a complete metric space X cannot be written as a countable union of nowhere dense sets in X*. Therefore nonempty, complete metric spaces are of the second category in themselves.

A function f from a topological space X into a topological space Y is *open* if $f(V)$ is an open set in Y whenever V is open in X.

D.9. Open Mapping Theorem. *Let X and Y be Banach spaces and let $T : X \to Y$ be a bounded linear operator.*

(i) *If $\delta B_Y = \{ y \in Y : \|y\| < \delta \} \subseteq \overline{T(B_X)}$ for some $\delta > 0$ then T is an open map.*

(ii) *If T is onto then the hypothesis of (i) holds. That is, every bounded operator from a Banach space onto a Banach space is open.*

D.10. Corollary. *If X and Y are Banach spaces and T is a continuous linear operator from X onto Y which is also one-to-one then $T^{-1} : Y \to X$ is a continuous linear operator.*

D.11. Closed Graph Theorem. *Let X and Y be Banach spaces. Suppose that $T : X \to Y$ is a linear mapping of X into Y with the following property: whenever $(x_n) \subset X$ is such that both $x = \lim x_n$ and $y = \lim T x_n$ exist, it follows that $y = Tx$. Then T is continuous.*

D.12. Uniform Boundedness Principle. *Suppose $(T_\gamma)_{\gamma \in \Gamma}$ is a family of bounded linear operators from a Banach space X into a normed linear space Y. If $\sup\{\|T_\gamma x\| : \gamma \in \Gamma\}$ is finite for each x in X then $\sup\{\|T_\gamma\| : \gamma \in \Gamma\}$ is finite.*

D.13. Banach-Steinhaus Theorem. *Let (T_n) be a sequence of continuous linear operators from a Banach space X into a normed linear space Y such that*

$$Tx = \lim_n T_n x$$

exists for each x in X. Then T is continuous.

D.14. Partial Converse of the Banach-Steinhaus Theorem. *Let (S_n) be a sequence of operators from a Banach space X into a normed linear space Y such that $\sup_n \|S_n\| < \infty$. Then, if $T : X \to Y$ is another operator, the subspace*

$$\{x \in X : \|S_n x - Tx\| \to 0\}$$

is norm-closed in X.

E

Convex Sets and Extreme Points

Let S be a subset of a vector space X. S is *convex* if $\lambda x + (1 - \lambda)y \in S$ whenever $x, y \in S$ and $0 \leq \lambda \leq 1$. Notice that every subspace of X is convex and if a subset S is convex so is each of its translates $x + S = \{x + y : y \in S\}$. If X is a normed space and S is convex then so is its norm-closure $\overline{S}$.

Given a real linear space X, let F and K be two subsets of X. A linear functional f on X is said to *separate F and K* if there exists a number α such that $f(x) > \alpha$ for all $x \in F$ and $f(x) < \alpha$ for all $x \in K$. As an application of the Hahn-Banach theorem we have:

E.1. Separation of convex sets. *Let X be a locally convex space and K, F be disjoint closed convex subsets of X. Assume that K is compact. Then there exists a continuous linear functional f on X that separates F and K.*

The *convex hull* of a subset S of a linear space X, denoted co(S), is the smallest convex set that contains S. Obviously, such a set always exists since X is convex and the arbitrary intersection of convex sets is convex, and can be described analytically by

$$\mathrm{co}(S) = \Big\{ \sum_{i=1}^{n} \lambda_i x_i \ : \ (x_i)_{i=1}^{n} \subset S, \ \lambda_i \geq 0 \text{ and } \sum_{i=1}^{n} \lambda_i = 1; n \in \mathbb{N} \Big\}.$$

If X is equipped with a topology τ, $\overline{\mathrm{co}}^\tau(S)$ will denote the *closed convex hull of S*, i.e., the smallest τ-closed, convex set which contains S (that is, the intersection of all τ-closed convex sets that include S). The closed convex hull of S with respect to the norm topology will be simply denoted by $\overline{\mathrm{co}}(S)$. Let us observe that, in general, $\overline{\mathrm{co}}^\tau(S) \neq \overline{\mathrm{co}(S)}^\tau$ but that the equality holds if τ is a vector topology on X.

If S is convex, a point $x \in S$ is an *extreme point of S* if whenever $x = \lambda x_1 + (1 - \lambda)x_2$ with $0 < \lambda < 1$, then $x = x_1 = x_2$. Equivalently, x is an extreme point of S if and only if $S \setminus \{x\}$ is still convex. $\partial_e(S)$ will denote the set of extreme points of S.

E.2. The Krein-Milman Theorem. *Suppose X is a locally convex topological vector space. If K is a compact convex set in X then K is the closed convex hull of its extreme points. In particular, each convex compact subset of a locally convex topological vector space has an extreme point.*

E.3. Milman's Theorem. *Suppose X is a locally convex TVS. Let K be closed and compact[1]. If u is an extreme point of $\overline{co}(K)$ then $u \in K$.*

E.4. Schauder's Fixed Point Theorem. *Let K be a closed convex subset of a Banach space X. Suppose $T : X \to X$ is a continuous linear operator such that $T(K) \subset K$ and $T(K)$ is compact. Then there exists at least one point x in K such that $Tx = x$.*

---

[1] Notice that we are not assuming that X has any topological separation properties. If X is Hausdorff then every compact subset of X is automatically closed.

F

The Weak Topologies

Let X be a normed vector space. The *weak topology of X*, usually denoted w-topology or $\sigma(X, X^*)$-topology, is the weakest topology on X such that each $x^* \in X^*$ is continuous. This topology is linear (addition of vectors and multiplication of vectors by scalars are continuous) and a base of neighborhoods of $0 \in X$ is given by the sets of the form

$$V_\epsilon(0; x_1^*, \ldots, x_n^*) = \left\{ x \in X : |x_i^*(x)| < \epsilon, \ i = 1, \ldots, n \right\},$$

where $\epsilon > 0$ and $\{x_1^*, \ldots, x_n^*\}$ is any finite subset of X^*. Obviously this defines a non-locally bounded, locally convex topology on X. One can also give an alternative description of the weak topology via the notion of convergence of nets: take a net (x_α) in X; we will say that (x_α) *converges weakly to $x_0 \in X$*, and we write $x_\alpha \xrightarrow{w} x_0$, if for each $x^* \in X^*$

$$x^*(x_\alpha) \to x^*(x_0).$$

Next we summarize some elementary properties of the weak topology of a normed vector space X, noting that it is in the setting of infinite-dimensional spaces that the different natures of the weak and norm topologies become apparent.

- *If X is infinite-dimensional, every nonempty weak open set of X is unbounded.*
- *A subset S of X is norm-bounded if and only if S is weakly bounded (that is, $\{x^*(a) : a \in S\}$ is a bounded set in the scalar field of X for every $x^* \in X^*$).*
- *If the weak topology of X is metrizable then X is finite-dimensional.*
- *If X is infinite-dimensional then the weak topology of X is not complete.*
- *A linear functional on X is norm-continuous if and only if it is continuous with respect to the weak topology.*
- *Let $T : X \to Y$ be a linear map. T is weak-to-weak continuous if and only if $x^* \circ T \in X^*$ for every $x^* \in X^*$.*

- *A linear map $T : X \to Y$ is norm-to-norm continuous if and only if T is weak-to-weak continuous.*

F.1. Mazur's Theorem. *If S is a convex set in a normed space X then the closure of S in the norm topology, $\overline{S}$, coincides with $\overline{S}^w$, the closure of S in the weak topology.*

F.2. Corollary. *If Y is a linear subspace of a normed space X then $\overline{Y} = \overline{Y}^w$.*

F.3. Corollary. *If S is any subset of a normed space X then $\overline{co}(S) = \overline{co}^w(S)$.*

F.4. Corollary. *Let (x_n) be a sequence in a normed space X that converges weakly to $x \in X$. Then there is a sequence of convex combinations of the x_n, $y_k = \sum_{i=k}^{N(k)} \lambda_i x_i$, $k = 1, 2, \ldots$, such that $\|y_k - x\| \to 0$.*

Let us turn now to the weak* topology on a dual space X^*. Let $j : X \to X^{**}$ be the natural embedding of a Banach space in its second dual, given by $j(x)(x^*) = x^*(x)$. As usual we identify X with $j(X) \subset X^{**}$. The weak* topology on X^*, denoted w^*-topology or $\sigma(X^*, X)$-topology, is the topology induced on X^* by X, i.e., it is the weakest topology on X^* that makes all linear functionals in $X \subset X^{**}$ continuous.

Like the weak topology, the weak* topology is a locally convex, Hausdorff linear topology and a base of neighborhoods at $0 \in X^*$ is given by the sets of the form

$$W_\epsilon(0; x_1, \ldots, x_n) = \left\{ x^* \in X^* : |x^*(x_i)| < \varepsilon \text{ for } i = 1, \ldots, n \right\},$$

for any finite subset $\{x_1, \ldots, x_n\} \in X$ and any $\epsilon > 0$. Thus by translation we obtain the neighborhoods of other points in X^*.

As before, we can equivalently describe the weak* topology of a dual space in terms of convergence of nets: we say that a net $(x_\alpha^*) \subset X^*$ *converges weak** to $x_0^* \in X^*$, and we write $x_\alpha^* \xrightarrow{w^*} x_0^*$, if for each $x \in X$

$$x_\alpha^*(x) \to x_0^*(x).$$

Of course, the weak* topology of X^* is no bigger than its weak topology and, in fact, $\sigma(X^*, X) = \sigma(X^*, X^{**})$ if and only if $j(X) = X^{**}$ (that is, if and only if X is reflexive). Notice also that when we identify X with $j(X)$ and consider X as a subspace of X^{**} this is not simply an identification of sets; actually

$$(X, \sigma(X, X^*)) \xrightarrow{j} (X, \sigma(X^{**}, X^*))$$

is a linear homeomorphism. Analogously to the weak topology, dual spaces are never w^*-metrizable or w^*-complete unless the underlying space is finite-dimensional. The most important feature of the weak* topology is the following compactness property, basic to modern functional analysis, which was discovered by Banach in 1932 for separable spaces and was extended to the general case by Alaoglu in 1940.

F.5. The Banach-Alaoglu Theorem. *If X is a normed linear space then the set $B_{X^*} = \{x^* \in X^* : \|x^*\| \le 1\}$ is weak*-compact.*

F.6. Corollary. *The closed unit ball B_{X^*} of the dual of a normed space X is the weak* closure of the convex hull of the set of its extreme points:*

$$B_{X^*} = \overline{co}^{w^*}\left(\partial_e(B_{X^*})\right)$$

If X is a non reflexive Banach space then X cannot be dense nor weak dense in X^{**}. However, it turns out that X must be weak* dense in X^{**}, as deduced from the next useful result, which is a consequence of the fact that the weak* dual of X^* is X.

F.7. Goldstine's Theorem. *Let X be a normed space. Then B_X is weak* dense in $B_{X^{**}}$.*

F.8. The Banach-Dieudonné Theorem. *Let C be a convex subset of a dual space X^*. Then C is weak*-closed if and only if $C \cap \lambda B_{X^*}$ is weak*-closed for every $\lambda > 0$.*

F.9. Proposition. *Let X and Y be normed spaces and suppose that $T : X \to Y$ is a linear mapping.*

(i) If T is norm-to-norm continuous then its adjoint $T^* : Y^* \to X^*$ is weak*-to-weak* continuous.

(ii) If $R : Y^* \to X^*$ is a weak*-to-weak* continuous operator then there is $T : X \to Y$ norm-to-norm continuous such that $T^* = R$.

F.10. Corollary. *Suppose X, Y are normed spaces. Then every weak*-to-weak* continuous linear operator from X^* to Y^* is norm-to-norm continuous.*

Let us point out here that the converse of Corollary F.10 is not true in general.

G

Weak Compactness of Sets and Operators

A subset A of a normed space X is said to be *[relatively]* *weakly compact* if [the weak closure of] A is compact in the weak topology of X.

G.1. Proposition. *If K is a weakly compact set of normed space X then K is norm-closed and norm-bounded.*

G.2. Proposition. *Let X be a Banach space. Then B_X is weakly compact if and only if X is reflexive.*

This proposition yields the first elementary examples of weakly compact sets, which we include in the next corollary.

G.3. Corollary. *Let X be a reflexive space.*

(i) If A is a bounded subset of X then A is relatively weakly compact.
(ii) If A is a convex, bounded, norm-closed subset of X then A is weakly compact.
(iii) If $T : X \to Y$ is a continuous linear operator then $T(B_X)$ is weakly compact in Y.

When X is not reflexive, in order to check if a given set is relatively weakly compact we can employ the characterization provided by the following result.

G.4. Proposition. *A subset A of a Banach space X is relatively weakly compact if and only if it is norm-bounded and the $\sigma(X^{**}, X^*)$-closure of A in X^{**} is contained in A.*

The most important result on weakly compact sets is the Eberlein-Šmulian theorem, which we included in Chapter 1 (Theorem 1.6.3). This is indeed a very surprising result; when we consider X endowed with the norm topology, in order that every bounded sequence in X have a convergent subsequence it is necessary and sufficient that X be finite-dimensional. If X is infinite-dimensional the weak topology is not metrizable, thus sequential extraction arguments would not seem to apply in order to decide whether a subset of X is weakly compact. The Eberlein-Šmulian theorem, oddly enough, tells us that a bounded subset A *is weakly compact if and only if every sequence in A has a subsequence weakly convergent to some point of A.*

A bounded linear operator $T : X \to Y$ is said to be *weakly compact* if the set $T(B_X)$ is relatively weakly compact, that is, if $\overline{T(B_X)}$ is weakly compact. Since every bounded subset of X is contained in some multiple of the unit ball of X, we have that T is weakly compact if and only if it maps bounded sets into relatively weakly compact sets. Using the Eberlein-Šmulian theorem one can further state that $T : X \to Y$ is weakly compact if and only if for every bounded sequence $(x_n)_{n=1}^\infty \subset X$ the sequence $(Tx_n)_{n=1}^\infty$ has a weakly convergent subsequence.

G.5. Gantmacher's Theorem. *Suppose X and Y are Banach spaces and let $T : X \to Y$ be a bounded linear operator. Then:*

(i) *T is weakly compact if and only if the range of its double adjoint T^{**} : $X^{**} \to Y^{**}$ is in Y, i.e., $T^{**}(X^{**}) \subset Y$.*

(ii) *T is weakly compact if and only if its adjoint $T^* : Y^* \to X^*$ is weak*-to-weak continuous.*

(iii) *T is weakly compact if and only if its adjoint T^* is.*

The next remarks follow easily from what has been said in this section:

- Let $T : X \to Y$ be an operator. If X or Y are reflexive then T is weakly compact;

- The identity map on a nonreflexive Banach space is never weakly compact;

- A Banach space X is reflexive if and only if X^* is.

List of Symbols

Blackboard bold symbols

$\mathbb{N}$ The natural numbers.

$\mathbb{Q}$ The rational numbers.

$\mathbb{R}$ The real numbers.

$\mathbb{C}$ The complex numbers.

$\mathbb{T}$ The unit circle in the complex plane, $\{z \in \mathbb{C} : |z| = 1\}$.

$\mathbb{P}$ A probability measure on some probability space $(\Omega, \Sigma, \mathbb{P})$ (Section 6.2).

$\mathbb{E}f$ The expectation of a random variable f (Section 6.2).

Classical Banach spaces

$L_\infty(\mu)$ The (equivalence class) of μ-measurable essentially bounded real-valued functions f with the norm $\|f\|_\infty := \inf\{\alpha > 0 : \mu(|f| > \alpha) = 0\}$.

$L_p(\mu)$ The (equivalence class) of μ-measurable real-valued functions f so that $\|f\|_p := (\int |f|^p \, d\mu)^{1/p} < \infty$.

$L_p(\mathbb{T})$ $L_p(\mu)$ when μ is the normalized Lebesgue measure on $\mathbb{T}$.

L_p $L_p(\mu)$ when μ is the Lebesgue measure on $[0,1]$.

$\mathcal{C}(K)$ The continuous real-valued functions on the compact space K.

$\mathcal{C}_{\mathbb{C}}(K)$ The continuous complex-valued functions on the compact space K.

$\mathcal{J}$ The James space (Section 3.4).

$\mathcal{J}\mathcal{T}$ The James tree space (Section 13.4).

$\mathcal{M}(K)$ The finite regular Borel signed measures on the compact space K.

ℓ_∞ The collection of bounded sequences of scalars $x = (x_n)_{n=1}^\infty$, with the norm $\|x\|_\infty = \sup_n |x_n|$.

ℓ_∞^n $\mathbb{R}^n$ equipped with the $\|\cdot\|_\infty$ norm.

ℓ_p	$L_p(\mu)$ when μ is the *counting measure* on $\mathcal{P}(\mathbb{N})$, that is, the measure defined by $\mu(A) = |A|$ for any $A \subset \mathbb{N}$. Equivalently, the collection of all sequences of scalars $x = (x_n)_{n=1}^\infty$ so that $\|x\|_p := (\sum_{n=1}^\infty |x_n|^p)^{1/p} < \infty$.
ℓ_p^n	$\mathbb{R}^n$ equipped with the $\|\cdot\|_p$ norm.
c	The convergent sequences of scalars under the $\|\cdot\|_\infty$ norm.
c_0	The sequences of scalars that converge to 0 endowed with the $\|\cdot\|_\infty$ norm.
c_{00}	The (dense) subspace of c_0 of finitely nonzero sequences.

Important constants

$C_q(X)$	The cotype-q constant of the Banach space X (Section 6.2).
K_G	The best constant in Grothendieck inequality (Section 8.1).
K_s	The suppression constant of an unconditional basis (Section 3.1).
K_u	The unconditional basis constant (Section 3.1).
$T_p(X)$	The type-p constant of the Banach space X (Section 6.2).

Operator-related symbols

T^*	The adjoint operator of T.
T^2	The composition operator of T with itself, $T \circ T$.
I_X	The identity operator on X.
j	The canonical embedding of X into its second dual X^{**}.
$\langle x, x^* \rangle$	The action of a functional x^* in X^* on a vector $x \in X$, also represented by $x^*(x)$.
$\ker T$	The null space of T; that is, $T^{-1}(0)$.
$T(X)$	The range (or image) of an operator T defined on X.
$T|_E$	The restriction of the operator T to the subspace E of the domain space.
$\pi_p(T)$	The p-absolutely summing norm of T (Section 8.2).

Distinguished sequences of functions

$(h_n)_{n=1}^\infty$	The Haar system (Section 6.1).
$(r_n)_{n=1}^\infty$	The Rademacher functions (Section 6.3).
$(\varepsilon_n)_{n=1}^\infty$	A Rademacher sequence (Section 6.2).

Sets and subspaces

B_X	The closed unit ball of a normed space X.
$\langle A \rangle$	The linear span of a set A.

$[A]$	The closed linear span of a set A; i.e., the norm-closure of $\langle A \rangle$.
$[x_n]$	The norm-closure of $\langle x_n : n \in \mathbb{N} \rangle$.
$\overline{S}$ or $\overline{S}^{\|\cdot\|}$	The closure of a set S of a Banach space in its norm topology.
$\overline{S}^w$ or $\overline{S}^{\text{weak}}$	The closure of a set S of a Banach space in its weak topology.
$\overline{S}^{w^*}$ or $\overline{S}^{\text{weak}^*}$	The closure of a set S of a dual space in its weak* topology.
$M^\perp$	The annihilator of M in X^*, i.e., the collection of all continuous linear functionals on the Banach space X which vanish on the subset M of X.
$\partial_e(S)$	The set of extreme points of a convex set S.
$\tilde{A}$ or $X \setminus A$	The complement of A in X.
$\mathcal{P}A$	The collection of all subsets of a (usually infinite) set A.
$\mathcal{P}_\infty A$	The collection of all infinite subsets of an A.
$\mathcal{F}A$	The collection of all finite subsets of an A.
$\mathcal{F}_r A$	The collection of all finite subsets of an A of cardinality r.

Abbreviations for properties

(BAP)	Bounded approximation property (Problems section of Chapter 1).
(DPP)	Dunford-Pettis property (Section 5.4).
(KMP)	Krein-Milman property (Section 5.4).
(MAP)	Metric approximation property (Problems section of Chapter 1).
(RNP)	Radon-Nikodym property (Section 5.4).
(u)	Pełczyński's property (u) (Section 3.5).
(UTAP)	Uniqueness of unconditional basis up to a permutation (Section 9.3).
wsc	Weakly sequentially complete space (Section 2.3).
(WUC)	Weakly unconditionally Cauchy series (Section 2.4).

Miscellaneous

sgn t	$= \begin{cases} t/|t| & \text{if } t \neq 0 \\ 0 & \text{if } t = 0. \end{cases}$
χ_A	The characteristic function of a set A, $\chi_A(x) = \begin{cases} 1 & \text{if } x \in A \\ 0 & \text{if } x \notin A. \end{cases}$
$X \approx Y$	X isomorphic to Y.
$|\cdot|$	The absolute value of a real number, the modulus of a complex number, the cardinality of a finite set, or the Lebesgue measure of a set, depending on the context.
δ_s	The Dirac measure at the point s, whose value at $f \in \mathcal{C}(K)$ is $\delta_s(f) = f(s)$.

δ_{jk}	The Kronecker delta: $\delta_{jk} = 1$ if $j = k$, and $\delta_{jk} = 0$ if $j \neq k$.
$X \oplus Y$	Direct sum of X and Y.
X^2	$= X \oplus X$.
$\ell_p(X_n)$	$= (X_1 \oplus X_2 \oplus \cdots)_p$, the infinite direct sum of the sequence of spaces $(X_n)_{n=1}^{\infty}$ in the sense of ℓ_p (Section 2.2).
$c_0(X_n)$	$= (X_1 \oplus X_2 \oplus \cdots)_0$, the infinite direct sum of the sequence of spaces $(X_n)_{n=1}^{\infty}$ in the sense of c_0 (Section 2.2).
$\ell_\infty^n(X)$	$= (X \oplus \cdots \oplus X)_\infty$, i.e., the space of all sequences $x = (x_1, \ldots, x_n)$ so that $x_k \in X$ for $1 \leq k \leq n$, with the norm $\|x\| = \sup_{1 \leq k \leq n} \|x_k\|_X$.
$\ell_\infty(X_i)_{i \in \mathcal{I}}$	The Banach space of all $(x_i)_{i \in \mathcal{I}} \in \prod_{i \in \mathcal{I}} X_i$ such that $(\|x_i\|)_{i \in \mathcal{I}}$ is bounded, with the norm $\|(x_i)_{i \in \mathcal{I}}\|_\infty = \sup_{i \in \mathcal{I}} \|x_i\|_{X_i}$.
$d(x, A)$	The distance from a point x to the set A in a normed space: $\inf_{a \in A} \|x - a\|$.
$d(X, Y)$	The Banach-Mazur distance between two isomorphic Banach spaces X, Y (Section 7.4).
d_X	The Euclidean distance of X (Section 12.1).
$\mathcal{E}$	The conditional expectation operator (Section 6.1), and also an ellipsoid (Section 12.1).
Δ	The Cantor set (Section 1.4).

References

[1] F. Albiac and N. J. Kalton, *A characterization of real $C(K)$-spaces*, to appear.
[2] D. J. Aldous, *Subspaces of L^1, via random measures*, Trans. Amer. Math. Soc. **267** (1981), 445–463.
[3] D. Alspach, P. Enflo, and E. Odell, *On the structure of separable $\mathcal{L}_p$ spaces* $(1 < p < \infty)$, Studia Math. **60** (1977), 79–90.
[4] D. Amir, *On isomorphisms of continuous function spaces*, Israel J. Math. **3** (1965), 205–210.
[5] J. Arazy and J. Lindenstrauss, *Some linear topological properties of the spaces C_p of operators on Hilbert space*, Compositio Math. **30** (1975), 81–111.
[6] H. Auerbach, *Sur les groupes linéaires bornés*, Studia Math. **5** (1935), 43–49.
[7] K. I. Babenko, *On conjugate functions*, Dokl. Akad. Nauk SSSR (N. S.) **62** (1948), 157–160. (Russian)
[8] S. Banach, *Théorie des opérations linéaires*, Warszawa, 1932.
[9] S. Banach and S. Mazur, *Zur Theorie der linearen Dimension*, Studia Math. **4** (1933), 100–112.
[10] B. Beauzamy, *Introduction to Banach spaces and their geometry*, North-Holland Mathematics Studies, vol. 68, North-Holland, Amsterdam, 1982, Notas de Matemática [Mathematical Notes], 86.
[11] Y. Benyamini and J. Lindenstrauss, *Geometric nonlinear functional analysis. Vol. 1*, American Mathematical Society Colloquium Publications, vol. 48, American Mathematical Society, Providence, RI, 2000.
[12] C. Bessaga and A. Pełczyński, *On bases and unconditional convergence of series in Banach spaces*, Studia Math. **17** (1958), 151–164.
[13] ———, *Spaces of continuous functions. IV. On isomorphical classification of spaces of continuous functions*, Studia Math. **19** (1960), 53–62.
[14] K. Borsuk, *Über Isomorphie der Funktionalraüme*, Bull. Int. Acad. Pol. Sci. (1933), 1–10.
[15] J. Bourgain, P. G. Casazza, J. Lindenstrauss, and L. Tzafriri, *Banach spaces with a unique unconditional basis, up to permutation*, Mem. Amer. Math. Soc. **54** (1985).
[16] J. Bourgain, D. H. Fremlin, and M. Talagrand, *Pointwise compact sets of Baire-measurable functions*, Amer. J. Math. **100** (1978), 845–886.
[17] J. Bourgain, H. P. Rosenthal, and G. Schechtman, *An ordinal L^p-index for Banach spaces, with application to complemented subspaces of L^p*, Ann. Math. (2) **114** (1981), 193–228.

[18] J. Bretagnolle and D. Dacunha-Castelle, *Application de l'étude de certaines formes linéaires aléatoires au plongement d'espaces de Banach dans des espaces* L^p, Ann. Sci. Ec. Norm. Sup. (4) **2** (1969), 437–480. (French)

[19] A. Brunel and L. Sucheston, *On B-convex Banach spaces*, Math. Syst. Theory **7** (1974), 294–299.

[20] D. L. Burkholder, *A nonlinear partial differential equation and the unconditional constant of the Haar system in* L^p, Bull. Amer. Math. Soc. (N.S.) **7** (1982), 591–595.

[21] _____, *A proof of Pełczynśki's conjecture for the Haar system*, Studia Math. **91** (1988), 79–83.

[22] M. Cambern, *A generalized Banach-Stone theorem*, Proc. Amer. Math. Soc. **17** (1966), 396–400.

[23] N. L. Carothers, *A short course on Banach space theory*, London Mathematical Society Student Texts, vol. 64, Cambridge University Press, Cambridge, 2005.

[24] P. G. Casazza, *Approximation properties*, Handbook of the Geometry of Banach Spaces, Vol. I, North-Holland, Amsterdam, 2001, pp. 271–316.

[25] P. G. Casazza, W. B. Johnson, and L. Tzafriri, *On Tsirelson's space*, Israel J. Math. **47** (1984), 81–98.

[26] P. G. Casazza and N. J. Kalton, *Uniqueness of unconditional bases in Banach spaces*, Israel J. Math. **103** (1998), 141–175.

[27] _____, *Uniqueness of unconditional bases in* c_0-*products*, Studia Math. **133** (1999), 275–294.

[28] P. G. Casazza and N. J. Nielsen, *The Maurey extension property for Banach spaces with the Gordon-Lewis property and related structures*, Studia Math. **155** (2003), 1–21.

[29] P. G. Casazza and T. J. Shura, *Tsirel'son's space*, Lecture Notes in Mathematics, vol. 1363, Springer-Verlag, Berlin, 1989, With an appendix by J. Baker, O. Slotterbeck, and R. Aron.

[30] J.A. Clarkson, *Uniformly convex spaces*, Trans. Amer. Math. Soc. **40** (1936), 396–414.

[31] H. B. Cohen, *A bound-two isomorphism between* $C(X)$ *Banach spaces*, Proc. Amer. Math. Soc. **50** (1975), 215–217.

[32] J. B. Conway, *A course in functional analysis*, Graduate Texts in Mathematics, vol. 96, Springer-Verlag, New York, 1985.

[33] D. Dacunha-Castelle and J. L. Krivine, *Applications des ultraproduits à l'étude des espaces et des algèbres de Banach*, Studia Math. **41** (1972), 315–334. (French)

[34] A. M. Davie, *The approximation problem for Banach spaces*, Bull. London Math. Soc. **5** (1973), 261–266.

[35] W. J. Davis, T. Figiel, W. B. Johnson, and A. Pełczyński, *Factoring weakly compact operators*, J. Funct. Anal. **17** (1974), 311–327.

[36] D. W. Dean, *The equation* $L(E, X^{**}) = L(E, X)^{**}$ *and the principle of local reflexivity*, Proc. Amer. Math. Soc. **40** (1973), 146–148.

[37] L. de Branges, *The Stone-Weierstrass theorem*, Proc. Amer. Math. Soc. **10** (1959), 822–824.

[38] R. Deville, G. Godefroy, and V. Zizler, *Smoothness and renormings in Banach spaces*, Pitman Monographs and Surveys in Pure and Applied Mathematics, vol. 64, Longman Scientific & Technical, Harlow, 1993.

[39] J. Diestel, *Sequences and series in Banach spaces*, Graduate Texts in Mathematics, vol. 92, Springer-Verlag, New York, 1984.

[40] J. Diestel, H. Jarchow, and A. Pietsch, *Operator ideals*, Handbook of the Geometry of Banach Spaces, Vol. I, North-Holland, Amsterdam, 2001, pp. 437–496.

[41] J. Diestel, H. Jarchow, and A. Tonge, *Absolutely summing operators*, Cambridge Studies in Advanced Mathematics, vol. 43, Cambridge University Press, Cambridge, 1995.

[42] J. Diestel and J. J. Uhl Jr., *Vector measures*, American Mathematical Society, Providence, RI, 1977.

[43] J. Dixmier, *Sur certains espaces considérés par M. H. Stone*, Summa Bras. Math. **2** (1951), 151–182. (French)

[44] L. E. Dor, *On sequences spanning a complex l_1 space*, Proc. Amer. Math. Soc. **47** (1975), 515–516.

[45] N. Dunford and B. J. Pettis, *Linear operations on summable functions*, Trans. Amer. Math. Soc. **47** (1940), 323–392.

[46] N. Dunford and J. T. Schwartz, *Linear operators. Part I*, Wiley Classics Library, John Wiley & Sons, New York, 1988.

[47] ———, *Linear operators. Part II*, Wiley Classics Library, John Wiley & Sons, New York, 1988.

[48] ———, *Linear operators. Part III*, Wiley Classics Library, John Wiley & Sons, New York, 1988.

[49] A. Dvoretzky, *Some results on convex bodies and Banach spaces*, Proc. Int. Symp. Linear Spaces (Jerusalem, 1960), Jerusalem Academic Press, Jerusalem, 1961, pp. 123–160.

[50] A. Dvoretzky and C. A. Rogers, *Absolute and unconditional convergence in normed linear spaces*, Proc. Natl. Acad. Sci. U. S. A. **36** (1950), 192–197.

[51] W. F. Eberlein, *Weak compactness in Banach spaces. I*, Proc. Natl. Acad. Sci. U. S. A. **33** (1947), 51–53.

[52] I. S. Èdel'šteĭn and P. Wojtaszczyk, *On projections and unconditional bases in direct sums of Banach spaces*, Studia Math. **56** (1976), 263–276.

[53] P. Enflo, *Banach spaces which can be given an equivalent uniformly convex norm*, Israel J. Math. **13** (1972), 281–288 (1973).

[54] ———, *A counterexample to the approximation problem in Banach spaces*, Acta Math. **130** (1973), 309–317.

[55] P. Enflo and T. W. Starbird, *Subspaces of L^1 containing L^1*, Studia Math. **65** (1979), 203–225.

[56] M. Fabian, P. Habala, P. Hájek, V. Montesinos Santalucía, J. Pelant, and V. Zizler, *Functional analysis and infinite-dimensional geometry*, CMS Books in Mathematics/Ouvrages de Mathématiques de la SMC, 8, Springer-Verlag, New York, 2001.

[57] W. Feller, *An introduction to probability theory and its applications. Vol. II*, Second edition, John Wiley & Sons, New York, 1971.

[58] H. Fetter and B. Gamboa de Buen, *The James forest*, London Mathematical Society Lecture Note Series, vol. 236, Cambridge University Press, Cambridge, 1997.

[59] T. Figiel and W. B. Johnson, *A uniformly convex Banach space which contains no l_p*, Compositio Math. **29** (1974), 179–190.

[60] T. Figiel, J. Lindenstrauss, and V. D. Milman, *The dimension of almost spherical sections of convex bodies*, Acta Math. **139** (1977), 53–94.

[61] I. Fredhom, *Sur une classe d'équations fonctionelles*, Acta Math. **27** (1903), 365–390.

[62] F. Galvin and K. Prikry, *Borel sets and Ramsey's theorem*, J. Symbolic Logic **38** (1973), 193–198.

[63] D. J. H. Garling, *Symmetric bases of locally convex spaces*, Studia Math. **30** (1968), 163–181.

[64] ———, *Absolutely p-summing operators in Hilbert space*, Studia Math. **38** (1970), 319–331 (errata insert).

[65] D. J. H. Garling and N. Tomczak-Jaegermann, *The cotype and uniform convexity of unitary ideals*, Israel J. Math. **45** (1983), 175–197.

[66] I. M. Gelfand, *Abstrakte Funktionen und lineare Operatoren*, Mat. Sb. **4(46)** (1938), 235–286.

[67] T. A. Gillespie, *Factorization in Banach function spaces*, Indag. Math. **43** (1981), 287–300.

[68] D. B. Goodner, *Projections in normed linear spaces*, Trans. Amer. Math. Soc. **69** (1950), 89–108.

[69] Y. Gordon, *Some inequalities for Gaussian processes and applications*, Israel J. Math. **50** (1985), 265–289.

[70] W. T. Gowers, *A new dichotomy for Banach spaces*, Geom. Funct. Anal. **6** (1996), 1083–1093.

[71] W. T. Gowers and B. Maurey, *The unconditional basic sequence problem*, J. Amer. Math. Soc. **6** (1993), 851–874.

[72] ———, *Banach spaces with small spaces of operators*, Math. Ann. **307** (1997), 543–568.

[73] L. Grafakos, *Classical and modern Fourier analysis*, Prentice Hall, Englewood Cliffs, NJ, 2004.

[74] A. Grothendieck, *Critères de compacité dans les espaces fonctionnels généraux*, Amer. J. Math. **74** (1952), 168–186. (French)

[75] ———, *Sur les applications linéaires faiblement compactes d'espaces du type C(K)*, Canad. J. Math. **5** (1953), 129–173. (French)

[76] ———, *Résumé de la théorie métrique des produits tensoriels topologiques*, Bol. Soc. Mat. São Paulo **8** (1953), 1–79. (French)

[77] M. M. Grunblum, *Certains théorèmes sur la base dans un espace du type (B)*, C. R. Dokl. Acad. Sci. URSS (N. S.) **31** (1941), 428–432. (French)

[78] S. Guerre-Delabrière, *Classical sequences in Banach spaces*, Monographs and Textbooks in Pure and Applied Mathematics, vol. 166, Marcel Dekker, New York, 1992, With a foreword by Haskell P. Rosenthal.

[79] J. Hoffmann-Jørgensen, *Sums of independent Banach space valued random variables*, Studia Math. **52** (1974), 159–186.

[80] R. C. James, *Bases and reflexivity of Banach spaces*, Ann. Math. (2) **52** (1950), 518–527.

[81] ———, *A non-reflexive Banach space isometric with its second conjugate space*, Proc. Natl. Acad. Sci. U. S. A. **37** (1951), 174–177.

[82] ———, *Separable conjugate spaces*, Pac. J. Math. **10** (1960), 563–571.

[83] ———, *Uniformly non-square Banach spaces*, Ann. Math. (2) **80** (1964), 542–550.

[84] ———, *Weak compactness and reflexivity*, Israel J. Math. **2** (1964), 101–119.

[85] ———, *Some self-dual properties of normed linear spaces*, Symposium on Infinite-Dimensional Topology (Louisiana State University, Baton Rouge, 1967), Princeton University Press, Princeton, NJ, 1972, pp. 159–175. Ann. Math. Studies, No. 69.

[86] _____, *Super-reflexive Banach spaces*, Canad. J. Math. **24** (1972), 896–904.

[87] _____, *A separable somewhat reflexive Banach space with nonseparable dual*, Bull. Amer. Math. Soc. **80** (1974), 738–743.

[88] _____, *Nonreflexive spaces of type 2*, Israel J. Math. **30** (1978), 1–13.

[89] F. John, *Extremum problems with inequalities as subsidiary conditions*, Studies and Essays Presented to R. Courant on His 60th Birthday, January 8, 1948, Interscience Publishers, New York, 1948, pp. 187–204.

[90] **W. B. Johnson and J. Lindenstrauss (eds.)**, *Handbook of the geometry of Banach spaces. Vol. I*, North-Holland, Amsterdam, 2001.

[91] W. B. Johnson and J. Lindenstrauss, *Basic concepts in the geometry of Banach spaces*, Handbook of the Geometry of Banach Spaces, Vol. I, North-Holland, Amsterdam, 2001, pp. 1–84.

[92] **W. B. Johnson and J. Lindenstrauss (eds.)**, *Handbook of the geometry of Banach spaces. Vol. 2*, North-Holland, Amsterdam, 2003.

[93] W. B. Johnson, B. Maurey, G. Schechtman, and L. Tzafriri, *Symmetric structures in Banach spaces*, Mem. Amer. Math. Soc. **19** (1979).

[94] W. B. Johnson, H. P. Rosenthal, and M. Zippin, *On bases, finite dimensional decompositions and weaker structures in Banach spaces*, Israel J. Math. **9** (1971), 488–506.

[95] W. B. Johnson and A. Szankowski, *Complementably universal Banach spaces*, Studia Math. **58** (1976), 91–97.

[96] P. Jordan and J. Von Neumann, *On inner products in linear, metric spaces*, Ann. Math. (2) **36** (1935), 719–723.

[97] M. I. Kadets and B. S. Mitjagin, *Complemented subspaces in Banach spaces*, Usp. Mat. Nauk **28** (1973), 77–94. (Russian)

[98] M. I. Kadets and A. Pełczyński, *Bases, lacunary sequences and complemented subspaces in the spaces L_p*, Studia Math. **21** (1961/1962), 161–176.

[99] _____, *Basic sequences, bi-orthogonal systems and norming sets in Banach and Fréchet spaces*, Studia Math. **25** (1965), 297–323. (Russian)

[100] M. I. Kadets and M. G. Snobar, *Certain functionals on the Minkowski compactum*, Mat. Zametki **10** (1971), 453–457. (Russian)

[101] J. P. Kahane, *Sur les sommes vectorielles $\sum \pm u_n$*, C. R. Acad. Sci. Paris **259** (1964), 2577–2580. (French)

[102] N. J. Kalton, *Bases in weakly sequentially complete Banach spaces*, Studia Math. **42** (1972), 121–131.

[103] _____, *The endomorphisms of L_p ($0 \leq p \leq 1$)*, Indiana Univ. Math. J. **27** (1978), 353–381.

[104] _____, *Banach spaces embedding into L_0*, Israel J. Math. **52** (1985), 305–319.

[105] N. J. Kalton and A. Koldobsky, *Banach spaces embedding isometrically into L_p when $0 < p < 1$*, Proc. Amer. Math. Soc. **132** (2004), 67–76 (electronic).

[106] M. Kanter, *Stable laws and the imbedding of L_p spaces*, Amer. Math. Mon. **80** (1973), 403–407 (electronic).

[107] S. Karlin, *Bases in Banach spaces*, Duke Math. J. **15** (1948), 971–985.

[108] Y. Katznelson, *An introduction to harmonic analysis*, Second corrected edition, Dover Publications, New York, 1976.

[109] J. L. Kelley, *Banach spaces with the extension property*, Trans. Amer. Math. Soc. **72** (1952), 323–326.

[110] A. Khintchine, *Über dyadische Brüche*, Math. Z. **18** (1923), 109–116.

[111] A. Khintchine and A. N. Kolmogorov, *Über Konvergenz von Reihen, dieren Glieder durch den Zufall bestimmt werden*, Mat. Sb. **32** (1925), 668–677.

[112] A. Koldobsky, *Common subspaces of L_p-spaces*, Proc. Amer. Math. Soc. **122** (1994), 207–212.

[113] _____, *A Banach subspace of $L_{1/2}$ which does not embed in L_1 (isometric version)*, Proc. Amer. Math. Soc. **124** (1996), 155–160.

[114] A. Koldobsky and H. König, *Aspects of the isometric theory of Banach spaces*, Handbook of the Geometry of Banach Spaces, Vol. I, North-Holland, Amsterdam, 2001, pp. 899–939.

[115] R. A. Komorowski and N. Tomczak-Jaegermann, *Banach spaces without local unconditional structure*, Israel J. Math. **89** (1995), 205–226.

[116] S. V. Konyagin and V. N. Temlyakov, *A remark on greedy approximation in Banach spaces*, East J. Approx. **5** (1999), 365–379.

[117] T. W. Körner, *Fourier analysis*, 2nd ed., Cambridge University Press, Cambridge, 1989.

[118] M. Krein, D. Milman, and M. Rutman, *A note on basis in Banach space*, Comm. Inst. Sci. Math. Méc. Univ. Kharkoff [Zapiski Inst. Mat. Mech.] (4) **16** (1940), 106–110. (Russian, with English summary)

[119] J. L. Krivine, *Sous-espaces de dimension finie des espaces de Banach réticulés*, Ann. Math. (2) **104** (1976), 1–29.

[120] _____, *Constantes de Grothendieck et fonctions de type positif sur les sphères*, Adv. Math. **31** (1979), 16–30. (French)

[121] J. L. Krivine and B. Maurey, *Espaces de Banach stables*, Israel J. Math. **39** (1981), 273–295. (French, with English summary)

[122] S. Kwapień, *Isomorphic characterizations of inner product spaces by orthogonal series with vector valued coefficients*, Studia Math. **44** (1972), 583–595, Collection of articles honoring the completion by Antoni Zygmund of 50 years of scientific activity, VI.

[123] H. Lemberg, *Nouvelle démonstration d'un théorème de J.-L. Krivine sur la finie représentation de l_p dans un espace de Banach*, Israel J. Math. **39** (1981), 341–348. (French, with English summary)

[124] P. Lévy, *Problèmes concrets d'analyse fonctionnelle. Avec un complément sur les fonctionnelles analytiques par F. Pellegrino*, Gauthier-Villars, Paris, 1951, 2d ed. (French)

[125] D. R. Lewis and C. Stegall, *Banach spaces whose duals are isomorphic to $l_1(\Gamma)$*, J. Funct. Anal. **12** (1973), 177–187.

[126] D. Li and H. Queffélec, *Introduction à l'étude des espaces de Banach*, Cours Spécialisés [Specialized Courses], vol. 12, Société Mathématique de France, Paris, 2004, Analyse et probabilités. [Analysis and probability theory]. (French)

[127] J. Lindenstrauss, *On a certain subspace of l_1*, Bull. Acad. Pol. Sci. Ser. Sci. Math. Astron. Phys. **12** (1964), 539–542.

[128] _____, *On extreme points in l_1*, Israel J. Math. **4** (1966), 59–61.

[129] _____, *On complemented subspaces of m*, Israel J. Math. **5** (1967), 153–156.

[130] _____, *On James's paper "Separable conjugate spaces"*, Israel J. Math. **9** (1971), 279–284.

[131] J. Lindenstrauss and A. Pełczyński, *Absolutely summing operators in L_p-spaces and their applications*, Studia Math. **29** (1968), 275–326.

[132] _____, *Contributions to the theory of the classical Banach spaces*, J. Funct. Anal. **8** (1971), 225–249.

[133] J. Lindenstrauss and H. P. Rosenthal, *The $\mathcal{L}_p$ spaces*, Israel J. Math. **7** (1969), 325–349.

[134] J. Lindenstrauss and C. Stegall, *Examples of separable spaces which do not contain ℓ_1 and whose duals are non-separable*, Studia Math. **54** (1975), 81–105.

[135] J. Lindenstrauss and L. Tzafriri, *On the complemented subspaces problem*, Israel J. Math. **9** (1971), 263–269.

[136] _____, *On Orlicz sequence spaces*, Israel J. Math. **10** (1971), 379–390.

[137] _____, *On Orlicz sequence spaces. II*, Israel J. Math. **11** (1972), 355–379.

[138] _____, *Classical Banach spaces. I*, Springer-Verlag, Berlin, 1977, Sequence spaces.

[139] _____, *Classical Banach spaces. II*, vol. 97, Springer-Verlag, Berlin, 1979, Function spaces.

[140] J. Lindenstrauss and M. Zippin, *Banach spaces with a unique unconditional basis*, J. Funct. Anal. **3** (1969), 115–125.

[141] J. E. Littlewood, *On bounded bilinear forms in n infinite number of variables*, Q. J. Math. (Oxford) **1** (1930), 164–174.

[142] G. Ja. Lozanovskiĭ, *Certain Banach lattices*, Sib. Mat. J. **10** (1969), 584–599. (Russian)

[143] B. Maurey, *Un théorème de prolongement*, C. R. Acad. Sci. Paris Ser. A **279** (1974), 329–332. (French)

[144] _____, *Théorèmes de factorisation pour les opérateurs linéaires à valeurs dans les espaces L^p*, Société Mathématique de France, Paris, 1974, With an English summary; Astérisque, No. 11. (French)

[145] _____, *Types and l_1-subspaces*, Texas Functional Analysis Seminar 1982–1983 (Austin, Tex.), Longhorn Notes, University of Texas Press, Austin, 1983, pp. 123–137.

[146] _____, *Type, cotype and K-convexity*, Handbook of the Geometry of Banach Spaces, Vol. 2, North-Holland, Amsterdam, 2003, pp. 1299–1332.

[147] B. Maurey and G. Pisier, *Séries de variables aléatoires vectorielles indépendantes et propriétés géométriques des espaces de Banach*, Studia Math. **58** (1976), 45–90. (French)

[148] C. A. McCarthy and J. Schwartz, *On the norm of a finite Boolean algebra of projections, and applications to theorems of Kreiss and Morton*, Commun. Pure Appl. Math. **18** (1965), 191–201.

[149] R. E. Megginson, *An introduction to Banach space theory*, Graduate Texts in Mathematics, vol. 183, Springer-Verlag, New York, 1998.

[150] A. A. Miljutin, *Isomorphism of the spaces of continuous functions over compact sets of the cardinality of the continuum*, Teor. Funkciĭ Funkcional. Anal. Priložen. Vyp. **2** (1966), 150–156. (1 foldout). (Russian)

[151] V. D. Milman, *Geometric theory of Banach spaces. II. Geometry of the unit ball*, Usp. Mat. Nauk **26** (1971), 73–149. (Russian)

[152] _____, *A new proof of A. Dvoretzky's theorem on cross-sections of convex bodies*, Funkcional. Anal. Priložen. **5** (1971), 28–37. (Russian)

[153] _____, *Almost Euclidean quotient spaces of subspaces of a finite-dimensional normed space*, Proc. Amer. Math. Soc. **94** (1985), 445–449.

[154] V. D. Milman and G. Schechtman, *Asymptotic theory of finite-dimensional normed spaces*, Lecture Notes in Mathematics, vol. 1200, Springer-Verlag, Berlin, 1986.

[155] L. Nachbin, *On the Han-Banach theorem*, An. Acad. Bras. Cienc. **21** (1949), 151–154.

[156] F. L. Nazarov and S. R. Treĭl', *The hunt for a Bellman function: applications to estimates for singular integral operators and to other classical problems of harmonic analysis*, Algebra i Analiz **8** (1996), 32–162. (Russian, with Russian summary)

[157] E. M. Nikišin, *Resonance theorems and superlinear operators*, Usp. Mat. Nauk **25** (1970), 129–191. (Russian)

[158] _____, *A resonance theorem and series in eigenfunctions of the Laplace operator*, Izv. Akad. Nauk SSSR Ser. Mat. **36** (1972), 795–813. (Russian)

[159] G. Nordlander, *On sign-independent and almost sign-independent convergence in normed linear spaces*, Ark. Mat. **4** (1962), 287–296.

[160] E. Odell and H. P. Rosenthal, *A double-dual characterization of separable Banach spaces containing l^1*, Israel J. Math. **20** (1975), 375–384.

[161] E. Odell and T. Schlumprecht, *The distortion problem*, Acta Math. **173** (1994), 259–281.

[162] W. Orlicz, *Beitrge zur Theorie der Orthogonalentwicklungen II*, Studia Math. **1** (1929), 242–255.

[163] _____, *Über unbedingte Konvergenz in Funktionenraümen I*, Studia Math. **4** (1933), 33–37.

[164] _____, *Über unbedingte Konvergenz in Funktionenraümen II*, Studia Math. **4** (1933), 41–47.

[165] R.E.A.C. Paley, *A remarkable series of orthogonal functions*, Proc. London Math., Soc. **34** (1932), 241–264.

[166] K. R. Parthasarathy, *Probability measures on metric spaces*, Probability and Mathematical Statistics, No. 3, Academic Press, New York, 1967.

[167] A. Pełczyński, *On the isomorphism of the spaces m and M*, Bull. Acad. Pol. Sci. Ser. Sci. Math. Astron. Phys. **6** (1958), 695–696.

[168] _____, *A connection between weakly unconditional convergence and weakly completeness of Banach spaces*, Bull. Acad. Pol. Sci. Ser. Sci. Math. Astron. Phys. **6** (1958), 251–253 (unbound insert). (English, with Russian summary)

[169] _____, *Projections in certain Banach spaces*, Studia Math. **19** (1960), 209–228.

[170] _____, *On the impossibility of embedding of the space L in certain Banach spaces*, Colloq. Math. **8** (1961), 199–203.

[171] _____, *Banach spaces on which every unconditionally converging operator is weakly compact*, Bull. Acad. Pol. Sci. Ser. Sci. Math. Astron. Phys. **10** (1962), 641–648.

[172] _____, *A proof of Eberlein-Šmulian theorem by an application of basic sequences*, Bull. Acad. Pol. Sci. Ser. Sci. Math. Astron. Phys. **12** (1964), 543–548.

[173] _____, *A characterization of Hilbert-Schmidt operators*, Studia Math. **28** (1966/1967), 355–360.

[174] _____, *Universal bases*, Studia Math. **32** (1969), 247–268.

[175] _____, *Any separable Banach space with the bounded approximation property is a complemented subspace of a Banach space with a basis*, Studia Math. **40** (1971), 239–243.

[176] _____, *Banach spaces of analytic functions and absolutely summing operators*, American Mathematical Society, Providence, RI, 1977, Expository lectures from the CBMS Regional Conference held at Kent State University, Kent, Ohio, July 11–16, 1976; Conference Board of the Mathematical Sciences Regional Conference Series in Mathematics, No. 30.

[177] A. Pełczyński and I. Singer, *On non-equivalent bases and conditional bases in Banach spaces*, Studia Math. **25** (1964/1965), 5–25.

[178] B. J. Pettis, *On integration in vector spaces*, Trans. Amer. Math. Soc. **44** (1938), 277–304.

[179] R. R. Phelps, *Dentability and extreme points in Banach spaces*, J. Funct. Anal. **17** (1974), 78–90.

[180] R. S. Phillips, *On linear transformations*, Trans. Amer. Math. Soc. **48** (1940), 516–541.

[181] A. Pietsch, *Absolut p-summierende Abbildungen in normierten Räumen*, Studia Math. **28** (1966/1967), 333–353. (German)

[182] G. Pisier, *Martingales with values in uniformly convex spaces*, Israel J. Math. **20** (1975), 326–350.

[183] _____, *Un théorème sur les opérateurs linéaires entre espaces de Banach qui se factorisent par un espace de Hilbert*, Ann. Sci. Ec. Norm. Sup. (4) **13** (1980), 23–43. (French)

[184] _____, *Counterexamples to a conjecture of Grothendieck*, Acta Math. **151** (1983), 181–208.

[185] _____, *Factorization of linear operators and geometry of Banach spaces*, CBMS Regional Conference Series in Mathematics, vol. 60, Published for the Conference Board of the Mathematical Sciences, Washington, DC, 1986.

[186] _____, *Factorization of operators through $L_{p\infty}$ or L_{p1} and noncommutative generalizations*, Math. Ann. **276** (1986), 105–136.

[187] _____, *Probabilistic methods in the geometry of Banach spaces*, Probability and Analysis (Varenna, 1985), Lecture Notes in Mathematics, vol. 1206, Springer, Berlin, 1986, pp. 167–241.

[188] _____, *The volume of convex bodies and Banach space geometry*, Cambridge Tracts in Mathematics, vol. 94, Cambridge University Press, Cambridge, 1989.

[189] H. R. Pitt, *A note on bilinear forms*, J. London Math. Soc. **11** (1932), 174–180.

[190] G. Plebanek, *Banach spaces of continuous functions with few operators*, Math. Ann., to appear.

[191] H. Rademacher, *Einige Sätze über Reihen von allgemeinen Orthogonalfuncktionen*, Math. Ann. **87** (1922), 112–138.

[192] F. P. Ramsey, *On a problem of formal logic*, Proc. London Math. Soc. **30** (1929), 264–286.

[193] T. J. Ransford, *A short elementary proof of the Bishop-Stone-Weierstrass theorem*, Math. Proc. Cambridge Philos. Soc. **96** (1984), 309–311.

[194] Y. Raynaud and C. Schütt, *Some results on symmetric subspaces of L_1*, Studia Math. **89** (1988), 27–35.

[195] H. P. Rosenthal, *On factors of $C([0, 1])$ with non-separable dual*, Israel J. Math. **13** (1972), 361–378 (1973); correction, ibid. **21** (1975), no. 1, 93–94.

[196] _____, *On subspaces of L^p*, Ann. Math. (2) **97** (1973), 344–373.

[197] _____, *A characterization of Banach spaces containing l^1*, Proc. Natl. Acad. Sci. U.S.A. **71** (1974), 2411–2413.

[198] _____, *On a theorem of J. L. Krivine concerning block finite representability of l^p in general Banach spaces*, J. Funct. Anal. **28** (1978), 197–225.

[199] _____, *The Banach spaces $C(K)$*, Handbook of the Geometry of Banach Spaces, Vol. 2, North-Holland, Amsterdam, 2003, pp. 1547–1602.

[200] H. L. Royden, *Real analysis*, 3rd ed., Macmillan Publishing Company, New York, 1988.

[201] W. Schachermayer, *For a Banach space isomorphic to its square the Radon-Nikodým property and the Kreĭn-Milman property are equivalent*, Studia Math. **81** (1985), 329–339.

[202] R. Schatten, *A theory of cross-spaces*, Annals of Mathematics Studies, no. 26, Princeton University Press, Princeton, NJ, 1950.

[203] J. Schauder, *Zur Theorie stetiger Abbildungen in Funktionalraumen*, Math. Zeit. **26** (1927), 47–65.

[204] G. Schechtman, *On Pełczyński's paper "Universal bases" (Studia Math. 32 (1969), 247–268)*, Israel J. Math. **22** (1975), 181–184.

[205] J. Schur, *Über lineare Transformationen in der Theorie der unendlichen Reihen*, J. Reine Angew. Math. **151** (1920), 79–111.

[206] I. Singer, *Basic sequences and reflexivity of Banach spaces*, Studia Math. **21** (1961/1962), 351–369.

[207] V. Šmulian, *Über lineare topologische Räume*, Rec. Math. [Mat. Sb.] N. S. **7 (49)** (1940), 425–448. (German, with Russian summary)

[208] A. Sobczyk, *Projection of the space (m) on its subspace (c_0)*, Bull. Amer. Math. Soc. **47** (1941), 938–947.

[209] C. Stegall, *A proof of the principle of local reflexivity*, Proc. Amer. Math. Soc. **78** (1980), 154–156.

[210] E. M. Stein, *On limits of sequences of operators*, Ann. Math. (2) **74** (1961), 140–170.

[211] A. Szankowski, *Subspaces without the approximation property*, Israel J. Math. **30** (1978), 123–129.

[212] S. J. Szarek, *A Banach space without a basis which has the bounded approximation property*, Acta Math. **159** (1987), 81–98.

[213] V. N. Temlyakov, *The best m-term approximation and greedy algorithms*, Adv. Comput. Math. **8** (1998), 249–265.

[214] ———, *Nonlinear methods of approximation*, Found. Comput. Math. **3** (2003), 33–107.

[215] N. Tomczak-Jaegermann, *The moduli of smoothness and convexity and the Rademacher averages of trace classes $S_p (1 \leq p < \infty)$*, Studia Math. **50** (1974), 163–182.

[216] ———, *Banach-Mazur distances and finite-dimensional operator ideals*, Pitman Monographs and Surveys in Pure and Applied Mathematics, vol. 38, Longman Scientific & Technical, Harlow, 1989.

[217] B. S. Tsirel'son, *It is impossible to imbed l_p of c_0 into an arbitrary Banach space*, Funkcional. Anal. Priložen. **8** (1974), 57–60. (Russian)

[218] L. Tzafriri, *Uniqueness of structure in Banach spaces*, Handbook of the Geometry of Banach Spaces, Vol. 2, North-Holland, Amsterdam, 2003, pp. 1635–1669.

[219] W. A. Veech, *Short proof of Sobczyk's theorem*, Proc. Amer. Math. Soc. **28** (1971), 627–628.

[220] R. J. Whitley, *Projecting m onto c_0*, Amer. Math. Mon. **73** (1966), 285–286.

[221] P. Wojtaszczyk, *Banach spaces for analysts*, Cambridge Studies in Advanced Mathematics, vol. 25, Cambridge University Press, Cambridge, 1991.

[222] ———, *Greedy algorithm for general biorthogonal systems*, J. Approx. Theory **107** (2000), 293–314.

[223] M. Zippin, *On perfectly homogeneous bases in Banach spaces*, Israel J. Math. **4** (1966), 265–272.

[224] ———, *A remark on bases and reflexivity in Banach spaces*, Israel J. Math. **6** (1968), 74–79.

[225] ———, *The separable extension problem*, Israel J. Math. **26** (1977), 372–387.

Index

Graduate Texts in Mathematics

(continued from p. ii)